The Urban Book Series

AF213265

Editorial Board

Margarita Angelidou, Aristotle University of Thessaloniki, Thessaloniki, Greece

Fatemeh Farnaz Arefian, The Bartlett Development Planning Unit, UCL, Silk Cities, London, UK

Michael Batty, Centre for Advanced Spatial Analysis, UCL, London, UK

Simin Davoudi, Planning & Landscape Department GURU, Newcastle University, Newcastle, UK

Geoffrey DeVerteuil, School of Planning and Geography, Cardiff University, Cardiff, UK

Jesús M. González Pérez, Department of Geography, University of the Balearic Islands, Palma (Mallorca), Spain

Daniel B. Hess, Department of Urban and Regional Planning, University at Buffalo, State University, Buffalo, NY, USA

Paul Jones, School of Architecture, Planning and Policy Development, Institute of Technology Bandung University (ITB), Bandung, Indonesia

Andrew Karvonen, Division of Urban and Regional Studies, KTH Royal Institute of Technology, Stockholm, Stockholms Län, Sweden

Andrew Kirby, New College, Arizona State University, Phoenix, AZ, USA

Karl Kropf, Department of Planning, Headington Campus, Oxford Brookes University, Oxford, UK

Karen Lucas, Institute for Transport Studies, University of Leeds, Leeds, UK

Marco Maretto, DICATeA, Department of Civil and Environmental Engineering, University of Parma, Parma, Italy

Ali Modarres, Tacoma Urban Studies, University of Washington Tacoma, Tacoma, WA, USA

Fabian Neuhaus, Faculty of Environmental Design, University of Calgary, Calgary, AB, Canada

Steffen Nijhuis, Architecture and the Built Environment, Delft University of Technology, Delft, The Netherlands

Vitor Manuel Aráujo de Oliveira, Porto University, Porto, Portugal

Christopher Silver, College of Design, University of Florida, Gainesville, FL, USA

Giuseppe Strappa, Facoltà di Architettura, Sapienza University of Rome, Rome, Roma, Italy

Igor Vojnovic, Department of Geography, Michigan State University, East Lansing, MI, USA

Claudia van der Laag Yamu, Oslo, Norway

Qunshan Zhao, School of Social and Political Sciences, University of Glasgow, Glasgow, UK

The Urban Book Series is a resource for urban studies and geography research worldwide. It provides a unique and innovative resource for the latest developments in the field, nurturing a comprehensive and encompassing publication venue for urban studies, urban geography, planning and regional development.

The series publishes peer-reviewed volumes related to urbanization, sustainability, urban environments, sustainable urbanism, governance, globalization, urban and sustainable development, spatial and area studies, urban management, transport systems, urban infrastructure, urban dynamics, green cities and urban landscapes. It also invites research which documents urbanization processes and urban dynamics on a national, regional and local level, welcoming case studies, as well as comparative and applied research.

The series will appeal to urbanists, geographers, planners, engineers, architects, policy makers, and to all of those interested in a wide-ranging overview of contemporary urban studies and innovations in the field. It accepts monographs, edited volumes and textbooks.

Indexed by Scopus.

Joaquim Farguell Pérez · Albert Santasusagna Riu
Editors

Urban and Metropolitan Rivers

Geomorphology, Planning and Perception

Springer

Editors
Joaquim Farguell Pérez
Department of Geography
University of Barcelona
Barcelona, Spain

Albert Santasusagna Riu
Institute of Water Research
University of Barcelona
Barcelona, Spain

ISSN 2365-757X ISSN 2365-7588 (electronic)
The Urban Book Series
ISBN 978-3-031-62643-2 ISBN 978-3-031-62641-8 (eBook)
https://doi.org/10.1007/978-3-031-62641-8

© The Editor(s) (if applicable) and The Author(s), under exclusive license to Springer Nature Switzerland AG 2024

This work is subject to copyright. All rights are solely and exclusively licensed by the Publisher, whether the whole or part of the material is concerned, specifically the rights of translation, reprinting, reuse of illustrations, recitation, broadcasting, reproduction on microfilms or in any other physical way, and transmission or information storage and retrieval, electronic adaptation, computer software, or by similar or dissimilar methodology now known or hereafter developed.
The use of general descriptive names, registered names, trademarks, service marks, etc. in this publication does not imply, even in the absence of a specific statement, that such names are exempt from the relevant protective laws and regulations and therefore free for general use.
The publisher, the authors and the editors are safe to assume that the advice and information in this book are believed to be true and accurate at the date of publication. Neither the publisher nor the authors or the editors give a warranty, expressed or implied, with respect to the material contained herein or for any errors or omissions that may have been made. The publisher remains neutral with regard to jurisdictional claims in published maps and institutional affiliations.

This Springer imprint is published by the registered company Springer Nature Switzerland AG
The registered company address is: Gewerbestrasse 11, 6330 Cham, Switzerland

If disposing of this product, please recycle the paper.

Glide, playful waves, and murmur softly!
No, rush quickly,
Making the banks and cliffs resound
frequently!
The joy that stirs our waters,
Which moves every wave to surge,
Rips through the barriers,
Set by astonishment and shyness.

Johann Sebastian Bach, Chorus from
Cantata BWV 206,
Glide playful waves, and murmur softly!
1735

Preface

The concept of the Anthropocene, conceived from the renewed, invigorating perspective afforded by geography, invites us to reflect on the role played by human societies in the history of the biosphere and our impact on its subsystems. The human footprint on the hydrosphere is increasingly apparent, and evidence of this is patent in the effects—both direct and indirect—on the quality, quantity and availability of the Earth's water. Today, and in the coming decades, our planet's natural phenomena, intensified by processes of climate change, will be subject to increasingly critical circumstances. The resulting global uncertainty needs to be carefully assessed and considered when deciding what measures need to be taken in all spheres, be it scientific, technical, social and political.

It is against this backdrop that the current book turns its attention to the urban rivers that flow through our cities and, at the broader scale, our metropolitan territories. Urban rivers have been defined from many different perspectives, but might best be characterised as those ecosystems of permanent flowing water that have been intensely altered by human transformations. These are rivers that have lost their so-called 'espace de liberté', their floodplains having been largely urbanised, their water courses channelised and transformed into linear spaces, and their riverbeds diverted and, even, eliminated. Urban rivers have been placed at the service of society's economic and extractive needs, and these impacts manifest themselves in a multiplicity of ways, dependent fundamentally on the technical-political vision of water afforded in each given place and moment in time.

The study of urban rivers poses challenges of great interest for geography. First, as a holistic and synthetic science, it allows us to address the relationship between society and nature from a range of different perspectives, which we bring together in this book by incorporating research conducted in the different branches of the discipline: including physical geography and geomorphology, urban geography, spatial planning and territorial organisation, regional geography, the geohistorical study of landscape and the geography of tourism. To these, we incorporate the confluent perspectives provided by the environmental sciences, ecology, hydrology, water engineering and urban planning. All these fields of study converge to offer, in the course of the book, a global vision of urban rivers.

Second, the very configuration of these spaces, hybrids of the natural and the artificial, shows the need to address the study of their complexity, overcoming sectoral visions that have historically prioritised the analysis of territories in isolation of their anthropic impacts. Urban rivers reflect the new reality of the Anthropocene and we cannot ignore this fact nor consider it devoid of scientific interest, no matter how transformed, urbanised and altered they may be. Urban rivers deserve to be reconsidered, revalued and rethought, above all as an integral part of the strategic planning conducted by government entities.

This book contains sixteen chapters that offer evidence of the need to consider urban rivers as an interdisciplinary field, in which geography has a leading role to play. To reflect this, the book adopts a transversal perspective and is divided into three main sections: **Geomorphology** (four chapters), **Planning** (six chapters) and **Perception** (five chapters). The studies of urban rivers concentrate mainly on southwestern Europe (France, Spain and Portugal), but cases are also included from Latin America. By way of introduction, chapter 1 (Olcina, Farguell & Santasusagna) summarises the particular problems in each of these three areas and brings to the table the main questions of debate that are then addressed more specifically in the following chapters.

The first section of the book, dedicated to the **Geomorphology** of urban rivers, emphasises the fact that river courses continue to erode, transport sediment and undergo variations in their flow regimes despite the reduction in their river space. It also describes and analyses interventions to mitigate, minimise or avoid these hydrogeomorphological processes. Thus, chapter 2 (Francos & Sánchez) describes the significance of the flood risk factor and the increasing vulnerability of urban areas to it, while chapter 3 stresses the effects of this hazard in urban areas subject to the growth brought by tourism, as illustrated by the case of the island of Mallorca (Rosselló & Grimalt). Chapter 4 (Farguell, Ochoa & Chavez) highlights the existence of these hidden or mitigated river processes when undertaking actions of river restoration, while chapter 5 (Yuste & Martínez) provides examples of how degraded river spaces can be recovered and reports the successful outcomes of such interventions.

The second section of the book examines the **Planning** processes carried out by urban, metropolitan and regional public entities when designating land uses and drawing up technical projects, sustainability policies and strategies for the revitalisation of urban river spaces from both a current and a historical evolutionary perspective. Chapter 6 (Rode) reports the experiences of small and medium river cities in France, to shed light on the processes of deurbanisation, re-naturalisation and adaptation to the challenges of global change. Chapter 7 (Santasusagna) examines the conception that city authorities have of urban rivers, especially when opting to implement the *gardenscape* model that has been so widely copied around the world. Chapter 8 (Valette & Hatvany) focuses specifically on several medium and large cities in the Garonne river valley and, from a geohistorical perspective, shows the evolution and adaptation of these cities to the river. Chapter 9 (Portugués) undertakes a case study of the city of Valencia and shows how a change in vision and awareness on the part of the city hall has guided the metamorphosis of the river Turia over the last half century. From a broader metropolitan scale, chapter 10 (Dournel) highlights the diversity of strategies adopted in the river cities that form part of Île-de-France region, with a particular concern for its wetlands. Chapter 11 (Rendón, Zúñiga & Santasusagna) turns its attention to the Huatanay in the Peruvian city of Cusco, a river whose extreme alterations and high levels of pollution mean its sustainable management faces an uncertain future.

Finally, the third section of the book explores society's **Perception** of the urban river and reflects on considerations of its importance and utility within the framework of the city. How is the passage of the river perceived? Does society consider it a hazard or an asset? And how does society perceive the river restoration measures along their courses in recent decades? Chapter 12 (Bonifácio) focuses on advances in neurourbanism and the consideration of river spaces as positive and necessary for the good mental health and quality of life of urban dwellers. Chapter 13 (Ollero, Albero, Boné, Díaz-Morlán, Pirchi & Marchioro) illustrates an example of the social perception of three rivers that converge in the city of Zaragoza and highlights the different impact society can have on each of them. Chapter 14 (Pavón, Benages-Albert, Vall-Casas, Garcia & Ribas) reflects on the transformations that have taken place in Barcelona's area of influence and which have facilitated a marked improvement in the characteristics and values of the river landscape. Chapter 15 (Cuello) makes a strong case for the need to use the river as an element of learning and perception in environmental education, in order to create a society that is aware of, and sensitised and committed to, the health of its urban rivers. Chapter 16 (Tort & López) turns its attention to hydronymy as a way of knowing and discovering society's perception of urban river environments.

In short, this book presents a series of concepts, ideas and case studies that facilitate an understanding of the geomorphological processes that characterise our cities' intensely transformed rivers; the plans, projects and policies of territorial and urban planning associated with these ecosystems; and, also, the way in which the urban river is perceived by citizens and river stakeholders.

It is our hope that this book will be useful not only for geographers interested in the river–city relationship and all the different approaches that might be adopted in its study, but also for those in other disciplines and professions that work in river environments. Our objective has been to leave a record of a very specific reality: urban rivers have enormous untapped potential, and if we can improve them and allow them to recover, then they can provide greater well-being in densely urbanised environments with their evident need for more open space.

Barcelona, Catalonia (Spain)
December 2023

Joaquim Farguell Pérez
Albert Santasusagna Riu

Acknowledgements

This book has benefitted from the scientific and economic support of Grant 2021SGR00859 awarded by the Agency for the Management of University and Research Grants (AGAUR) of the Catalan Government of the *Generalitat* (SGR 2021–2024).

The editors want to acknowledge all author's contributions.

Contents

Contributors

Laura Albero Department of Geography and Regional Planning (Climate, Water, Global Change, and Natural Systems Research Group), University of Zaragoza, Zaragoza, Spain

Marta Benages-Albert UIC Barcelona School of Architecture, Universitat Internacional de Catalunya, Barcelona, Spain

Pedro Boné TYPSA Aragón, Zaragoza, Spain

Ana Bonifácio Associate Laboratory TERRA, Centre of Geographical Studies, Institute of Geography and Spatial Planning, University of Lisbon, Lisbon, Portugal;
Institute of Physiology, Lisbon School of Medicine, University of Lisbon, Lisbon, Portugal

Jhesibel Chavez Department of Geography, GRAM (Grup de Rercerca Ambiental Mediterrània), University of Barcelona (UB), Barcelona, Spain

Agustín Cuello Gijón Fundación Nueva Cultura del Agua (FNCA), Education Commission, Cádiz, Spain

Jaime Díaz-Morlán Department of Urban and Territorial Planning, Engineering and Architecture School, University of Zaragoza. Atalaya Territorio, S.L, Zaragoza, Spain

Sylvain Dournel CEDETE Laboratory, University of Orléans (FR), Orléans, France

Joaquim Farguell Pérez Department of Geography, GRAM (Grup de Rercerca Ambiental Mediterrània), Barcelona, Spain;
Water Research Institute (IdRA), University of Barcelona (UB), Barcelona, Spain

José Anastasio Fernández-Yuste Department of Forest and Environmental Engineering and Management, School of Forest Engineering and Natural Resources, Universidad Politécnica de Madrid, Madrid, Spain

Marcos Francos Department of Geography, Faculty of Geography and History, University of Salamanca, Cervantes S/N, 37002 Salamanca, Spain

Xavier Garcia Catalan Institute for Water Research (ICRA-CERCA), University of Girona (UdG), Girona, Spain

Miquel Grimalt-Gelabert Departament de Geografia, Universitat de Les Illes Balears (UIB), Palma de Mallorca, Spain

Matthew Hatvany Department of Geography, Laval University, Quebec, Canada

César López Leiva School of Forest Engineering and Natural Resources (ETSI), Polytechnic University of Madrid (UPM), Madrid, Spain

Eberval Marchioro Department of Geography, Federal University of Espírito Santo (UFES), Vitória, Brasil

Carolina Martínez Santa-María Department of Forest and Environmental Engineering and Management, School of Forest Engineering and Natural Resources, Universidad Politécnica de Madrid, Madrid, Spain

Lucero Ochoa Department of Geography, GRAM (Grup de Rercerca Ambiental Mediterrània), University of Barcelona (UB), Barcelona, Spain

Jorge Olcina Cantos Department of Regional and Physical Geography. Climate and Regional Planning Research Group, University of Alicante (UA), Alicante, Spain

Alfredo Ollero Department of Geography and Regional Planning (Climate, Water, Global Change, and Natural Systems Research Group), University of Zaragoza, Zaragoza, Spain

David Pavón Department of Geography, SAMBI (Grup de Recerca en Canvi Socioambiental), University of Girona (UdG), Girona, Spain

Valeria N. Pirchi Department of Geography and Regional Planning (Climate, Water, Global Change, and Natural Systems Research Group), University of Zaragoza, Zaragoza, Spain

Iván Portugués Mollà Department of Geography. Section of Physical Geography, University of Valencia (UV), Valencia, Spain

Sisko Fernando Rendón Cusi Faculty of Engineering and Architecture, Andean University of Cusco, Cusco, Peru

Anna Ribas Department of Geography, SAMBI (Grup de Recerca en Canvi Socioambiental), University of Girona (UdG), Girona, Spain

Sylvain Rode Department of Geography and Planning, UMR 5281 ART-Dev, University of Perpignan Via Domitia (UPVD), Perpignan, France

Joan Rosselló-Geli Estudis d'Arts I Humanitats, Universitat Oberta de Catalunya (UOC), Barcelona, Spain

Carlos Sánchez-García Department of Geography, Autonomous University of Madrid, Calle Francisco Tomas y Valiente, 1. 28049, Madrid, Spain

Albert Santasusagna Riu Department of Geography. GRAM (Grup de Rercerca Ambiental Mediterrània) and Institute of Water Research (IdRA), University of Barcelona (UB), Barcelona, Spain

Joan Tort Donada Department of Geography. GRAM (Grup de Rercerca Ambiental Mediterrània) and Water Research Institute (IdRA), University of Barcelona (UB), Barcelona, Spain

Philippe Valette Department of Geography, Planning and EnvironmentGeode UMR 5602 CNRS, University Toulouse Jean Jaurès, Toulouse, France

Pere Vall-Casas UIC Barcelona School of Architecture, Universitat Internacional de Catalunya, Barcelona, Spain

Juan José Zúñiga Negrón Faculty of Engineering and Architecture, Andean University of Cusco, Cusco, Peru

Chapter 1
Urban and Metropolitan Rivers: Current Processes, Trends and Challenges

Jorge Olcina Cantos, Joaquim Farguell Pérez, and Albert Santasusagna Riu

Abstract What is an urban river? What physical factors characterise it? What social implications does it have, above all, for urban planning and risk management? And, what historical, heritage and cultural values can we ascribe to it? These and other questions are specifically addressed in this chapter, with the aim of responding in a concise, structured manner to the objectives that this book sets itself. The urban river presents a quite specific geomorphology, one subject to notable alterations as the result of human transformations in the course of history. Thus, to the changes wrought by the civil works that sought the domestication of river channels as they run through the urban fabric have been added urban-territorial planning actions that seek the most appropriate form of technical-political management, measures that have marked much of the evolution of urban rivers. Social perceptions of the urban river, moreover, are critical for understanding the attitudes, values and influences that derive from their cultural representation at each moment in history. Urban rivers today are an object of study of great geographical interest, best analysed by adopting an integrative perspective that concerns itself with the physical, social, economic and cultural factors that come together in the river territory.

Keywords Hydro-geomorphological dynamics · Structural works · Urban and territorial planning · Historical change · Social perceptions · Heritage value

J. Olcina Cantos (✉)
Department of Regional and Physical Geography. Climate and Regional Planning Research Group, University of Alicante (UA), Alicante, Spain
e-mail: jorge.olcina@ua.es

J. Farguell Pérez · A. Santasusagna Riu
Department of Geography, GRAM (Grup de Rercerca Ambiental Mediterrània), Barcelona, Spain

Water Research Institute (IdRA), University of Barcelona (UB), Barcelona, Spain

J. Farguell Pérez
e-mail: jfarguell@ub.edu

A. Santasusagna Riu
e-mail: asantasusagna@ub.edu

© The Author(s), under exclusive license to Springer Nature Switzerland AG 2024
J. Farguell Pérez and A. Santasusagna Riu (eds.), *Urban and Metropolitan Rivers*, The Urban Book Series, https://doi.org/10.1007/978-3-031-62641-8_1

1.1 Introduction

Providing a definition of an *urban river* is far from straightforward, despite numerous studies of the effects that urbanisation has on rivers and on fluvial systems in general (Wolman 1967; Paul and Meyer 2001; Schumm 2005; Gregory 2006). Indeed, Durán et al. (2020), in seeking to define the urban river, highlight the absence of any clear concept of the phenomenon but, at the same time, identify a number of different perspectives from which a definition might usefully be established. Thus, they propose a conceptualisation of the urban river based on the union of its physical and geomorphological features—including the river channel and adjacent areas—with those aspects derived, in the specific case of Spain, from regulations governing the *dominio público hidráulico* or public water domain, and which specify what should be understood by the river environment, including its floodplains. In short, the authors provide a definition that combines aspects of a river's physical geography with those related to its management and prevailing legislation (Durán et al. 2020).

This definition holds in the case of Spain, thanks to the legal definition of the *dominio público hidráulico*; however, in the English-speaking world, rivers are deemed *natural* up to the point at which the deforestation of part of a river basin for its subsequent urbanisation changes the land cover and, in so doing, transforms the behaviour of the river, modifying the basin's hydrological and geomorphological processes—thus, a *natural* river channel becomes *urbanised*. In short, the changes made in the basin have an impact on the fluvial system downstream of the point of modification, precisely because the basin conditions have been modified (Wolman 1967; Gregory and Chin 2002; Gregory 2006).

What the two definitions have in common is the importance conceded to hydro-geomorphological processes, given that the specific morphology of a channel depends on the slope of the terrain, the fluvial regime and the occurrence of extreme hydrological events (floods) and the magnitude and characteristics of the sediment transported, all of which is determined by the characteristics of the basin's climate and lithology, as well as its land cover (Schumm 2005).

Urban rivers are, in conclusion, rivers that have experienced a transformation of their channel and adjacent land cover as a result of processes of construction and urbanisation that alter their catchment. Such a transformation is characterised by intense modifications of the erodible corridor as a result of major civil works that seek to prevent the river channel from behaving naturally: that is, from forming its own riverbed, adapting to the valley slope and shaping the floodplains that make up the channel's migration zone (Kondolf et al. 2012).

1.2 The Main Alterations Made to Rivers in Urban Areas

Urban or *urbanised* rivers have been subject to significant alterations so that any similarities with *natural* rivers are few, in terms of both their morphology and function, which include given features of the river ecosystem and the hydro-geomorphological processes inherent to the river's work. The urbanisation of the geographical space simplifies the complex river structure to that of a drainage channel, normally to facilitate a series of urban advantages, but in the process the characteristics of the fluvial system are destroyed (Kondolf et al., 2012).

In terms of their morphology, urban rivers are tamed and channelised so as to segregate them from their own floodplain, and they are artificially straightened in order to reduce the number and length of meanders (i.e., their sinuosity) and, thus, increase the channel slope (Chin 2006). These far-reaching modifications are carried out with the goal of avoiding the flooding of the spaces colonised for urban and industrial expansion, but also of accelerating the flow so that, during an eventual flood, the water circulates as quickly as possible (Kondolf et al. 2012). Apart from these modifications to the channel, the width of an urban river tends to increase (Chin and Gregory 2005; Chin 2006; Kondolf 2012); however, in some cases and for reasons of space, the width of a river has been known to decrease, the case, for example, of the river Congost in Granollers (Barcelona, Spain) (Farguell et al. 2022).

But channelisation is not solely responsible for the loss of shape and the disruption of the hydro-geomorphological dynamics suffered by urban rivers; there are other factors that influence this degradation. Wolman (1967) was the first to describe how the processes of urbanisation—which he broke down into distinct phases—impact this dynamic. Thus, starting from a supposed steady-state equilibrium, the construction or urbanisation phase is characterised by increased erosion rates due to exposure of a bare surface, which results in an increase in the sediment yield accumulating in the channel. Once construction is complete, there is an obvious increase in impervious surfaces, which means infiltration is virtually eliminated. This leads to an excess of runoff water that is quickly drained into the river via a system of gutters and drains, reducing the lag between the precipitation episode and the river's peak discharge (Dunne and Leopold 1978; Gregory and Chin 2011). As a result, a change is wrought in the magnitude and frequency of flood events, and in the flood return period, which also has consequences for the river ecosystem (Poff et al. 1997).

These alterations can also be related to changes in the fluvial regime and in the quality of the circulating water by changing the frequency of low flows, as wastewater treated in plants is poured into the river.

The runoff from impervious urban areas introduces large amounts of nutrients, pollutants and heavy metals into rivers. In heavy rainfall events, effluents overwhelm the sewage system and solid waste is discharged and transported in the rivers. This not only impacts water quality but also the quality of the river sediments, which may contain organic contaminants. These are transported and, once sedimented, can modify the quality of the floodplains and beaches (Kevin et al. 2008).

Fig. 1.1 Undercutting of flood protection walls on the river Congost in Granollers (Barcelona, Spain). *Source* Joaquim Farguell (October) 2022

It is also common in urban areas that the bed of the river is eroded due to the low sediment load. An increase in the water flow energy tends to result in the erosion and incision of the waterbed, posing an obvious hazard to the infrastructure built along the channel—most typically bridges—but also the channel infrastructure itself, which is at risk of caving in during repeated flood episodes (Fig. 1.1).

Despite the significant alterations suffered by river channels in the urban environment, measures can be taken to improve their hydro-morphology and the quality of their waters. Current understanding of the hydro-geomorphological dynamics of river systems, as furnished by the geographical discipline of fluvial geomorphology, has led to numerous river restoration actions (Gregory et al. 2008; García et al. 2021). And although such actions are unlikely to restore the river space to its condition prior to human intervention (Dufour and Piégay 2009), restoration based on the recovery of hydro-geomorphological processes is deemed the best form of management to recover the space's functions (Fig. 1.2), as well as river habitats and the fluvial ecosystem in general (Bernhart and Palmer 2007), and also the most likely to guarantee success (Wohl 2015). The recovery of hydro-geomorphological processes and the restitution of a river regime in accordance with natural dynamics should lead to the river ecosystem *healing itself*, in a process whereby the river acts as the agent that restores its space and maintains it over time (Kondolf 2011).

Fig. 1.2 Images of the river Congost (Granollers, Barcelona). In **a** the river is channelised and immobilised by parallel roads, while in **b** the hydro-geomorphological processes and the channel's morphology have been recovered following the elimination of these roads. *Source* Images taken from Google Earth (2021)

Apart from the recovery of these fluvial processes, improvements to drainage systems have also been studied with the aim of mitigating alterations to the channel caused by the excess runoff attributable to the impervious nature of urban areas. The reduction in channel discharge should favour the recovery of river habitats to conditions comparable to those of non-urban rivers (Anim et al. 2018). Similarly, recommendations have been made to adopt river restoration measures both for the recovery of natural river regimes and for seeking to recover a river morphology that is as natural as possible so as to restore the habitats and ecology of the river ecosystem (Anim et al. 2019).

Studies of this type are crucial for determining which elements or processes contribute to improving a river ecosystem given that a consensus has yet to be reached as to what *river restoration* actually means. For some, it is a form of river space management, while for others it means creating or maintaining some of the river's landscape qualities, or the building of structures to reduce flood risk, or improving the fish habitat and water quality or creating recreational areas (Wohl et al. 2015). Whatever the target pursued, river restoration only acquires sense when society demands an improvement of the river space and its environment, actions that have repercussions for an improvement in the quality of life of urban dwellers (Wohl et al. 2005).

This means that river restoration is perhaps best approached by adopting a holistic, transversal perspective of the river basin (Gregory and Chin 2002). Moreover, as human societies increasingly tend to concentrate in urban areas, urban ecology and the cultural and socioeconomic characteristics of urban dwellers are rapidly establishing themselves as a consolidated branch of research (Gurnell et al. 2007). Likewise, river restoration is also emerging as a branch of scientific research within geography, in which hydrology and fluvial geomorphology, together with spatial planning and management, have a leading role to play.

1.3 Urban Planning as a Tool: Recent River Transformations

1.3.1 The Historical Absence of Planning: The Achilles' Heel of Urban Rivers

Rivers have been exploited as a basic resource since historical times. Indeed, being able to access water both for personal consumption and for farming has meant urban areas, and their associated agricultural activities, have grown up adjacent to their courses. Yet, while rivers have been a fundamental resource in human development, they also represent a hazard when their rate of discharge is not as expected, be it because of an excess or deficit of water.

Until the second half of the twentieth century, any actions implemented in urban river channels were fundamentally structural in nature, a response to episodes of flooding that could be particularly destructive of human communities. After that date, the expansion or diversion of rivers, the building of levees and dykes and the creation of urban parks and river walks along their banks all became common actions in the urban rivers of developed countries. Indeed, regulations governing the planning of water, land and the wider territory all recognise the need to incorporate analyses of the associated flood hazard, analyses which today form part of a city's hydrological and urban-territorial plan.

As societies have advanced, attitudes towards urban river channels have undergone significant changes. From a resource deemed essential for supplying the water required by the city, urban rivers became a dumping ground for the waste generated by the urban environment, the economic activities that grew up along its banks and the river transport that plied their trade on its waters. In countries and regions without the sufficient technological or economic capacity, urban rivers quickly become vile channels of pollution. If, in addition, the fluvial behaviour of the urban channel presents frequent hydrological extremes (especially floods), then they become hazardous spaces devoid of any urban attraction. Only when river risk management practices and the quality of the river's waters can guarantee the existence of a healthy, safe river environment can river courses be revalued as quality urban spaces and riverfront programmes be deployed that include leisure spaces, river walks and urban reforms. Yet, to date, this only occurs in cities in advanced countries that have adopted the requisite environmental and urban regulations to implement the sustainable management of their rivers.

In a city's urban plan, the authorities have traditionally considered rivers as a complementary element—one of embellishment—of the city's system of spaces free from urban development. Only since the end of the twentieth century, with the adoption of sustainability as a guiding principle of spatial planning, has the urban river course been considered an important part of the urban fabric's green and blue infrastructure. As a result, riverfront renewal projects have been developed, with the creation of ecological corridors that value the importance of the geomorphological and biogeographic wealth of the river channel.

An additional concern is those river channels without water that occupy significant spaces in arid and semi-arid zones of the planet and in relation to which urban planning has been scarce. Indeed, on occasions, their existence has been ignored and they have been incorporated directly into the street map, as streets and avenues that are exposed to all the risks of the flood hazard. Such environments have been allowed to form owing to the absence of adequate protective regulations (generally the case in developing countries) or non-compliance with existing norms (generally the case in more advanced countries) (Fig. 1.3).

Urban planning has as its goal the definition of a spatial model that establishes present and future land uses in a city and its surrounding territory (or municipal area). In developed countries, this is a legally regulated process in which the objectives and requisites are first identified for the establishment of this spatial plan. In short, it is a regulated act where the land typologies defined by legislation (Land Use and Urban Planning Law) are clearly laid down. In these land categories, river channels form part of the system deemed free from urban development, that is, space that must be protected or whose uses—especially residential uses—are restricted due to its ecological value or the dangers posed by evident hazards.

Cities can, therefore, regulate the banks of a river course that crosses the urban fabric, but they cannot intervene in the public water domain, which is regulated by the water authorities. River channels, by definition, form part of this public domain and this space is subject to strict measures of environmental or hazard protection. What can be managed in a city's urban plan are the riverside areas, albeit taking into consideration any restrictions in use that may derive from the flood risk maps and the risk management measures provided for under law. In the EU, the water policy (60/2000) and the assessment and management of flood risk (60/2007) directives establish guidelines for the management of river courses in the territory, including urban areas. The quality of circulating waters and the reduction of flood risks by limiting the occupation of riverside spaces underpin all territorial and urban planning actions. In less advanced countries, the absence of norms regulating territorial processes condemns river courses to be spaces of pollution and results in the hazardous occupation of river banks and of the channel itself; in short, insalubrious areas highly vulnerable to flooding. And this despite initiatives and projects carried out in recent years by international organisations aimed at implementing the UN's sustainable development goal 6—that is, access to clean water and sanitation, and investment to ensure the management of hydrological extremes.

Fig. 1.3 Contrasting examples of territorial planning and the management of urban river channels: Above, a planned urban stretch of a river course in the river Segre park (Lleida, Spain); below, a heavily polluted, unplanned stretch of the river in the suburb of Dharavi (Mumbai, India). *Source* (top) https://urbanisme.paeria.cat/sostenibilitat/aigua/pla-del-riu-segre-a-lleida. Universal access; (bottom) https://es.123rf.com/photo_149854255_barrios-pobres-y-empobrecidos-de-dharavi-en-la-ciudad-de-mumbai.html. Universalaccess

1.3.2 Managing Extreme Hydrological Events: The Adaptation of Cities to River Floods

Flooding in the urban environment converts the water resource into a hazard. The flood risk can be quantified in terms of the specific hazard level represented by a river course and the degree of human occupation of its river environment. Flood management has evolved through various stages with the emergence of new techniques and methods for analysing the spatial risk. Thus, we have seen an evolution from the celebration of the flood in ancient societies, where river flooding was a method of natural fertilisation of the fields occupying the floodplain, to the dread generated by such catastrophic events among traditional societies—until the end of the Modern Age—with insufficient knowledge of engineering techniques that might *contain* the flood and guarantee human lives and property. The advances in hydraulic knowledge that occurred in the nineteenth and, above all, twentieth centuries paved the way for the development of high-impact structural actions in urban river courses, especially in developed countries. Floods are, in the words of White (1945), always "watery marauders which do no good, and against which society wages a bitter battle" to control. This, in short, is the old paradigm of flood management that places all its confidence in the effectiveness of large hydraulic engineering works (dam construction, channelisation, diversion, flood walls and dykes).

Since the end of the twentieth century, however, new theoretical and methodological approaches have emerged to manage the flood risk in urban areas that call into question the blind trust placed in such infrastructure. Moreover, catastrophic flood events—in terms of their economic toll and the number of lives claimed—in river channels controlled by civil works highlight the need for alternative risk reduction policies. The celebration of the Earth Summit in Rio de Janeiro in 1992, which led to the adoption of Agenda 21, a series of actions implemented eventually at the local level and the deployment of the European Territorial Strategy (1999), at the European level, to promote sustainability in land use planning have opened the door to the incorporation of risk management in territorial plans. This strategy has been adopted by developed and developing countries and regions alike, in the case of the latter with the help of international organisations and actions (United Nations, World Bank, development aid programmes) (Fig. 1.4).

In Europe, in the aftermath of the floods of the large central European rivers in 1997 and 2002, a new way of addressing the flood risk was adopted in the light of the ineffectiveness of the continent's major hydraulic works. The occupation of spaces at high risk of flooding by urban uses, services and infrastructure had resulted in considerable economic losses and numerous fatalities. The European Commission, in response, issued a flood risk management directive 60/2007 requiring European countries to draw up flood risk maps and management plans as part of their territorial planning measures. The Member States have assumed their responsibilities in this regard, while at the state level, the norms regulating the use of water supplies in the public domain have been improved and flood risk reduction plans have been introduced within territorial plans. In Spain, for example, the regions or autonomous

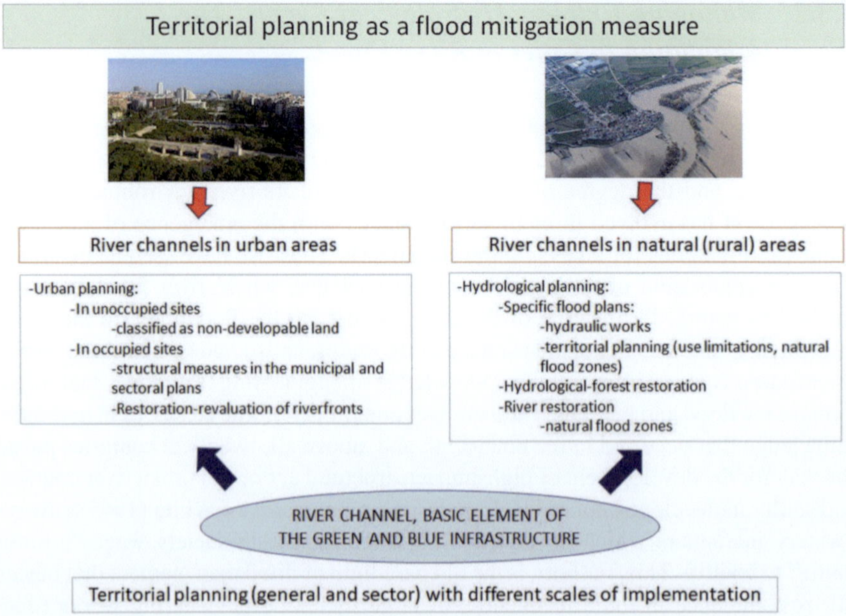

Fig. 1.4 Territorial planning as a flood mitigation measure. *Source* Authors

communities (e.g., Catalonia, Basque Country, Valencian Country, Balearic Islands, Andalusia) have played a leading role in the development of such policies. At the same time, the country's municipalities employ their urban planning tools to reduce the likelihood of the flood risk, while the maintenance of the river course as it flows through the urban centre is the responsibility of the town or city council. Land use legislation should provide precise regulations prohibiting the implementation of permanent uses in high-risk flood areas. In developed countries, urban planning regulations require that land at risk of flooding is not zoned as suitable for development, with official accreditation of this risk being demonstrated by the flood risk map, which has to be presented in any planning process. Here, it should be borne in mind that a hierarchy applies to the use of flood risk maps within prevailing territorial planning regulations, given that mapping may be undertaken by different tiers of government in accordance with the powers assigned to them. In Spain, for example, priority is given to the national SNCZI (*Sistema Nacional de Cartografía de Zonas Inundables*) maps over those of the autonomous communities (CC.AA. in their Spanish abbreviation) for the same mapped area (Fig. 1.5).

Territorial planning tools for flood risk mitigation have incorporated the idea of creating natural flood zones, or providing *room for rivers*, in urban centres to reduce the magnitude of floods. And, more recently, sustainable urban drainage systems (SuDS) have been built, adapted to cope with the intense rainfall events that can cause such damage to towns and cities. This design and construction of SuDS has accelerated in recent decades against the backdrop of climate change attributable to

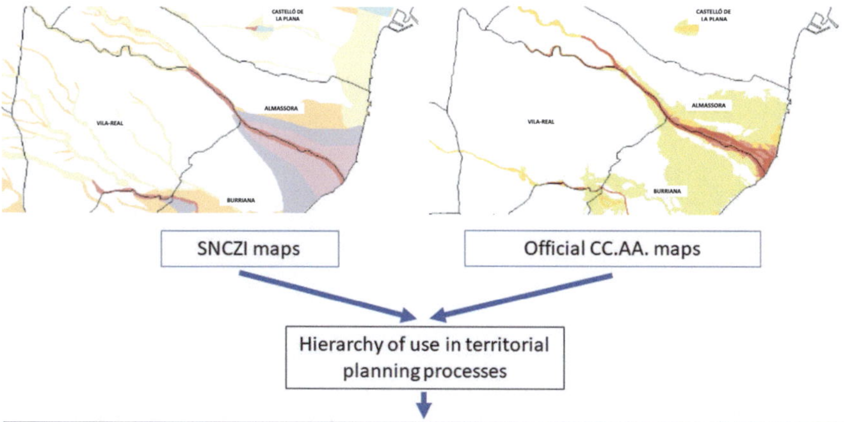

AREA OF APPLICATION	PRIORITY	COMPLEMENTARY USE
In areas mapped by both entities	SNCZI	Official CC.AA. maps (if available)
In areas not mapped by SNCZI	Official CC.AA. maps (if available)	Other official maps (civil protection)
In areas not mapped by any entity	Flood studies (regulations of CC.AA. sectoral plans, if available)	Other official maps (civil protection)

Fig. 1.5 Official flood risk maps in Spain for territorial planning. *Source* Authors

the anthropogenic greenhouse effect, which in many regions of the world is modifying the type of precipitation that falls, with marked increases in intensity. Some cities have built stormwater tanks and floodable parks to store excess flood water, thus reducing the vulnerability of certain urban areas. These waters, moreover, can also be incorporated into the urban water management cycle, following their purification and regeneration (Fig. 1.6).

Flood management in many urban areas of advanced countries is today based on hydrological information systems that use sensors for the continuous measurement of rainfall and river discharge data. With this information and real-time, small-scale weather predictions (using downscaling methods), flood warning systems can be designed for populations, together with emergency management plans. In Europe, for example, the EU-Alert system has recently been activated, based on the transmission of territorial warnings from official civil protection agencies via mobile telephone operators.

Finally, a measure that has a great impact but which has been employed little to date, even in advanced areas, is flood risk education based on the teaching in schools of disaster management processes and guidelines for action against natural hazards. Flood risk education programmes combined with the effective communication of such risks (using media and government sources) are essential for reducing the human risk factor, especially in hazardous geographical areas.

Fig. 1.6 Stormwater tank (SuDS) in Barcelona for the storage of urban surface runoff from rainfall events. *Source* Barcelona City Council. Stormwater tank plan (Universal access)

1.3.3 The Rediscovery of Riverfronts and the Social Connectivity of Urban Rivers. A New Opportunity for Green and Blue Infrastructure

Territorial and urban planning incorporating flood risk management measures has made possible the renewal and regeneration of riverfronts in many cities around the world. River channels, with a regular hydrology and steady flow, have become much prized green (or blue) ecological spaces that have to be protected in accordance with hydrological, environmental and urban planning regulations. These urban riverfront renewal programmes have generally been launched following extraordinary catastrophic flood episodes and, while precedents can be found in the modern age, most projects (of urban reform and expansion) have been developed in cities of the developed world since the mid-nineteenth century.

In Spain, for example, the main projects undertaken include the channelisation and beautification of the river Segura as it flows through Murcia (1785), and, later, in the twentieth century, the actions to channelise and divert the Guadalquivir in Seville, the Turia in Valencia, and various river channels in the Basque Country following the floods of August 1983 and the reaches of the river Segura between Murcia and its mouth in southern Valencia. In each case, the urban–river front was redeveloped as a space for city leisure and recreation. In some instances, these river regeneration programmes in urban environments were not a direct response to the flood risk, but rather to the need to reform and adapt these spaces to a city's new social needs as regards leisure and green areas. This is the case, for example, of the *Madrid Rio* project on the course of the river Manzanares (Fig. 1.7).

In this way, the river channel has become a basic element of a territory's green and blue infrastructure. In many territories, spatial planning today actually centres on the design of its green infrastructure, as a planning tool, dependent on the development of a territorial information system, that incorporates all those elements to be protected, or which have a limited use, as determined by the significance of their ecology and heritage, their socio-cultural importance and their risk value. Examples of best practices in the incorporation of the river channel into the green and blue infrastructure of a spatial plan can be found in the *Directrices de Ordenación del Territorio* (Territorial Planning Guidelines) of the Basque Country (2019) and the *Plan de Acción Territorial* (Territorial Action Plan) for flood risk reduction in the Valencian Country (2014) (Fig. 1.8).

1.4 The Perception of Urban Rivers as Social and Cultural Phenomena

As we have sought to stress, rivers occupy a privileged position in the development of human societies. Water is at the root of the organisation of humanised space, and rivers have served as flows of connection and communication between peoples and cultures; yet, as complex, diverse ecosystems subject to constant change, they have also been responsible for negative, at times, catastrophic events. Around the globe, the urban phenomenon rose up and was consolidated near bodies of water, and this has given rise to significant risks in the relationship between rivers and urban areas.

Yet, urban rivers, in addition to being considered interesting objects of study for geomorphology, can also be understood from a social perspective and, therefore, observed as cultural phenomena. The river–city interaction has given rise to an endless number of distinct scenarios: historical facts, artistic expressions, heritages, mythologies and memories, scenarios and values that contribute to forming a multiplicity of diverse views and perceptions of a given urban river landscape. Both from the perspective of their historical evolution and from that of the present day, urban rivers have been perceived as complex, attractive elements from a range of different viewpoints. In this brief section of the chapter, we outline the basic ideas that have sought to relate urban rivers and society's changing perception of them.

Fig. 1.7 Transformation of the Turia river channel in the city of Valencia: Above, the new channel in the south of the city (South Plan); below, recovery of the former Turia riverbed as it passes through the urban heart of Valencia and transformation into a green and leisure space for citizens. *Source* (top) Valencia City Council (Universal access); (bottom) José Luis Filpo (published under Wikimedia Commons licence)

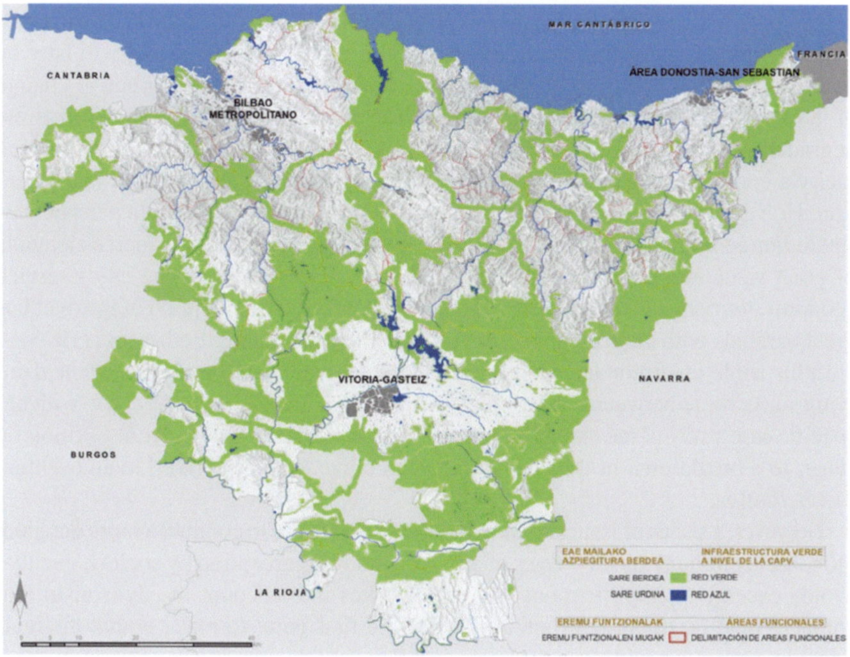

Fig. 1.8 Green and blue infrastructure in the autonomous community of the Basque Country. *Source Directrices de Ordenación del Territorio. Lurraldea 2040* (2019). Available at: https://www.euskadi.eus/directrices-de-ordenacion-territorial-dot/web01-a2lurral/es/

1.4.1 How Urban Rivers Have Been Perceived in the Course of Their History?

The passage of history has had a varied impact on the evolution of the river–city relationship, not solely from a utilitarian or productive perspective, referring exclusively to the way rivers have been transformed and used for specific purposes in urban contexts, but also in terms of society's perception of them. Indeed, perceptions of what a river is and what it is not, the purposes it does and does not serve, and the advantages and drawbacks of living near it are what have marked various stages of socialisation. Saraiva (1999, 2009) proposes an evolution in the human perception of urban rivers that develops through different phases, albeit without a fixed chronology, since it is by no means a globally shared process.

In Fig. 1.9, we seek to adapt and complement the ideas of this author, emphasising three fundamental questions that, in our view, ought to be highlighted in any such analysis: that is, the *attitude* of society towards a river (be it one of respect, adaptation, subordination and awareness), its *social value* (which, for practical purposes, derives from specific beliefs, uses and techniques, which are manifest in a technical-political agenda) and, finally, the *impacts* on the hydrosystem derived from this social perception. Broadly speaking, we identify a shifting, pendular relationship, which begins with the river's sacralisation, typical of pre-modern, pre-industrial societies, with environmental alterations of a much lower impact than those that will later be manifest with the emergence of industrialisation. The profound disruptive change of the transition from a traditional to an industrial society marks the transformation, degradation and abandonment of urban rivers. Rivers shift from being considered a city's commercial gateway to become marginal receptacles of waste and pollution. However, in the last half century, urban rivers have, from multiple perspectives (including ecological, social and cultural), once again come to be considered elements of great value and have been integrated into the development of the city's service economy, tourism and urban leisure. This worldview of the perception of the river has great parallels with contributions from other authors, including Bethemont (1993)—with his triple evolutionary vision of the *natural*, *engineered* and *discordant* river, lying somewhere between ecology and technology—and Gérardot (2004)—with her specific case study of the city of Lyon and the river Rhône, evolving from a symbiotic stage, to a breakdown in the contact between citizens and the river, to an eventual *reconciliation*.

However, a series of issues need to be stressed if we are to conduct a more detailed, global reflection on the city–river relationship and its perception by society, one that avoids excessive simplifications and which takes into account the dynamism that characterises the urban river environment. The first point to make is that the relationship cannot be understood from a rigid, inflexible standpoint. Although there is a tendency to see the relationship as entirely harmonious before industrialisation, the reality is that both positive and negative impacts have tended to coincide throughout history. The supposed duality of cities working *with* their rivers and cities working *against* their rivers is not exclusively marked by the technical and engineering improvements of the nineteenth century. As Carré and Deutsch (2015) point out, we need to take into account the means available, local needs, collective decisions and the multiple representations of the relationship with water and with nature. In short, the specific geographies of each river system, with their rich histories and with their changing stages, which lead to advances in some cases and, regrettably, to setbacks in others, whether or not they occupy the same territory (Penning-Rowsell 1997; Dournel et al. 2011).

Likewise, industrialisation should not be considered synonymous with unintelligent water management. The social hygiene movement and the management measures derived from the development of hydrologic engineering sought to respond to the actual needs of city populations, either by limiting stagnant bodies of water—one of the main sources of serious disease—or by discharging urban surface waters. Needs and their perception change to the point that, today, we speak of *river quality*. This concept is, clearly, a modern invention that directly impacts river management and identifies new objectives centred on the protection of the natural environment and its resources. Although from a present-day perspective it seems logical to strive for the conservation and proper functioning of the river ecosystem—as advocated by the scientific community—it has not always been like this. And, in fact, it is not a concept that is always taken into account today, since the transition from sectoral

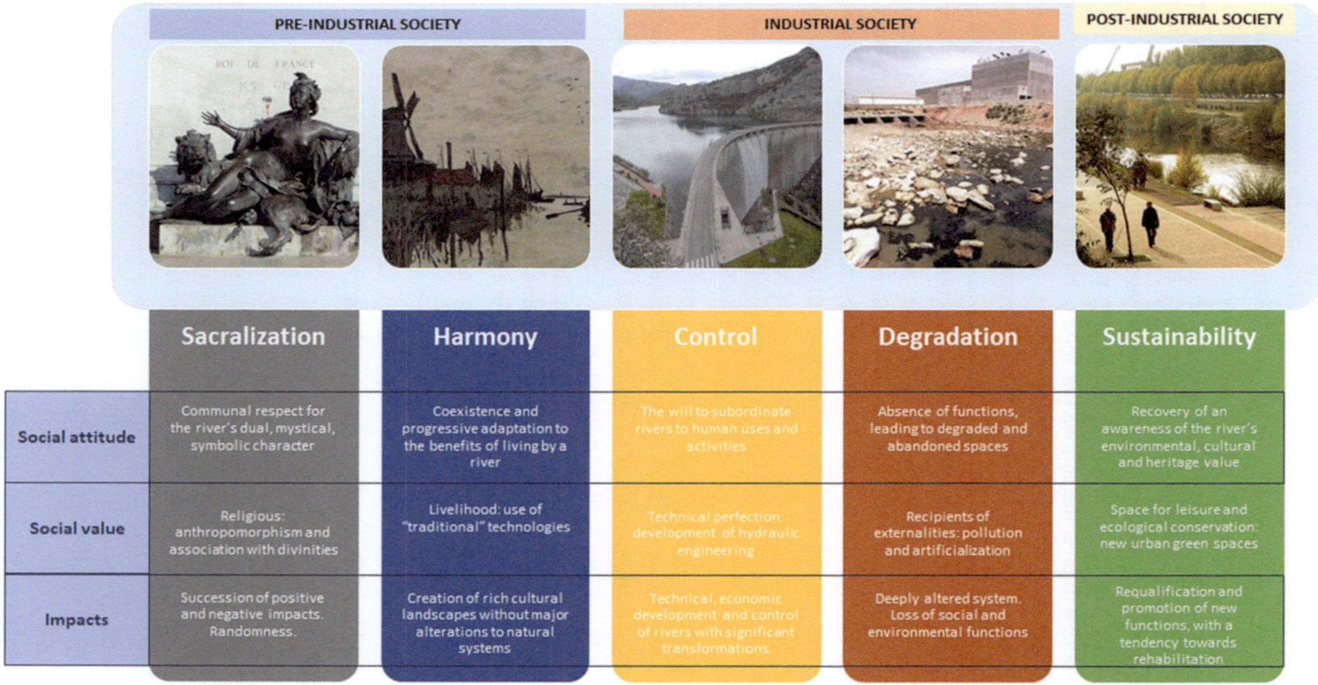

Fig. 1.9 Phases of socialisation in the historical evolution of the city–river relationship. *Source* Authors, based on Saraiva (1999, 2009) and Carré and Deutsch (2015)

to comprehensive policies has not occurred at the same time or in the same way throughout the world.

Moreover, the *sustainability* phase, attributed broadly to present-day society, needs to be examined critically, especially as regards the actual effectiveness of river restoration processes in convergence (or divergence) with the so-called social renewal or social revitalisation of urban rivers. Does the Anthropocene really represent a phase of sustainability for urban rivers? The answer is probably ambivalent with some processes promoting a re-naturalisation and the creation of green infrastructure and others that impoverish the landscape, prioritising economic interests (Rode 2023). There will be some actions that tend to repair critical scenarios derived from poor decisions taken by management entities, as well as new anthropogenic artificialisations and homogenisations of the landscape that are the result of impositions, negligence and/or urgent necessity. Looking back in time allows us to realise that this ambivalence has, to a greater or lesser degree, always existed. But what is different with respect to the past, and something that we should take very much into account today, is human society's capacity to transform the river environment.

What is required, therefore, is a non-linear approach to the study of rivers and cities that facilitates an understanding of the inherent complexity of the system itself (Bravard 2004), and to be sure that the means, needs and actors are always different, influential and dynamic. We must understand the relationship between rivers and cities as a system of interconnected elements, with continuities and discontinuities that directly influence human perception. And that, even today, this perception differs around the world, subject as it is to the specific interpretations afforded by the cultural filter, the historical heritage, (limited) decision-making power and the economic context. Only in this way can we understand, for example, that the sacredness of rivers is not exclusive to pre-industrial societies, and that the respect, admiration and veneration for rivers continue to be expressed to this day. A good example of this in Spain is the popular name given to the river Ebro: *padre Ebro* or father Ebro, a heartfelt, poetic way of referring to the largest river in the Iberian Peninsula.

1.4.2 Can Urban Rivers Be Managed on the Basis of Social Perceptions?

The social perception of a given object of study is a question that is of both interest and concern for researchers in the field of geography—that is, the social perception of geographical phenomena matters, and increasingly so, especially when this perception is at the root of the behaviour and responses of society to a specific reality. What perception do riverside dwellers have of their own river? Does the perception of the inhabitants of small river towns differ from that of the inhabitants of large river cities? From the citizens' perspective, how should an urban river be planned? Is the river viewed as *dirty*? Should it be subject to interventions that can improve its aesthetics? And what perceptions do citizens have of the flood risk?

These and similar questions are what, for several decades now, teams of researchers from all over the world have sought to answer by undertaking specific case studies (see, among many others, Llasat et al. 2008; Bodoque et al. 2016; Adomah Bempah and Olav Øyhus 2017; Ruiz-Villanueva et al. 2018; Arsénio et al. 2020). And, while it is necessary to ensure citizen participation and involvement in decisions that affect river management, from a strictly technical and scientific perspective, it is not always the case that a river needs to be re-naturalised or its ecosystem needs to be enhanced. The concept of the Anthropocene is not without its discordances and contradictions: just when mankind can be considered another engine of planetary development, with an enormous capacity to initiate change, societies are emerging that appear to be *disconnected* from the basic notions that might allow them to understand natural cycles. Urban rivers tend to be seen today as just another public amenity at the service of the city. By definition, such amenities should be functional and meet a series of requirements as identified by the population, including those of an aesthetic nature. A library, a hospital or a market should dispose of the necessary safety measures, offer convenience and provide a diversity of functions. But, as a rule, they do not incorporate a natural space, be it one that has been subject to varying degrees of transformation or urbanisation. The public amenity that most resembles an urban river is a park or a garden, but in most cases we are dealing with an area of domesticated nature, having been created within the city for this express purpose. For this reason, the management of an urban river is usually much more complex and requires cooperation between various public bodies, even if it is considered a specific type of park.

The fact of being able to enjoy a natural space within (or in the close vicinity of) a city can contribute to generating a feeling of constancy, permanence and an absence of risk, which could not be more removed from the reality of fluvial dynamics. Although the urban river in question may have undergone a total transformation and have been channelised, in episodes of rain and rising channel levels it can suffer quite significant changes, be they in water turbidity, biodiversity, smell or organic matter. Each of these changes may have an impact on the day-to-day lives of the *users* of urban rivers, resulting in a distorted perception of the processes, quality and reality of a river ecosystem. Likewise, interventions in urban rivers that favour simpler, more aesthetically pleasing transitions can generate a positive perception among citizens but, depending on the case, may well constitute significant alterations of the ecosystem and an increased flood risk.

However, cases have also been reported in which the response of citizens has been essential in avoiding extreme alterations and transformations of urban rivers. Disposing of accurate scientific data is basic to ensure the critical involvement of citizens in questions of an environmental nature. A good example from Spain, and one which can be considered representative of many around the world, is provided by the small city of Orihuela, in the Valencian province of Alicante (Fig. 1.10).

Following a serious episode of flooding in 1987, an exceptional measure was passed as part of the city's flood defence plan, to create a *bypass* for the river Segura as it runs through the historic centre. Eventually, however, and largely thanks to political and social opposition, the intervention was rejected, opting instead for new

Fig. 1.10 Channelisation of the river Segura in Orihuela (Alicante, Valencian Country, Spain). This intervention involved a process of urban reform—with the façades of the buildings constructed on the banks of the new river channel now facing the river—as well as an aesthetic overhaul. *Source* Jorge Olcina (2022)

river channels (Martí et al. 2013) which, in turn, has allowed the aesthetic and architectural renewal of the urban façade. Similarly, the preservation of the former Turia riverbed in Valencia and its conversion into urban parkland in the city river diversion that took place between the 1960s and 1970s is a further example of citizen pressure successfully attaining certain environmental goals (Portugués-Mollá 2017; Guia 2023). This begs the question: Should urban rivers be managed in line with citizen perceptions? The answer is clearly dependent on many factors. But one thing is certain, the opportunity that urban rivers represent for the environmental education of citizens is irrefutable. The fact of having a *proximity ecosystem* within the city itself contributes, beyond doubt, to improving the citizens' appreciation of their environment, thanks to the role played in this regard by schools, museums and the cultural entities of the local authorities.

1.5 Conclusions

Urban river courses are unique because of the route they must take as they run through the urban fabric. Yet, urban river channels are subject to alterations that seek their adaptation to a city's needs, and this often results in their transformation into spaces of risk, especially during episodes of extreme flooding. However, at the same time, their presence in the urban area means they have come to constitute part of both the natural and cultural heritage of the inhabitants of that urban centre.

Yet, the urbanisation of rivers has certain consequences for both its channel and fluvial regime. Higher rates of runoff resulting from the increase in the impervious nature of the urban fabric lead to alterations in the river regime, changes in the magnitude and frequency of floods and modifications in the type and magnitude of the river's sediment load. Moreover, the straightening of the channel and the disconnection of the river from its floodplain have undesired effects on urban rivers.

An understanding of the hydro-geomorphological processes that occur in a river is, as such, key for being able to manage these spaces, as well as for implementing successful restoration projects.

Perceptions of urban river channels have undergone marked changes as societies have advanced. From being viewed an essential resource supplying a city with its water, many urban rivers became a receptacle for the rubbish and waste generated in the urban environment. Traditionally, cities have considered their rivers as elements of the urban zoning system free from urban development, but only since the end of the twentieth century has the urban river course been deemed a relevant part of the green and blue infrastructure of the urban fabric, and have riverfront renewal programmes been implemented. Yet, rivers can become hazards due to extreme increases in their rates of discharge. The earlier flood management paradigm entrusted all its efforts to large hydraulic engineering works (dam construction, channelisation, diversions, flood walls, and dykes, etc.); but, since the end of the twentieth century, new theoretical and methodological approaches have emerged for flood management in urban areas that call into question the blind trust placed in such infrastructure. In this scenario, territorial and urban planning has become an essential risk mitigation tool, by controlling the types of land use that can be developed in the vicinity of the river course. The consideration of urban river channels as key elements of the city's green and blue infrastructure allows the land they occupy to be treated as protected spaces that have to be managed in accordance with the principles of sustainability and respect for the river. This approach to urban river channels seeks, above all, to highlight the significance of their ecology and heritage, their socio-cultural importance and their risk value.

If we look back in time—beyond the events of the last century and a half—it is possible to gain a geohistorical perspective of urban rivers that allows us to understand how the city–river relationship has evolved from earliest times. As the urban phenomenon has been consolidated, winning space and resources from ecological systems in the process, rivers have come to play a leading cultural role in their cities, becoming integrated (to varying degrees) into urban life. Urban rivers have served as commercial gateways, providing economic opportunities between peoples and cultures; as borders between nations and territories; as mystical and religious symbols; and even as artistic elements, veritable symbols of the landscape. This is how urban rivers have been seen, valued and perceived around the world—in every country, nation and region—from the multiple perspectives of coexisting societies. Broadly speaking, our understanding and cultural representations of urban rivers have evolved from an initial phase characterised by a scarce knowledge of the physical and ecological functioning of the river system to the adoption of radical processes of transformation that sought their domestication and adaptation to human needs, especially in Europe during its period of industrialisation. Technical and scientific advances have led to alterations of, and imbalances in, the city–river relationship, as well as, fortunately, to sound, environmentally based actions since the end of the twentieth century. Planners today question the social perception of urban rivers and the impact such attitudes might have on their efficient, sustainable management, and conclude that they may well be a double-edged sword. Thus, while thanks to increased citizen awareness, much progress has been made in the re-naturalisation of urban rivers, at the same time, a social perception devoid of technical and scientific information tends to simplify and reduce rivers to mere *gardens* at the service of consumer society.

Finally, the main conclusion to be drawn from this chapter is the very evident need for contemporary societies to respect urban river channels and their sustainable management. Urban rivers today constitute a fundamental element in the green and blue infrastructure of cities, and territorial planning should consider them critical for implementing programmes of conservation. Moreover, those reaches of rivers that run through the urban fabric should be subjected to as few alterations as possible. These are fundamental aspects in the respect demanded for riverside territories in a context that is today characterised by the need for sustainability and adaptation to climate change and its associated extremes.

References

Adomah Bempah S, Olav Øyhus A (2017) The role of social perception in disaster risk reduction: Beliefs, perception, and attitudes regarding flood disasters in communities along the Volta River, Ghana. Int J Disas Risk Reduct 23:104–108

Anim DO, Fletcher TD, Pasternack GB, Vietz GJ, Duncan HP, Burns MJ (2019) Can catchment-scale urban stormwater management measures benefit the stream hydraulic environment? J Environ Manage 233:1–11. https://doi.org/10.1016/j.jenvman.2018.12.023

Anim DO, Fletcher TD, Vietz GJ, Pasternack GB, Burns MJ (2018) Effects of urbanization on stream hydraulics. River Res Appl 34:661–674. https://doi.org/10.1002/rra

Arsénio P, Rodríguez-González PM, Bernez I, Dias FS, Bugalho MN, Dufour S (2020) Riparian vegetation restoration: does social perception reflect ecological value? River Res Appl 36(6):907–920

Bernhart E, Palmer M (2007) Restoring streams in an urbanizing world. Freshw Biol 52(4):738–751. https://doi.org/10.1111/j.1365-2427.2006.01718.x

Bethemont J (1993) La société au miroir du fleuve. In Piquet F (ed) Le fleuve et ses métamorphoses. Actes du colloque international tenu à l'Université Lyon 3—Jean Moulin les 13, 14 et 15 mai 1992. Lyon, Didier Erudition, 13–18

Bodoque JM, Amérigo M, Díez-Herrero A, García JA, Cortés B, Ballesteros-Cánovas JA, Olcina J (2016) Improvement of resilience of urban areas by integrating social perception in flash-flood risk management. J Hydrol, 541(A):665–676

Bravard JP (2004) Le façonnement du paysage fluvial de Lyon: Choix urbanistiques et héritages de l'histoire hydro-morphologique. Boletín De La Asociación De Geógrafos Españoles 37:3–16

Carré C, Deutsch JC (2015) L'eau dans la ville. Une amie qui nous fait la guerre. La Tour d'Aigues, Éditions de l'Aube

Chin A (2006) Urban transformation of river landscapes in a global context. Geomorphology 79:460–487. https://doi.org/10.1016/j.geomorph.2006.06.033

Chin A, Gregory KJ (2005) Managing urban river channel adjustments. Geomorphology 69:28–45. https://doi.org/10.1016/j.geomorph.2004.10.009

Dournel S, Franchomme M, Sajaloli B (2011) Géohistoire d'une résurgence d'eaux troubles: Les zones humides urbaines et les cités fluviales du Nord de la France (début XIXe-fin XXe s.). Revue Du Nord 26:169–188

Dufour S, Piégay H (2009) From the myth of a lost paradise to targeted river restoration: forget natural references and focus on human benefits. River Res Appl 25(5):568–581. https://doi.org/10.1002/rra.1239

Dunne T, Leopold L (1978) Water in environmental planning. 15th edition, Freeman (ed), 818

Durán, F.; Pons, J.J.; Serrano, M. (2020): ¿Qué es un río urbano? Propuesta metodológica para su delimitación en España. ACE: Architecture, City Environ 15(44), 9035. https://doi.org/10.5821/ace.15.44.9035

Farguell J, Chavez J, Ochoa L (2022) Anàlisi geomorfològica i ambiental del riu Congost al seu pas per Granollers. Unpublished technical report

García H, Ollero A, Ibasate A, Fuller IA, Death R, Piégay H (2021) Promoting fluvial geomorphology to "live with rivers" in the Anthropocene Era. Geomorphology, 380. https://doi.org/10.1016/j.geomorph.2021.107649

Gérardot C (2004) Les élus lyonnais et leurs fleuves: Une reconquête en question. Géocarrefour 79(1):75–84

Gregory KJ (2006) The human role in changing river channels. Geomorphology 79:172–191. https://doi.org/10.1016/j.geomorph.2006.06.018

Gregory KJ, Chin A (2002) Urban stream channel hazards. Area 34(3):312–321

Gregory KJ, Benito G, Downs P (2008) Applying fluvial geomorphology to river channel management: background FOR progress towards a palaeohydrology protocol. Geomorphology 98(1–2):153–172. https://doi.org/10.1016/j.geomorph.2007.02.031

Gregory KJ (2011) Wolman MG (1967) A cycle of sedimentation and erosion in urban river channels. Goegrafiska Annaler 49A:385–395; Progress in Physical Geography 35(6):831–841. https://doi.org/10.1177/0309133311414527

Guia A (2023) La rebel·lió dels vianants. El Jardí del riu Túria al centre d'una nova València. Valencia, Bromera

Gurnell A, Lee M, Souch C (2007) Urban rivers. Geogr Compass 1(5):1118–1137. https://doi.org/10.1111/j.1749-8198.2007.00058.x

Kevin G, Philip N, Batalla RJ, García C (2008) Sediment and contaminant sources and transfers in river basins. Sustain Manage Sediment Resour 4:83–135. https://doi.org/10.1016/S1872-1990(08)80006-2

Kondolf M, Podolak K, Grantham ET (2012) Restoring mediterranean-climate rivers. Hydrobiologia 719:527–545. https://doi.org/10.1007/s10750-012-1363-y

Kondolf M (2011) Setting goals in river restoration: when and where can the river "Heal Itself"? In Simon A, Bennet SJ, Castro JM (eds) Stream restoration in dynamics fluvial systems. Geopress, AGU, 29–43. ISBN: 978–0–87590–483–2

Llasat MC, López L, Barnolas M, Llasat-Botija M (2008) Flash-floods in Catalonia: the social perception in a context of changing vulnerability. Adv Geosci 17:63–70

Martí Ciriquián P, García Mayor C, Nolasco Cirugeda A (2013) Transformación urbana, espacio público y su percepción social. El río Segura a su paso por Orihuela y Rojales. Polígonos. Revista De Geografía 25:185–214

Paul MJ, Meyer JL (2001) Streams in the urban landscape. Annu Rev Ecol Syst 32:333–365

Penning-Rowsell E (1997) Rius i ciutats: amenaces i potencialitats. Documents D'anàlisi Geogràfica 31:23–34

Poff N, LeRoy, Allan D, Bain BM, Karr RJ, Prestegaard KL, Richter DB, Sparks ER, Stromberg CJ (1997) The natural flow regime. Bioscience 47(11):769–784.

Portugués-Mollá I (2017) La metamorfosis del río Turia en Valencia (1897–2016): de cauce torrencial urbano a corredor verde metropolitano. Departamento de Geografía de la Universidad de Valencia. Tesis doctoral, Valencia

Rode S (2023) Écologiser l'urbanisme. Pour un ménagement de nos milieux de vie partagés. Le Bord de l'Eau, Lormont

Ruiz-Villanueva V, Díez-Herrero A, García JA, Ollero A, Piégay H, Stoffel M (2018) Does the public's negative perception towards wood in rivers relate to recent impact of flooding experiencing? Sci Total Environ 635(1):294–307

Saraiva M (1999) O rio como paisagem. Lisboa, Fundaçao Calouste Gulbenkian, Fundaçao para a Ciência e Tecnologia
Saraiva M (2009) Cidades e rios. Problemas e desafios. In Saraiva M (ed) Cidades e Rios. Perspectivas para uma relaçao sustentável. Lisboa, Parque EXPO 98, 17–32
Schumm SA (2005) River variability and complexity. Cambridge University Press, Cambridge, United Kingdom, 220. ISBN: 978–0–521–84671–4
White GF (1945) Human adjustment to floods. A geographical approach to the flood problem in the United States. Research Paper nº 29. Chicago, University of Chicago
Wohl E, Lane SN, Wilcox CA (2015) The science and practice of river restoration. Water Resour Res 51:5974–5997. https://doi.org/10.1002/2014WR016874
Wohl E, Angermeier PL, Bledsoe B, Kondolf M, MacDonnel L, Merrit MD, Palmer AM, Poff L, Tarboton D (2005) River restoration. Water Resour Res 41:W10301. https://doi.org/10.1029/2005WR003985
Wolman MG (1967) A cycle of sedimentation and erosion in Urban River channels. Geografiska Annaler. Series A, Physical Geography, 49(2/4), Landscape and Processes: Essays in Geomorphology, 385–395

Part I
Geomorphology

Chapter 2
The Role of Floods in Urban Environments

Carlos Sánchez-García and Marcos Francos

Abstract Floods are the natural hazard that affects the largest population in the world. In addition, the demographic evolution moves from a population majority located in rural areas to urban areas and changes toward the city. Many of the main European cities, such as Paris, Vienna and Budapest, are located in fluvial plains alongside more or less mighty rivers. This geographical situation contributes to floods taking on a more localized impact, no longer solely determined by river discharge. Instead, the evolving flood patterns are increasingly influenced by the interaction between the discharge resulting from torrential rainfall and the soil's sealing, leading to a greater runoff. This means that the potential risk for loss of life is much higher than it was just a few decades ago. In this chapter, we analyze the evolution of urban flooding, and how the effects of large river floods have changed, and which mitigation measures have been taking place, together with the demographic increase that has occurred in these populations in recent centuries.

Keywords Urban flooding · Extreme rainfall · Flood risk · Fluvial geomorphology · Nature-based solutions

2.1 Introduction

Floods can cause a very high impact in urban environments, affecting people, economy and environment. That is why the public administrations in charge of territorial planning must take the potential risk of flooding in nearby areas into account. Exposure to floods has increased in recent decades and this is confirmed by studies

C. Sánchez-García (✉)
Department of Geography, Autonomous University of Madrid, Calle Francisco Tomas y Valiente, 1. 28049, Madrid, Spain
e-mail: carlos.sanchezg@uam.es

M. Francos
Department of Geography, Faculty of Geography and History, University of Salamanca, Cervantes S/N, 37002 Salamanca, Spain
e-mail: mfq@usal.es

© The Author(s), under exclusive license to Springer Nature Switzerland AG 2024
J. Farguell Pérez and A. Santasusagna Riu (eds.), *Urban and Metropolitan Rivers*, The Urban Book Series, https://doi.org/10.1007/978-3-031-62641-8_2

focused on the impact of floods at a historical level (Sánchez-García et al. 2019; Barriendos et al. 2019; Ribas and Saurí 2021; Sánchez-García and Schulte 2023). Economic damage has also increased on a global scale (EEA 2020), and yet vulnerability to risk is experiencing a decreasing trend because of a significant increase in the capacity to adapt to the risk (Angelakis et al. 2020). The economic damages caused by floods are being debated because, since in relative terms, one can speak of an increase in economic losses, in normalized way, the losses have stabilized or even decreased (Barredo 2009). On the other hand, already in the late 1970s, Burton et al. (1978) argued that economic damages are related to the level of development of a country or region. Taking this premise into account, Jongman et al. (2015) suggested that, indeed, the period between 1980 and 2010 coincided with an increase in the quality of life at the European average level and a decrease in mortality and a drop in economic losses, normalized in relation to GDP. Therefore, both the losses and the economic well-being of a region are intrinsically related (Schumacher and Strobi 2011).

This increase in exposure and decrease in vulnerability jointly with mortality must be explained in a context in which the European population has shifted from residing 50% in rural areas in 1950 to less than 15% in 2020 (EEA 2017). Thus, urban demography in Europe has increased along with exposure, in population terms: at the beginning of the nineteenth century, just over 169 million people lived in Europe (McEvedy and Jones 1978); while currently, more than 700 million inhabitants lives. At least 80% of them live in medium-sized cities (Lutz et al. 2019). In this way, Europe has become urbanized, with all the consequences that this entails. Thus, many of the cities were built around a river to take advantage of the benefits it provided, like energy, food, transportation and others (Elleder 2015; Wetter 2017), and now some of them need to change the way to watch the rivers its relationship.

Several studies have analyzed human motivation to occupy areas close to rivers in nineteenth century (Lowry and McCardle 1891) and more in depth from the end of twentieth century to the present (Griffin 1996; Munzwa and Wellington 2010; Biginagwa and Ichumbaki 2018; Manning et al. 2021). These studies address the history of human occupation of areas near rivers using different case studies in Europe, Africa and America. Among the most important motivations are access to water for domestic purposes (e.g., hygiene, washing, drinking, cooking), product manufacturing (e.g., steam production, cooling machinery) and agriculture (e.g., irrigation, soil fertility). In these occupations, the risk associated with these areas was underestimated, prioritizing the benefit of the proximity to rivers (Utami and Bisri 2014; Yang et al. 2022). The utilization of river resources by various societies was influenced not only by the sediments introduced through the rise in discharge but also by the periodic damage caused by floods (Brázdil et al. 2006; Sánchez-García and Francos 2022). Some studies indicate that floods will become more frequent and more seasonal, so adaptation to them must be greater (Blöschl et al. 2020). In this sense, the areas of greatest concern are those occupied by humans, with special emphasis on floods occurring in urban environments (Merz et al. 2021). Currently, the sealing of much of the soil in urban areas has changed the hydrological dynamics in many places. Areas that were traditionally affected by flash floods now suffer even more damaging consequences in particular areas, such as ravines, streams or *ramblas* (Barrera et al. 2006). This has led to increased attention over the years to the extreme events that cause these kind of floods and their trends (e.g., torrential rains) and some buildings surrounding the location of cities (e.g., illegal buildings) (Juma et al. 2021; Qiang et al. 2021). Therefore, an increasing number of studies have been carried out to measure flood risk and to establish protective measures to best manage urban rivers (Morita 2008; Chia et al. 2020).

The process of human occupation and the development of cities have been closely related to river courses (Bellintani and Saracino 2015; Grana et al. 2016). This has generated positive and negative synergies in both, leading to constant interactions between them and the coining of terms such as "urban rivers" (Everard and Moggridge 2012). The growth of the world's population and, the increase of cities, has resulted in many case in an increase of exposure. Furthermore, the natural environment of rivers has been disturbed and construction has taken place in areas where the river has a great influence and impact (Yamanouchi and Ishikawa 2011; Wei et al. 2017). This has given rise to the existence of cities whose dynamic distribution and structure cannot be understood without the existence of an important river, making the city and the river appear symbiotically united on many occasions (e.g., Vienna-Danube, Paris-Seine, London-Thames) (Sutton 2007; Hohensinner et al. 2013). These constructions modify the runoff characteristics producing that, in events of intense precipitation, the return periods are even shorter than those previously established in studies that did not contemplate the construction of these areas (Yao et al. 2016; Ciupa and Suligowski 2020; Patowary et al. 2020). This is even more relevant in environments such as the Mediterranean, where the ephemeral nature of river courses and the increasingly frequent torrential rains can lead to events of great danger for the population (Kosmas et al. 1997; Tuset et al. 2016). For this reason, chapters like this one, which highlight an international problem in the Mediterranean environment due to its climatic characteristics, are even more relevant and necessary.

Therefore, the aims of this chapter are to compile the latest studies on urban rivers, mainly in European Mediterranean basins. Three main objectives are established: (I) understand the relationship between cities and the rivers that flow through them; (II) analyze the possible flood risks associated with these urban settlements; and (III) synthesize the most appropriate management tools in each case, using the current existing bibliography.

2.2 Urban Rivers

Urban settlements have traditionally been located close to watercourses, but in locations that allow the use of the resource and, at the same time, safety from flooding (Calvo García-Tornel 1997). Over the years, the increase in the size of cities and the peaks in the maximum discharge rates of rivers are leading to a spatial confluence of urban phenomena and flood zones (Tierolf et al. 2021). This phenomenon occurs most notably in the rivers of the Mediterranean region where anthropic pressure is greatest (Vian et al., 2021). The evolution of cities and river dynamics has often resulted in a symbiotic relationship, creating clear interdependence and influencing the socio-economic dynamics of the regional system (Mahoney 1985).

Socio-economic interests are often given priority over respect for and preservation of the natural course of the river, that means that construction can take place in areas where the law does not allow it (e.g. floodplain) (Hohensinner et al. 2013). Several studies evidenced this new situation pointing out the violation of such legislation and the risk assessment carried out by specialists (de Rodrigues and de Faria 2009; Anisimov and Ryzhenkov 2016). These are not only huts, vegetable gardens or even permanent building structures, but also agricultural activities, gravel pits, buildings, etc. The occupation of the watercourse, which is the land covered by water at ordinary high-water levels of public domain, apart from being illegal in most cases, also causes serious problems for the river ecosystem, preventing its conservation or recovery (Janauer 2016).

The occupation of the river channel affects several river properties such as: the flow of water discharge, the level reached by the water sheet, the material transported by the river and the chemistry of the water (Bridgland 2000). River ecosystems are extremely vulnerable to the effects of urban development, being irreversible to recover in many cases, although they also have a great capacity for natural regeneration once the affectations and obstacles imposed by human intervention have been eliminated (Mondal and Satpati 2019). Furthermore, urban areas alter the environment through their effects on geomorphology, hydromorphology, hydrology, energy flows and biological communities, creating landscapes that are highly modified according to the conditions of equilibrium in the natural environment (Mount 1995). When urban development areas are inadequately planned, they can significantly alter the flow regime of rivers for different reasons such as water consumption needs and land sealing (Sala and Inbar 1992). Likewise, the shape of rivers can be altered by urbanization and land sealing processes, construction of hydraulic works (Chin and Gregory 2005).

Soils occupation causes serious risks due to the possibility of flooding. Thus, it can have important consequences for the population and the buildings in these areas (Munawar et al. 2021). Floods will become more frequent over the years (Rubinato et al. 2019; Nguyen et al. 2021) and due to the human intervention of these areas they will be increasingly catastrophic. These events must be considered when carrying out regional planning, ensuring that the rules are complied with and not taken up, and trying to ensure that the actions carried out to mitigate the risk are respectful with the environment. The protection of these areas must be much stronger at the legislative level, trying to avoid breaches of the law and taking into account studies carried out by experts. Regarding urban planning, to incorporate professionals and technicians from areas related to hydrology is essential, so that, based on general international consensus measures, they can adapt the standards to each specific location. To this end, they should consider various factors such as the rainfall regime of each area, the recurrence of floods and their magnitude, possible modifications of the upstream channel and thus adapt the area that may be flooded and delimit the areas of potential risk (Watson et al. 2020). To achieve this, it may be a good option to define the area around the river as a special protection area in which occupation is impossible (Junk et al. 2021).

2.3 Floods and Cities

In the past, urban centers were located in higher areas and outside the flood plain, leaving these for agricultural use. However, we currently find ourselves with a totally anthropized and altered environment, in which exposure to flood risk is higher than ever (Ribas and Saurí 2021). The cities, therefore, have gained surface from the river, and the river, in moments of torrential rains, follows old meanders that it already flowed before the sealing of the area and urbanization (Barrera et al. 2006; Benito et al. 2012; Martínez-Gomariz et al. 2019). In addition, the consequences of these floods are increased by, in many cases, inefficient river regulation measures, deforestation and the null planning of emergency plans for flood risks (Tomar et al. 2021). Floods caused by torrential rains exceed the capacity of drainage systems (Guerreiro et al. 2021). For that reason, to model the rainfall intensity necessary to create a torrential flow is being a study object in the last decades in Europe (Guerreiro et al. 2021).

Urban flooding is growing for numerous reasons. Apart from the increase in the population living in urban areas, the consequent increase in population density is a factor to consider (Remondia et al. 2016). For this reason, floods do not only occur in rivers where they were traditionally generated, but also in areas with ephemeral rivers or boulevards or streams that flow through the streets of the cities, are susceptible to being flooded (Martínez-Gomariz et al. 2019). The morphology of some cities in the Mediterranean area of Spain, Italy and France presents areas with high gradients and flat areas near the Mediterranean Sea. Moreover, the land has been strongly urbanized during the last decades in many cases. These characteristics, together with the Mediterranean rainfalls characterized by high intensity and short duration, leave some cities in a flood-prone situation.

Given that the risk of flooding, in recent decades, is being reduced due to mitigation measures, floodplains continue to be a very useful resource for urban, agricultural and/or industrial development (Hey 1997). Therefore, flood-mitigation measures such as channeling or the construction of dams directly affect the dynamic processes of transport and deposit of fluvial sediments throughout the river basin (Heckmann et al. 2018). Rivers have an unstable development in time series (hundreds of years), which is why the works dedicated to historical hydrology in urban areas are increasingly developed, using methods for calculating return periods in which the uncertainty increase with time (Herget and Meurs 2010; Gaál et al. 2010; Gaume 2018). In this way, the use of historical information can be helpful to understand the evolution of a territory, particularly in urban settings, since it allows us to know the behavior of the river on a much larger time scale (Sánchez-García et al. 2019). One of the positive aspects of using historical documentation is the large number of data sources available (technical reports, hydraulic works projects, meteorological warnings, handwritten notes, maps or photographs), and the study time, which in some cases can be studied from the twelfth century (Audisio and Turconi 2011).

Currently, the line to follow in many cities is to reduce the environmental impact they are having. However, the trend continues to be to increase population density in areas close to flood-risk areas, therefore, another of the objectives of some cities is to reduce this density and convert areas close to rivers into arable areas (UN 2014). The current situation is so alarming in some cities that 15% of the world's population lives in flood-risk areas, both fluvial and marine (Ligtvoet et al. 2014). Recently, severe flooding hit highly developed cities like Prague, Dresden and several other cities (2002), Bern and several other cities (2005), New Orleans (2005), Copenhagen (2010, 2011 and 2014) and New York (2012), as well as areas like Queensland (2010), South-western England (2013–2014) and the French Riviera (2015). The societal consequences are severe. In Europe only, the average cost of flood damage between 2000 and 2012 has been estimated to be about 4.9 billion euros per year. It is estimated that this figure may increase to about 23.5 billion per year by 2050, i.e. with almost 400% (Jongman et al. 2014).

The economic cost of floods is an argument to keep studying their genesis. Some recent studies on flash floods have allowed progress in some research areas. In the context of urban flooding, some authors have developed and improved flood and flash flood mapping methods, using GIS, as an alternative to hydraulic modeling (Popa et al. 2019), since it presents more inconsistencies in some areas (Samela et al. 2020). Flood modeling in Europe, and more specifically in European urban areas, has advanced a lot in recent decades, and some revisions have been carried out considering many of the data at the European level of flows and flood zones (Guerreiro et al. 2021). These challenges are relevant for the study of daily observed floods at the continental level as well as pluvial floods.

A good example of bad regional planning is the southeast years of the Iberian Peninsula, in the Province of Almería. In 2012, the Antas River overflowed its banks as a result of torrential rains at the end of September. The historical hydrological context of the Antas River Basin is that in 1973 and 1989 catastrophic floods occurred. The hydrological and paleohydrological records show that these types of events are relatively frequent, as observed in the stratigraphy of the paleoflood deposits, and the fact that there have been three events (1973, 1989 and 2012) of equal magnitude or higher in the last 40. During them, fluvial systems are activated both in ramblas (channels and flood plains) and in alluvial fans. Bad planning of urban areas has been denounced by various researchers (Sánchez-García et al. 2019; Sánchez-García and Schulte 2023). In this case, it is constituted by the towns of Puerto del Rey and Costa Laguna that expand near the flooding area of the Antas River in its section close to the end. In this area, there is an artificial dike that borders the riverbed and that has been exceeded by the flood by at least half a meter. The houses in the urbanization whose streets are 3m below the maximum level of the dike were flooded up to their first floor, with streets that functioned as drainage channels (Fig. 2.1).

Fig. 2.1 Flooding through one of the streets of the coastal urbanization of Vera Playa, in the Antas River mouth (28th September 2012) (*Source* El Ideal newspaper)

Adaptive and multifunctional infrastructure in combination with water sensitive urban design is seen as means to reinforce resilience against climate change (Ashley et al. 2013). However, incorporation of mitigation measures into decision-making and ways to handle integrative and multi-criteria aspects in the legal and organizational system are still largely undeveloped (Brown et al. 2009).

2.4 Flood-Risk Perception

Social perception of flood risk in urban areas is key role of hydrological studies (Tripathi et al. 2014; Yin et al. 2021). Despite this, these studies are very recent, the oldest with clear sources being carried out by Bruen and Gebre (2001) and the most recent by Elum and Lawal (2022). Thus, this is an unresolved problem for which there are still many to be carried out. Society, as previously mentioned, has been congregating around river areas for years. The growth of cities and the increase of the world's population mean that cities are becoming denser, and the potential risks are increasing (Rana et al. 2020). To avoid the catastrophe and potential risk that floods pose in densely populated urban environments, authors such as Rufat and Howe (2022) and Yin et al. (2022) propose risk management plans supported by maps to help plan cities and in case of an extreme event to carry out evacuation of potentially affected areas. Multi-criteria analysis can also help to manage this type of extreme events (Hong and Chang 2020). This study shows how the experiences lived by citizens are fundamental when it comes to knowing how they interpret a risk such as flooding. Other authors point out that the solution to mitigate the impact of urban flooding is to invest in Nature-based Solutions (NbS) and generate more protected and sustainable areas with the environment (Li et al. 2022) and the incorporation of social perception on flash flood-risk management (Bodoque et al. 2016).

Some authors have gone a step further and have estimated the urban damage that can be caused by flooding. In this context, Martínez-Gomariz et al. (2021) analyze the possible damage that a flood can produce in Barcelona. The authors show that they could predict the damage estimated from a comprehensive model of pluvial flood damage (SFLOOD). This type of study is extremely necessary in areas such as the Mediterranean, where floods and extreme events are going to be increasingly recurrent. These studies should also be focused on areas where the population density is higher and therefore society may be more affected. These kinds of studies should be considered in risk analyses and in urban planning and development, as well as in the EU Flood Directive in Spain, to manage the territory in the most sustainable way, considering the drift of events such as floods that are expected. Society's awareness of the need to respect the natural environment and specifically the riverside area is essential. Changes in the floodplain are not only caused by construction in the area near the riverbed, but also because it is full of all kinds of materials and organic or construction waste, often turning it into a dam during a flood. For this reason, to respect current legislation and carrying out appropriate urban planning by forming multidisciplinary teams, to make society aware of the need to respect these areas is important (Sutejo et al. 2019). This respect must be transmitted to the population, involving everyone in the possible disasters that may occur because of their actions. In some cases, many of these areas near rivers are rubbish dumps and this is punishable by fines and the obligation to repair the damage; therefore, the administrations are looking for ways to manage the waste that reaches the rivers so that it does not affect the morphology of the riverbeds (He et al. 2023).

2.5 New Tools for Flood-Risks Management in Urban Environments

After a flood event which causes catastrophic damage in an urban area, the population quickly experiences a lack of security and demands mitigation measures against the risk of another catastrophic event. Historically, this situation has been increasing in areas where urbanization and population pressure are positively correlated with the increase in floods, for this reason many resources have been allocated with the intention of reducing the risk of flooding (Pérez-Morales et al. 2021). Mitigation measures included the construction of dikes, walls or channeling of rivers in dangerous areas such as Murcia (Lemeunier and Picazo 1988) or Almería (Sánchez-García et al. 2019), the transfer of the final sections of the rivers that pass through the center of big cities, such as the case of the Túria River as it passes through the city of Valencia (Spain), the embankment program that extends along the entire Dutch coast (Wesselink 2007) or in the coast in front of New Orleans in the USA (Galloway 2004).

After the damage, especially economic, that urban floods continue to produce during catastrophic episodes and without taking into account the current climate change, which they will become more frequent, mitigation measures based on structural changes are not enough in the face of the increase of exposure to flood risk that continues to grow (López-Martínez et al. 2020). Structural measures such as the construction of a reservoir make the population and administrations consider that the area is safe, and the perception of flood risk is considerably reduced. That is why, after a measure of this type, risk areas are once again considered suitable for occupation based on the false security generated by the new infrastructure (Pérez-Morales et al. 2021).

Therefore, the increased incidence of urban flooding over recent years highlights the need for reliable flood-mitigation strategies, and hence flood-risk management is increasingly recognized as a global challenge (Shah et al. 2018). Thus, there is a shift from large-scale structural measures toward mitigation strategies based on reducing vulnerability in flood-prone urban areas from a more holistic perspective (Patowary and Sarma 2020). The combination of variables such as the depth of the water table and the velocity of flow, through damage-vulnerability curves, can be used to estimate the effects of a flood in urban areas, as well as the effect that a flood would have on infrastructures, buildings, people or vehicles (Cea and Costabile 2022).

Urbanization considerably influences hydrological processes, by lowering infiltration, increasing runoff and triggering higher and more rapidly occurring peak streamflow, and higher recurrence of floods (Jongman 2014). However, the magnitude of the hydrological changes is affected by several biophysical parameters, such as soil type, topography, the percentage of sealing and the spatial heterogeneity of the urban features (Leandro et al. 2016). Urban floods are difficult to control, so a strategy of "living with floods" is considered more reasonable (Leandro et al. 2016). Depending on the level of risk, political decisions and funding, a suitable combination of measures should be adopted to mitigate rising flood risk driven by urbanization, but also climate change (Collentine and Futter 2016). Structural measures as mentioned above are not enough for flood control, so NbS are increasingly being applied in flood management (Jongman et al. 2015). By using flexible and cost-effective solutions which mimic natural processes, NbS have the potential to build urban resilience and provide a number of ecosystem services associated with environmental, social and economic benefits (Ferreira et al. 2020).

Distinct NbS strategies have been used as mitigation measures trying to increase water infiltration (to reduce runoff), surface water retention and evapotranspiration, breaking flow connectivity over the landscape (Brown et al. 2009). The application of the NbS can be in a context of micro-scale techniques (e.g., green roofs and detention basins), usually implemented near runoff sources, to the context of large-scale areas which are allowed to be temporarily flooded (Sharifi 2019). Although several studies address the usefulness and potential of NbS in water management, very limited information regarding their effectiveness on flood mitigation is available (Ferreira et al. 2020). Furthermore, the performance of different NbS in flood protection is strongly influenced by specific local conditions, including urban distribution, pavement type and NbS design, which bring additional challenges to understand the impact of specific NbS strategies (Kalantari et al. 2019).

Application of NbS demands comprehensive knowledge and evaluation. So far, research has mostly focused on technical aspects of NbS and evaluation of a few benefits (Caparros-Martinez et al. 2020), while the link between this kind of measures and the wider water system and comprehensive methodology remains to be determined (Ferreira et al. 2020). Despite strong evidence that NbS are a sustainable solution for reducing urban flooding, their full effect potential is still largely unexploited. The need for quantitative assessments of the effectiveness of NbS on flood mitigation has been stressed by several authors, particularly for the Mediterranean region given the limited number of studies available (Collentine and Futter 2016).

2.6 Final Remarks

With the current climate change, catastrophic floods will be more and more recurrent. However, there has been no increase in personal damage due to this increase, and this has been due to the change in land planning in areas affected by extreme hydrological events. Historically, hydrological regimes have behaved differently, with floods becoming more frequent, but also less damaging to people, contrary to what happened in past centuries. Urban areas have increased exponentially and that is why a plan of mitigation measures toward flood-prone areas is necessary. The measures related to the construction of infrastructures or pharaonic channelizing (case of the Túria River in Valencia) are less frequent since their double consequence has been verified: false citizen security and affectation in the hydrological dynamics of the river. That is why measures such as Nature-based Solutions are becoming the most widely used and recommended. On the other hand, the measures that are currently taken in urban areas in developed countries are far from those that continue to be taken in underdeveloped countries. Flood-risk mitigation in urban areas is related to the purchasing power of the neighborhood, the city and ultimately, the country. While in Europe the path is clearly directed toward more sustainable measures, with green spaces that improve infiltration or water retention capacity, in South America or Africa, measures based on the construction of infrastructures are still being carried out.

Structural measures have not stopped growing in number and size in urban areas. Throughout the twentieth century and up to the present, an ascending trend line in the number of flood events has been demonstrated by several studies (Blöschl et al. 2020; Merz et al. 2021; Sánchez-García and Schulte 2023). In the first half of the century, the actions were mainly limited to dams and reservoirs capable of regulating channels, to supply a growing population with water resources and defend it from flood waves. In the second half of the century, this population increase, which had been making greater demands in terms of space and water for both agricultural and urban use, accelerated the progressive occupation of the riverbeds with the consequent increase in flood events and associated economic losses, although proportionally they are not increasing (comparing the GDP of 1990 to the present). At this point, large supply projects ceased to be important due to, among other reasons, a lack of space for the location of profitable dams.

Flood-risk assessment tools and flood-risk management strategies need to account for the fact that we are currently in an exceptional flood-rich period (and it will increase with climate change) in terms of timing of flood occurrence, magnitudes and spatial extent within urban areas, mainly in Europe. Process-based models that capture the physical mechanisms in the atmosphere and rainfall–runoff transformation on the land surface, including the role of precipitation, soil moisture, snowmelt and seasonality in flood generation in both recent and historical times, will be an essential component of flood-risk assessment tools in a changing climate.

Acknowledgements This research has received funding, in the case of CSG from the European Union-Next Generation EU, the Spanish Ministry of Universities under the Horizon 2020 research and innovation program through Margarita Salas fellowship (ref. MARSALAS21-22). CSG belongs to the FluVAlps Research Group (PID2020-113664RB-I00) supported by the Spanish Ministry of Science and Innovation.

References

Angelakis AN, Antoniou G, Voudouris K et al (2020) History of floods in Greece causes and measures for protection. Nat Hazards 101:833–852

Anisimov AP, Ryzhenkov AJ (2016) Environmental disaster in the Volga-Akhtuba floodplain in the context of global climate change: legal approaches and methods to decrease the gravity of the problem. Euro Online J Natl Soc Sci 5(3):724

Ashley RM, Lundy L, Ward S (2013) Water-sensitive urban design: opportunities for the UK. Proc Inst Civ Eng Munic Eng 166:65–76

Audisio C, Turconi L (2011) Urban floods: a case study in the Savigliano area (North-Western Italy). Nat Hazard 11:2951–2964

Barredo JI (2009) Normalised flood losses in Europe: 1970–2006. Nat Hazard 9:97–104

Barrera A, Llasat MC, Barriendos M (2006) Estimation of extreme flash flood evolution in Barcelona County from 1351 to 2005. Nat Hazards Earth Syst Sci 6:505–518

Barriendos M, Gil-Guirado S, Pino D et al (2019) Climatic and social factors behind the Spanish Mediterranean flood event chronologies from documentary sources (14th-20th centuries). Global and Planetary Change 102997

Bellintani P, Saracino M (2015) Rivers, human occupation and exchanges around the Late Bronze Age settlement of Frattesina (NE Italy). Rivers in Prehistory, Archaeopress, Oxford, 77–87

Benito G, Machado MJ, Rodríguez X (2012) Las crecidas del 28 de septiembre de 2012 en el SE de España: Cómo un evento moderado puede causar un impacto extremo. Enseñanza de las Ciencias de la Tierra 20(3):207–209.

Biginagwa TJ, Ichumbaki EB (2018) Settlement history of the islands on the Pangani River, northeastern Tanzania. Azania: Archaeol Res Africa 53(1):63–82

Blöschl G, Kiss A, Viglione A et al (2020) Current European flood-rich period exceptional compared with past 500 years. Nature 583(7817):560–566

Bodoque JM, Amérigo M, Díez-Herrero A et al (2016) Improvement of resilience of urban areas by integrating social perception in flash-flood risk management. J Hydrol 541:665–676

Brázdil R, Kundzewicz ZW, Benito G (2006) Historical hydrology for studying flood risk in Europe. Hydrol Sci J 51:739–764

Bridgland DR (2000) River terrace systems in north-west Europe: an archive of environmental change, uplift and early human occupation. Quatern Sci Rev 19(13):1293–1303

Brown RR, Keath N, Wong THF (2009) Urban water management in cities: historical, current and future regimes. Water Sci Technol 59:847–855
Bruen M, Gebre FA (2001) Worldwide public perception of flood risk in urban areas and it's consequences for hydrological design in Ireland. In Flood Risk Management Proc. Irish National Hydrology Seminar, Tullamore
Burton I, Kates RW, White GF (1978) The environment as hazard. Oxford University Press, New York
Calvo García-Tornel F (1997) Las transformaciones de los espacios urbanos fluviales en zonas áridas: lecciones de la cuenca del Segura. Documents D'anàlisi Geogràfica 31:0103–0116
Caparros-Martinez JL, Milan-Garcia J, Rueda-Lopez N et al (2020) Green infrastructure and water: and analyses of global research. Water 12:1760
Cea L, Costabile P (2022) Flood risk in urban areas: modelling, management and adaptation to climate change. A Rev Hydrol 9:50
Chia B, Wang Y, Chen Y (2020) Flood resilience of urban river restoration projects: case studies in Hong Kong. J Manag Eng 36(5):05020009
Chin A, Gregory KJ (2005) Managing urban river channel adjustments. Geomorphology 69:28–45
Ciupa T, Suligowski R (2020) Impact of the city on the rapid increase in the runoff and transport of suspended and dissolved solids during rainfall—the example of the Silnica River (Kielce, Poland). Water 12(10):2693
Collentine D, Futter MN (2016) Realising the potential of natural water retention measures in catchment flood management: trade-offs and matching interests. J Flood Risk Manag. 11:76–84
de Rodrigues NM, de Faria ALL (2009) Use of SIG tools in the urban area illegal occupation of part of the Sao Bartolomeu river–Vicosa, Minas Gerais, Brazil/utilizacao de ferramentas SIG na area urbana: ocupacao ilegal de um trecho do ribeirao Sao Bartolomeu-Vicosa (MG). Revista Geográfica Acadêmica 3(1):18–28
EEA (European Environment Agency) (2017) Urban and rural population in developed and less developed regions. https://www.eea.europa.eu/data-and-maps/figures/urban-and-rural-population-in Accessed on 10 Jan 2023
EEA (European Environment Agency) (2020) Floodplains: a natural system to preserve and restore. https://www.eea.europa.eu/publications/floodplains-a-natural-system-to-preserve-and-restore Accessed on 10 Jan 2023
Elleder L (2015) Historical changes in frequency of extreme floods in Prague. Hydrol Earth Syst Sci 19:4307–4315
Elum ZA, Lawal O (2022) Flood risk perception, disaster preparedness and response in flood-prone urban communities of Rivers State. Jàmbá: J Disas Risk Stud 14(1):10
Everard M, Moggridge HL (2012) Rediscovering the value of urban rivers. Urban Ecosyst 15(2):293–314
Ferreira CSS, Mourato S, Kasanin-Grubin M et al (2020) Effectiveness of nature-based solutions in mitigating flood hazard in a Mediterranean peri-urban catchment. Water 12:2893
Gaál L, Szolgay J, Kohnová S et al (2010) Inclusion of historical information in flood frequency analysis using a Bayesian MCMC technique: a case study for the power dam Orlík, Czech Republic. Contribut Geophys Geodesy 40(2):121–147
Galloway G (2004) Integrated flood management, case study, USA: flood management—Mississippi River. WMO/GWP Associated Programme on Flood Management
Gaume E (2018) Flood frequency analysis: the Bayesian choice. Wiley Interdisciplinary Reviews: Water, Wiley, 5(4):23
Grana L, Tchilinguirian P, Hocsman S et al (2016) Paleohydrological changes in Highland Desert rivers and human occupation, 7000-3000 Cal. Yr BP, South-Central Andes, Argentina. Geoarchaeology, 31(5):412–433
Griffin D (1996) A culture in transition: a history of acculturation and settlement near the mouth of the Yukon River, Alaska. Arctic Anthropol, 98–115
Guerreiro SB, Glenis V, Dawson RJ et al (2021) Pluvial flooding in European cities—a continental approach to urban flood modelling. Water 9:296

He X, Xu M, Cui C et al (2023) Evaluating the social license to operate of waste-to-energy incineration projects: a case study from the Yangtze River Delta of China. J Clean Prod, 135966
Heckmann T, Cavalli M, Cerdan O et al (2018) Indices of sediment connectivity: opportunities, challenges and limitations. Earth-Sci Rev 187:77–108
Herget J, Meurs H (2010) Reconstructing peak discharges for historical flood levels in the city of Cologne Germany. Glob Planetary Change 70(1–4):108–116
Hey RD (1997) River engineering and management in the 21st century, in: Applied fluvial geomorphology for river engineering and management, Thorne CR, Hey RD, Bathrust MD (eds). Wiley, London, England, 3–11
Hohensinner S, Sonnlechner C, Schmid M et al (2013) Two steps back, one step forward: reconstructing the dynamic Danube Riverscape under human influence in Vienna. Water History 5(2):121–143
Hong CY, Chang H (2020) Residents' perception of flood risk and urban stream restoration using multi-criteria decision analysis. River Res Appl 36(10):2078–2088
Janauer GA (2016) Ecohydrology, floodplain water bodies, European legal provisions and the future. Ecohydrol Hydrobiol 16(1):58–65
Jongman B (2014) Effective adaptation to rising flood risk. Nat Commun 9:1986
Jongman B, Hochrainer-Stigler S, Feyen L (2014) Increasing stress on disaster-risk finance due to large floods. Nat Clim Chang 4:264–268
Jongman B, Winsemius HC, Aers JC et al (2015) Declining vulnerability to river floods and the global benefits of adaptation. Proc Natl Acad Sci 112(18):E2271–E2280
Juma B, Olang LO, Hassan M et al (2021) Analysis of rainfall extremes in the Ngong River Basin of Kenya: towards integrated urban flood risk management. Phys Chem Earth, Parts A/B/C 124:102929
Junk WJ, da Cunha N, Thomaz SM et al (2021) Macrohabitat classification of wetlands as a powerful tool for management and protection: the example of the Paraná River floodplain Brazil. Ecohydrol Hydrobiol 21(3):411–424
Kosmas C, Danalatos N, Cammeraat LH et al (1997) The effect of land use on runoff and soil erosion rates under Mediterranean conditions. CATENA 29(1):45–59
Leandro J, Schumann A, Pfister A (2016) A step towards considering the spatial heterogeneity of urban key features in urban hydrology flood modelling. J Hydrol 535:356–365
Lemeunier G, Picazo MT (1988) La sociedad murciana frente a las inundaciones (1450–1900). In Gil Olcina A, Morales Gil A (Dir.). *Avenidas fluviales e inundaciones en la cuenca del mediterráneo.* Universidad de Alicante, Instituto Interuniversitario de Geografía, 365–373. Alicante
Li J, Nassauer JI, Webster NJ, Preston SD et al (2022) Experience of localized flooding predicts urban flood risk perception and perceived safety of nature-based solutions. Frontiers in Water 4:210
Ligtvoet W, Hilderink H, Bouwman A (2014) Towards a World of Cities in 2050. An Outlook on Water-Related Challenges; Background Report to the UN-Habitat Global Report; PBL Netherlands Environmental Assessment Agency: The Hague, The Netherlands
López-Martínez F, Pérez-Morales A, Illán-Fernández EF (2020) Are local administrations really in charge of flood risk management governance? The Spanish Mediterranean coastline and its institutional vulnerability issues. J Environ Planning Manage 63(2):257–274
Lowry R, McCardle WH (1891) A History of Mississippi: From the Discovery of the Great River by Hernando DeSoto, Including the Earliest Settlement Made by the French Under Iberville, to the Death of Jefferson Davis. RH Henry and Company
Lutz W, Amran G, Belanger A et al (2019) Demographic Scenarios for the EU: migration, population and education. Publications Office of the European Union, Luxembourg
Mahoney TR (1985) Urban history in a regional context: river towns on the upper Mississippi, 1840–1860. J Am Hist 72(2):318–339
Manning SW, Lorentzen B, Hart JP (2021) Resolving Indigenous village occupations and social history across the long century of European permanent settlement in Northeastern North America: the Mohawk River Valley~ 1450–1635 CE. PLoS ONE 16(10):e0258555

Martínez-Gomariz E, Locatelli L, Guerrero M (2019) Socio-economic potential impacts due to urban Pluvial floods in Badalona (Spain) in a context of climate change. Water 11:2658

Martínez-Gomariz E, Forero-Ortiz E, Russo B et al (2021) A novel expert opinion-based approach to compute estimations of flood damage to property in dense urban environments. Barcelona case study. J Hydrol 598:126244

McEvedy C, Jones R (1978) Atlas of world population history. Penguin Books Ltd., Harmondsworth, Middlesex, England

Merz B, Blöschl G, Vorogushyn SD et al (2021) Causes, impacts and patterns of disastrous river floods. Nat Rev Earth Environ 2(9):592–609

Mondal M, Satpati L (2019) Human intervention on river system: a control system—a case study in Ichamati River, India. Environ Dev Sustain 22:5245–5271.

Morita M (2008) Flood risk analysis for determining optimal flood protection levels in urban river management. J Flood Risk Manage 1(3):142–149

Mount JF (1995) California rivers and streams: The conflict between fluvial processes and land use. University of California Press, Berkeley, California

Munawar HS, Khan SI, Anum N et al (2021) Post-flood risk management and resilience building practices: a case study. Appl Sci 11(11):4823

Munzwa K, Wellington J (2010) Urban development in Zimbabwe: a human settlement perspective. Theor Empirical Res Urban Manage 5(14):120–146

Nguyen HD, Fox D, Dang DK et al (2021) Predicting future urban flood risk using land change and hydraulic modeling in a river watershed in the central Province of Vietnam. Remote Sens 13(2):262

Patowary S, Sarma AK (2020) Projection of urban settlement in eco-sensitive hilly areas and its impact on peak runoff. Environ Dev Sustain 22(6):5833–5848

Pérez-Morales A, Romero-Díaz MA, Gil Guirado S (2021) Structural measures against floods on the Spanish Mediterranean coast. Evidence for the persistence of the "Escalator Effect." Geograph Res Lett 47:33–50

Popa MC, Peptenatu D, Drâghici CC et al (2019) Flood hazard mapping using the flood and flash-flood potential index in the Buzâu River catchment Romania. Water 11:2116

Qiang Y, Zhang L, He J et al (2021) Urban flood analysis for Pearl River Delta cities using an equivalent drainage method upon combined rainfall-high tide-storm surge events. J Hydrol 597:126293

Rana IA, Jamshed A, Younas ZI et al (2020) Characterizing flood risk perception in urban communities of Pakistan. Int J Disas Risk Reduct 46:101624

Remondia F, Burlando P, Vollmer D (2016) Exploring the hydrological impact of increasing urbanisation on a tropical river catchment of the metropolitan Jakarta, Indonesia. Sustainable Cities Society 20:210–221

Ribas Palom A, Saurí Pujol D (2021) What can we learn from the past? A century of Changes in vulnerability to floods in the Ter River basin. Geograph Res Lett 47:51–72

Rubinato M, Nichols A, Peng Y et al (2019) Urban and river flooding: comparison of flood risk management approaches in the UK and China and an assessment of future knowledge needs. Water Sci Eng 12(4):274–283

Rufat S, Howe PD (2022) Small-area estimations from survey data for high-resolution maps of urban flood risk perception and evacuation behavior. Ann Am Assoc Geograph, 1–23

Sala M, Inbar M (1992) Some hydrologic effects of urbanization in Catalan rivers. CATENA 19:363–378

Samela C, Persiano S, Bagli S et al (2020) Safer_RAIN: a DEM-based hierarchical filling-&-Spilling algorithm for pluvial flood hazard assessment and mapping across large urban areas. Water 12:1514

Sánchez-García C, Schulte L, Carvalho F et al (2019) A 500-year flood history of the arid environments of southeastern Spain. The case of the Almanzora River. Glob Planetary Change 181:102987

Sánchez-García C, Francos M (2022) Human-environmental interaction with extreme hydrological events and climate change scenarios as background. Geogr Sustain 3:232–236
Sánchez-García C, Schulte L (2023) Historical floods in the southeastern Iberian Peninsula since the 16th century: trends and regional analysis of extreme flood events. Global and Planetary Change 231:104317.
Schumacher I, Strobi E (2011) Economic development and losses due to natural disasters: the role of hazard exposure. Ecol Econ 72:97–105
Shah MAR, Rahman A, Chowdhury SH (2018) Challenges for achieving sustainable flood risk management. J. Flood Risk Manag 11:352–358
Sharifi A (2019) Urban form resilience: a meso-scale analysis. Cities 93:238–252
Sutejo A, El Chidtian ASCR, Nisa DA (2019) Free of waste river concept with social campaign creative strategy. In IICACS: International and Interdisciplinary Conference on Arts Creation and Studies, vol 1, 135–142
Sutton JE (2007) The human past: world prehistory and the development of human societies. J Glob Hist 2(1):113–118
Tierolf L, de Moel H, van Vliet, J (2021) Modeling urban development and its exposure to river flood risk in Southeast Asia. Comp. Envir. and Urban Syst 87:101620.
Tomar P, Singh SK, Kanga S et al (2021) GIS-based urban flood risk assessment and management—a case study of Delhi National Capital Territory (NCT) India. Sustainability 13:12850
Tripathi R, Sengupta SK, Patra A et al (2014) Climate change, urban development, and community perception of an extreme flood: a case study of Vernonia, Oregon, USA. Appl Geogr 46:137–146
Tuset J, Vericat D, Batalla RJ (2016) Rainfall, runoff and sediment transport in a Mediterranean mountainous catchment. Sci Total Environ 540:114–132
United Nations (2014) World urbanization prospect; The Revision 2014. Highlights. UN, New York, NY, USA
Utami S, Bisri M (2014) Disaster risk and adaptation of settlement along the River Brantas in the context of sustainable development, Malang, Indonesia. Procedia Environ Sci 20:602–611
Vian F, Izquierdo J J, Serrano-Martínez M (2021) River-city recreational: interaction: A classification of urban riverfront parks and walks. Urban Forestry and Urban Greening 59:127042. https://doi.org/10.1016/j.ufug.2021.127042
Watson F, Guzmán-Arias I, Villagra-Mendoza K (2020) Case study: design parameter analysis for a hydraulic modeling of a floodplain protection of La Estrella River, Limón Costa Rica. Revista Tecnología En Marcha 33(1):44–54
Weeelink AJ (2007) Flood safety in the Netherlands: the Dutch response to Hurricane Katrina. Technol Soc 29(2):239–247
Wei C, Taubenböck H, Blaschke T (2017) Measuring urban agglomeration using a city-scale dasymetric population map: a study in the Pearl River Delta, China. Habitat Int 59:32–43
Wetter O (2017) The potential of historical hydrology in Switzerland. Hydrol Earth Syst Sci 21:5781–5803
Yamanouchi T, Ishikawa S (2011) Relationships between water quality, river course modification history and distribution of Nuphar population-the case in the Koda River, Kochi City. Japanese J Conserv Ecol 16(2):169–179
Yang F, Xiong S, Ou J et al (2022) Human settlement resilience zoning and optimizing strategies for river-network cities under flood risk management objectives: taking Yueyang city as an example. Sustainability 14(15):9595
Yao L, Wei W, Chen L (2016) How does imperviousness impact the urban rainfall-runoff process under various storm cases? Ecol Ind 60:893–905
Yin Q, Ntim-Amo G, Ran R et al (2021) Flood disaster risk perception and urban households' flood disaster preparedness: the case of Accra Metropolis in Ghana. Water 13(17):2328
Yin Q, Ntim-Amo G, Xu D et al (2022) Flood disaster risk perception and evacuation willingness of urban households: the case of Accra, Ghana. Int J Disas Risk Reduct 78:103126

Chapter 3
Tourism-Related Urbanization and Flooding: Some Examples in Mallorca

Joan Rosselló-Geli and Miquel Grimalt-Gelabert

Abstract The growth of impervious surfaces caused by urbanization processes on the coast means an increase in flooding episodes affecting urban tourist centers. The island of Mallorca is a clear example of mass tourism, and the territorial impacts it has caused. These include the massive construction of hotels and buildings on the coasts, creating urbanizations that are often uncontrolled by the public administration, thus not following regulations on risk and safety issues, which are usually more recent in time. This chapter presents the impacts that this tourist urbanization has caused in Mallorca in the form of flooding episodes that can be classified as urban as they affect flood areas that had not previously been occupied by man. The proliferation of such spaces, together with the increase in natural episodes of intense rainfall due to climate change, suggests that this type of flooding in tourist areas will increase in the future, with the consequent economic and societal impacts.

Keywords Mallorca · Floods · Urbanization · Tourism · Hazard

3.1 Introduction

Flooding has become one of the main hazards worldwide as every year thousands of persons and millions in economic losses are recorded, 299 billion $ between 2018 and 2022 (Munich Re 2023). Despite its natural origin, floods are also the result of an anthropic intervention, the use by man of flood-prone areas to develop economic activities and, more frequently, lodging. The urban growth of the latest decades of the twentieth century mainly occupies flood-prone territories, thus increasing the affectation of natural events, often with damaging results (Rentscher et al. 2023).

J. Rosselló-Geli (✉)
Estudis d'Arts I Humanitats, Universitat Oberta de Catalunya (UOC), Barcelona, Spain
e-mail: jrosselloge@uoc.edu

M. Grimalt-Gelabert
Departament de Geografia, Universitat de Les Illes Balears (UIB), Palma de Mallorca, Spain
e-mail: miquel.grimalt@uib.es

© The Author(s), under exclusive license to Springer Nature Switzerland AG 2024
J. Farguell Pérez and A. Santasusagna Riu (eds.), *Urban and Metropolitan Rivers*,
The Urban Book Series, https://doi.org/10.1007/978-3-031-62641-8_3

This growing trend is happening around the world and has impacts on both high-income and low-income countries (Rentscher et al. 2023). Europe is an example of developed countries affected by flood events, which are related to urbanization (Kaspersen et al. 2017), as shown by floods in 2021 (Yang et al. 2021) or recently in 2023 (Stamataki 2023). The expansion of urban areas can be linked to several factors but, in Mediterranean coastal spaces, it is mostly a result of the tourism industry expansion since the second half of the twentieth century. The Spanish coast is the paradigm of the uncontrolled urbanization, with a 30% increase of impervious surfaces between 1990 and 2000 with more than 2 million dwellings built (Fernández Muñoz and Barrado Timón 2011).

Such growth has not considered environmental factors, namely the occupation of flood-prone areas (López Martínez and Pérez Morales 2017) and the risks involved (Gallegos-Reina and Perles-Roselló 2022).

Because of the intense urbanization process, the number of flood events affecting coastal locations has increased in the past decades. Examples are found on the Canary Islands (Máyer 2002; Máyer et al. 2006; López et al. 2019; Dorta Antequera et al. 2020), the Spanish southeast coast (Gallegos-Reina 2023) or northwest coast (Toubes et al. 2017). An especial interest is located on the Eastern coast from Alicante (Olcina-Cantos et al. 2010; Sánchez-Almodovar et al. 2023; García Botella and Ramon Morte 2023) to Catalonia (Lara et al. 2010; Llasat et al. 2014; Ribas et al. 2020).

Another case of intense urbanization related to the tourism industry is found in the Balearic Islands, an archipelago located off the Spanish Eastern coast. Composed of five islands, its main economic activity is, since the 1960s, the tourism industry, mostly related to sun and sand offer, currently changing to another tourist offers (Minguez 2012).

Despite the climatic trends, often related to sunny weather, Mallorca is affected by heavy rainfall episodes, which can reach amounts over 200 mm in 24 h (Grimalt et al. 2006). Such events cause floods, with records dating back to the fifteenth century, affecting historically agricultural plains and some urban centers of the island, built around streams (Grimalt and Rosselló 2020). While it was common to avoid flood-prone areas, such behavior changed at the start of the twentieth century, with the development of urban extensions and especially, since the 1960s, when, with the start of tourism, the coastline has been extensively urbanized, occupying the mouths of the island's torrential watersheds and even the stream channels. A result has been an increase of flood events and its associated impacts but unlike in other areas of Spain, no studies have been carried out on the relationship between tourist urbanization and flooding.

The research herein presented aims to identify the location of tourism-related flood events affecting the coasts of Mallorca. From a general review to some specific examples, we will present the impact of urbanization over the island's catchments in the past 60 years.

3.2 Research Area

The island of Mallorca is the largest of the Balearic archipelago with a surface of 3640 km^2 and a population of 914.564 in 2022 (IBESTAT 2023). The morphology of the landscape is structured around two mountainous areas, the Serra de Tramuntana in the northwest and the Serres de Llevant in the southeast, while lying in between is a relatively flat area, known as Es Pla de Mallorca, composed by hilly areas separated by small valleys. Around those three main features, a carbonate Miocene platform and some alluvial plains shape the physiographic units of the island (Fig. 3.1a).

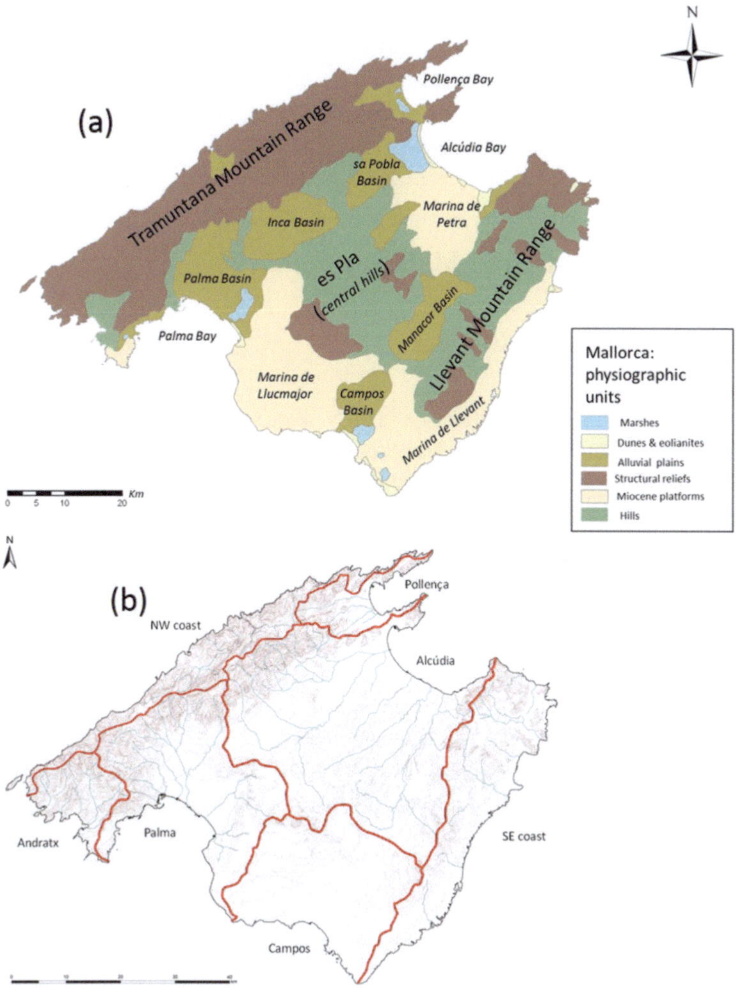

Fig. 3.1 Physiographic units of Mallorca (a) and basin location (b) (*Source* Grimalt and Rosselló 2020)

The climate regime of Mallorca is Mediterranean, with mild temperatures during the year while rainfall is characterized by the summer drought and a concentration of events during the fall, with yearly averages of 600 mm. Nonetheless, heavy rainfall events are common, usually in autumn, with rainstorms reaching over 400 mm in 24 h affecting mostly the north and the eastern parts of the island.

Regarding the surface runoff system, the island's small size and the precipitation regime led to the existence of an irregular network, formed by ephemeral streams, locally called torrents, which are organized by basins (Fig. 3.1b). The runoff is caused by rainfall and lasts hours and, in some events, even days but the streams are usually dry. Despite its irregularity, extreme rainfall episodes led to large flooding events as extreme peaks exceed the streambed surface and the overflow affects areas more or less populated.

The economy of Mallorca was focused on agricultural activities until the twentieth century but, since the 1960s, the main activity is related with tourism, an industry that has shaped the societal and economical structure of Mallorca, leading to an increase of urbanization and a population growth (Salvà 2008).

3.3 Flood Affectation in Mallorca

3.3.1 Historical Floods

Flooding has been a part of Mallorca's history since, at least, the Middle Ages. The first knowledge of a flood impact was the 1403 event, which destroyed the city of Palma, killing close as 5000 inhabitants according to the chronicles (Alcantara Peña 1891). Historical floods affected urban centers and agricultural plains (Fig. 3.2). Three cities have streams running within the urban center (Manacor, Palma and Sóller) while two others (Esporles and Sant Llorenç) have the main channel of the surface water system close to the city. As for agricultural plains, the areas of Sa Pobla and those of Campos are located in alluvial plans and the urban sprawl is formed over paleo channels because of the presence of alluvial fans, which lead to the flooding of streets, houses and farming land when a large flood occurs.

This very specific location of flooded areas is related to human adaptation measures, which avoided risky areas to locate populations on high ground wherever possible and only use flood-prone zones for farming purposes.

3.3.2 Modern Floods

At the start of the twentieth century, a change regarding the location of flooded area is identified. There was an urban expansion toward flood-prone areas, mostly located in non-coastal zones but close to the largest cities of the island. Such process

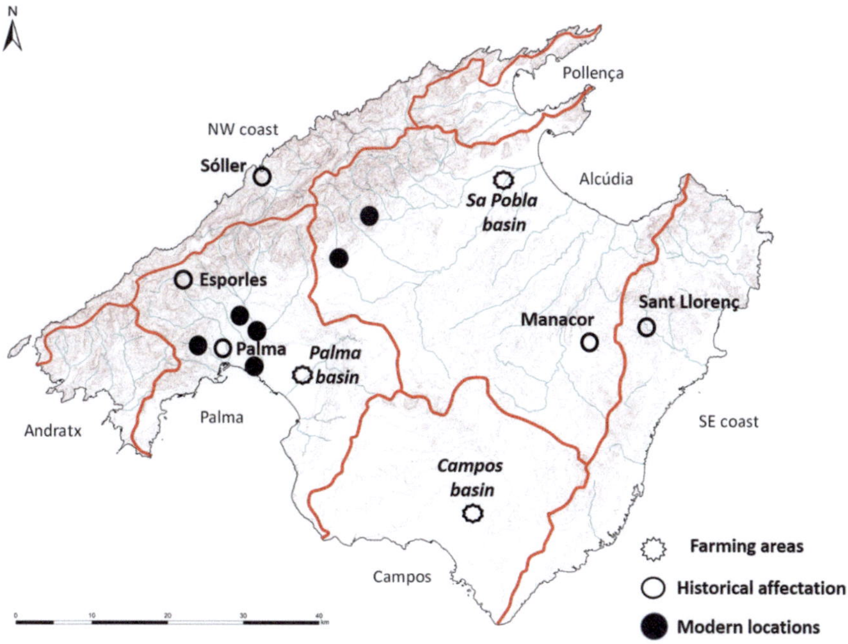

Fig. 3.2 Flood events affecting non-tourist locations (*Source* Authors)

affects heavily the surroundings of Palma, the regional capital. which increased its population, attracting inhabitants from the other towns of the island, people that urgently needed housing arrangements. The increase of urbanization appears as well in municipalities close to Palma, such as Marratxí, or cities with a growing economic activity, like Inca (Fig. 3.2).

Urban growth was directed toward areas previously unoccupied by man, due to the risk of flooding, but the lack of free spaces, together with the infrequency of flooding episodes, favored this process. The result was a large increase of the occurrence of flood events as the case of Palma illustrates, with more than 20 episodes between 1930 and 2020 affecting those new urban areas (Rosselló-Geli and Grimalt-Gelabert 2021).

3.4 Tourism-Related Flood Location and Classification

Since the 1960s, there is an urbanization development toward the coast, because of the arrival of mass tourism (Salvà 1990). Those tourism development areas were built in spaces that can be included in different floodable morphologies.

(a) The first typology corresponds to those spaces that are beach-dune systems. In this model, the constructions are generally located on the dunes that separate the lagoon from the sea or around the lagoon. These conditions are found in the urbanizations located on the first coastline at the bottom of the main bays and gulfs of the island, such as Arena or Palma.

In the case of flooding, these affect large areas, although the water levels are not very noticeable, as they are slow water flows, which do not cause a risk of fatalities but can generate serious economic losses.

(b) A second typology includes the cove-bottom environments (a coastal formation in the form of an enclosed bay typical of the limestone coastlines of Mallorca, characterized by being narrow and enclosed by high cliffs on both sides), which flow into a small lagoon. This typology presents an increased risk due to the custom of draining these wetlands to build or construct swimming pools or solariums. Most of the tourist areas on the east coast of Mallorca, as well as areas located on other stretches of coastline, fall within this model. The consequences of floods can be severe, due to the sudden rise in water levels and very high flow velocities, capable of sweeping away vehicles and pedestrians as well as damaging buildings and public works. In general. the watercourses that flow into these areas are short and steep, with a significant erosive dynamic and a high capacity for clogging the bottom of the cove (Estrany and Grimalt 2014).
(c) The third type of flood zone comprises those watercourses that have, in their final stretch, a relatively wide valley, in which a small flood plain and even a system of fluvial terraces develop. In this case, these areas are potentially dangerous because high flow levels and velocities affect them, especially when the capacity of the channels is exceeded, and the flow encroaches on the flood plains and terraces.
(d) A fourth typology is made up of a limited number of areas where the torrential river course opens up into different branches at its mouth, functioning as an alluvial fan. In the event of flooding, the course is distributed through the different channels, which join if the flood level is high and create wide flood fronts. The most outstanding example is the central part of the bay of Palma, at the mouth of the Gros torrent, which is interconnected with the nearby Bàrbara torrent, and which represents a potential danger for the tourist areas of Molinar and Ciudad Jardín.
(e) Finally, a fifth type is formed by those courses that end directly at the coast as the mouth of a valley formed by mountainous sectors close to the coast. The outlet is in an area without any lagoon formation that would allow the floods to be regulated, thus increasing the risk for many coastal urbanizations, such as those located on the coast of Calvià or the eastern part of the municipality of Palma.

The occurrence of floods affecting the above-mentioned locations led to man-made actions to prevent the risk. The so-called societal answers are diverse and can be classified as follows.

(a) Open channelization of the streambed. Engineering works reinforce and limit the sides of the course, usually with concrete, which is transformed into an open channel. The capacity of these canals is usually low, especially in relation to the possible peak flows of large storms, because they are designed only for normal rainfall events or because the calculations use models that underestimate the generation of runoff in Mediterranean areas.
(b) Streambed enclosure. The final part of the watercourse flows in a concrete box that forms an underground bed, so that the water cannot occupy its natural space in the event of a flood. In addition, the space above the canal becomes a street or a promenade and is often built on. Moreover, the calculation of the underground channel section usually underestimates the potential flood peak that can circulate within the streambed.
(c) Course deviation. An alternative passage for the course of the torrent is constructed, overriding the original course. This action is common when the mouth coincides with the central part of a beach. As in the previous cases, the new channel suffers from a lack of capacity to contain possible floods.
(d) Clogging of lacustrine spaces. Located at the mouths of the river, these lagoons were filled in at the beginning of the twentieth century to gain building space near the coastal areas and to avoid the presence of stagnant water that favored the presence of insects and to offer a safe image. What used to be an estuary was thus modified, in some cases leaving a reduced drainage channel or diverting the channel away from the beach.
(e) Occupation of the streambed. The original channel is forgotten, usually due to the lack of runoff events, and is built directly into it, leaving no passage for the water. On other occasions, the bed is converted into a route such as a road or street.
(f) Transformation of the lagoons at the bottom of a creek into a false estuary. The lagoons are artificially dredged and are connected to the sea by the destruction of the retaining wall that used to separate them. In this false estuary, marinas are normally built which, in the event of major floods, can be affected by phenomena such as the dragging of boats or the destruction of buildings or annexed installations.

On more than one occasion, human actions are carried out in combination on the same watercourse. An example is the case of S'Illot, on the east coast of Mallorca. Its cove-bottom lagoon was filled in 1973 (action type d), a small artificial channel was left in a side area of the beach (action type a), and the final stretch was diverted so that it did not end at the beach (action type c) using a closed channel (action type b).

Table 3.1 includes a list of tourist coastal locations affected by flood events since the 1960s. For each case the flooding morphology, the human actions and the date of the episodes are included. To emphasize the hazard of these events, the damage level is also included, according to a three-level classification, from ordinary (1) to catastrophic floods (3) being extraordinary (2) a medium level of damages (Llasat et al. 2013). Ordinary floods cause damages to the watercourse and nearby areas while the extraordinary are events, affecting the streambed too but exceeding its banks. It can cause damages to farming or human-related activities and disturbances of the daily life of the population, such as road closures or traffic delays. Finally, catastrophic events are those where the overflowing of the stream banks can cause serious damages or even the destruction of infrastructure such as bridges, walls or roads as well as buildings located relatively away from the stream, thus being the cause of significant material damage.

It should be noted that all the previous classification could provoke fatalities as the levels only consider the direct material damage because of the flood. In that sense, a flood classified as ordinary can led to a fatality if a person is in the streambed and is swept away by the water flow, thus being the effect of an imprudent behavior.

Other table captions are as follows: regarding the morphologies of flood, five levels (1–5) are listed, according to the previously stated types, while the human answers to flood risk and the changes made in the dangerous areas are described in letters in order from a to g, also following the explanation given above. Finally, an area identifier with a number helps to locate the affected places in Fig. 3.3.

To illustrate the impact of flooding on tourist coastal areas, two examples have been selected, both representative of the different morphologies of tourist areas located in flood-prone zones. On the one hand, the event of November 1990 in Santa Ponça, located within the municipality of Calvià, on the southwestern side of the island. The hotel facilities were built there on the lower terrace of the torrent de Santa Ponça (Fig. 3.4), a channeled course with a concrete bed ten meters wide and two meters high for a basin of just over 70 km^2. A significant number of hotels and commercial establishments are located at elevations below 1.5 m above the base of the watercourse and were affected by torrential events, such as the flood of 8 November 1990. In addition, there is evidence of other events of great intensity in periods prior to the tourist occupation, the case of 25 September 1962 being particularly noteworthy, when the overflowing waters reached the limit of the lower terrace, an event that, if it were to happen today, could cause serious impacts on tourist activity and its users. A second example is found in S'Illot, a tourist resort on the east coast of Mallorca, located in a cove with a lagoon at the back of the beach where the torrent flows into a 70-km^2 basin (Fig. 3.5). The natural conditions were extensively modified when hotels were built behind the beachfront, the lagoon was drained, and the bed of the course was modified into an artificial channel of tiny dimensions (6 m wide and 1 m high), with a covered section in the last 100 m.

The flood of 6 September 1989 caused a large flooded area, with water levels reaching 2 m high inside the hotels located at the beachfront (Ortega-Mclear 2019). Subsequent improvements were the re-excavation of the lagoon and the increase in the section of the channel to 22 m wide and 2 m high. The outcome was that another major storm event, the flood of 9 October 2018, with higher flows than those of 1989, flooded a smaller area, although upstream it caused serious damage in Sant Llorenç des Cardassar and 10 fatalities (Grimalt et al. 2021).

Table 3.1 Flood events on coastal tourist areas

Date	Affected area	Id number	Stream	Damage level	Flood morphologies	Human actions
15/09/1962	s'Arenal	12	Torrent des Jueus	3	5	a
19/09/1962	Port de Pollença	1	La Gola	2	1	g
25/09/1962	s'Arenal	12	Torrent des Jueus	3	5	a
25/09/1962	Passeig Marítim Palma	14	Torrent Sant Magí	3	5	a
25/09/1962	Camp de Mar	18	Torrent des Camp de Mar	4	5	a
25/09/1962	Ciudad Jardín	13	Torrent Gros	3	4	g
25/09/1971	Sant Elm	20	Torrent de Sant Elm	3	5	e
25/09/1971	Port Andratx	19	Torrent des Saluet	2	2	f
01/10/1973	s'Arenal	12	Torrent des Jueus	2	5	a
29/03/1974	Port de Sóller	21	Torrent Major	3	3	a
18/10/1974	Cala Major	15	Cala Major	3	5	e
18/10/1974	s'Arenal	11	Son Verí	3	5	b
19/09/1975	Canyamel	3	Torrent de Canyamel	3	2	g
18/10/1978	Port de Sóller	21	Torrent Major	2	3	a
25/08/1983	Cala Millor	4	Estanyol de Son Moro	2	1	d
25/08/1983	Port de Manacor	6	Torrent de ses Talaioles	3	2	f
16/09/1988	Peguera	17	Torrent de sa Coma	3	5	c
06/09/1989	Portocolom	9	Torrent des Corso	4	2	e
06/09/1989	Port de Manacor	6	Torrent de ses Talaioles	3	2	f
06/09/1989	Cala Murada	8	Torrent de sa Plana	3	2	d
06/09/1989	S'Illot	5	Torrent de ca n'Amer	3	2	d,c, a, b

(continued)

Table 3.1 (continued)

Date	Affected area	Id number	Stream	Damage level	Flood morphologies	Human actions
08/10/1990	Cala Sant Vicenç	22	Torrent de Cala Molins	3	5	c,b
08/10/1990	Port de Pollença	1	La Gola	3	1	g
09/10/1990	Port d'Alcúdia	2	Albufera	3	1	f
08/11/1990	Santa Ponça	17	Torrent de Santa Ponça	3	3	a
08/11/1990	Peguera	18	Torrent de sa Coma	3	5	c
13/10/1994	Port de Manacor	6	Torrent de ses Talaioles	2	2	f
13/10/1994	Estany den Mas	7	Torrent des Morts	3	2	d
17/10/1994	Port de Sóller	21	Torrent de sa Figuera	3	5	a
11/11/2001	Cala d'Or	10	Torrent de Cala Llonga	2	2	f
08/08/2007	Magaluf	16	Prat de Magaluf	2	1	d
15/12/2008	Ciudad Jardín	13	Torrent Gros	2	4	a
22/09/2009	Port Andratx	19	Torrent des Saluet	2	2	f
09/10/2018	Canyamel	3	Torrent de Canyamel	4	3	g
09/10/2018	S'Illot	5	Torrent de ca n'Amer	4	2	a

Source Authors

3.5 Conclusion

The research identifies the spatial location of flood events as the result of the extensive urbanization of Mallorca's coastlines. The urbanization process linked to tourism has led to the occupation of flood zones, historically unattractive for use by Mallorcan society, which prevented the existence of a risk in the event of flooding.

Several typologies of floodable areas are identified along with the human answer to the flooding risk, measures that, sometimes, increase the risk faced by population as the technical solutions underestimate the runoff capacity of the island torrential streams. Particularly noteworthy is the case of creek-bottom lagoons, which were

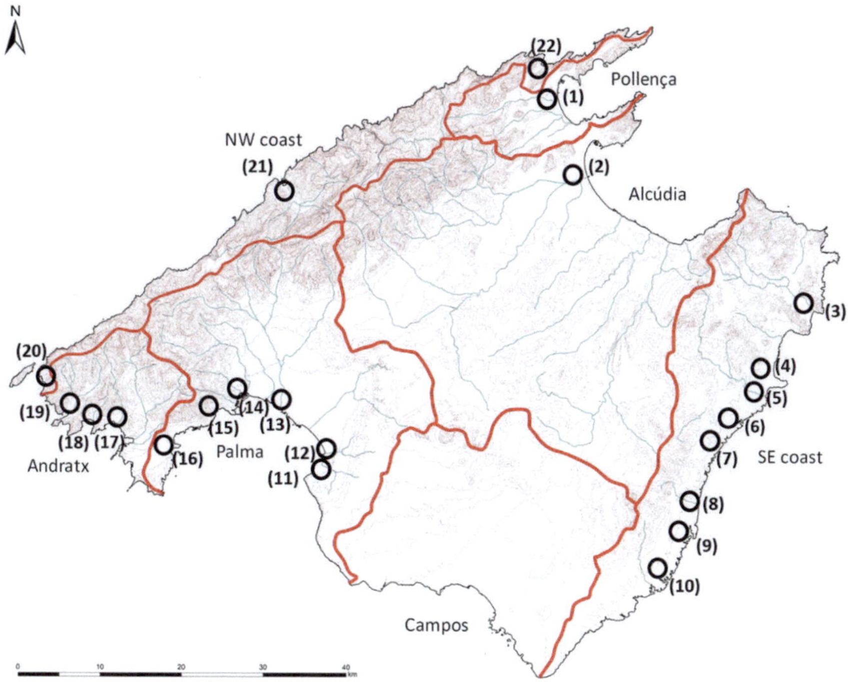

Fig. 3.3 Spatial location of significant tourist-related flood events (*Source* Authors)

dredged or filled, a process that, combined with the presence of buildings and marinas, increases the floodable surfaces and the affectations in case of large runoff events.

Two examples in different island locations highlight the results of that urban pressure on flood-prone areas in terms of damages and fatalities. Despite the risks, there are no clear solutions to the problem as the tourism industry of the island does not allow large modifications of coastal zones to provide safety to inhabitants and visitors.

The probable increment of extreme rainfall events affecting the Mediterranean basin can lead to a worsening of the problem, increasing the danger and intensity of the flood phenomenon in the island. This problem can be extended to other Mediterranean island areas, with physical characteristics and tourist development like the Mallorca case.

Fig. 3.4 Evolution of Santa Ponça's urbanization and floodable areas (*Source* Authors)

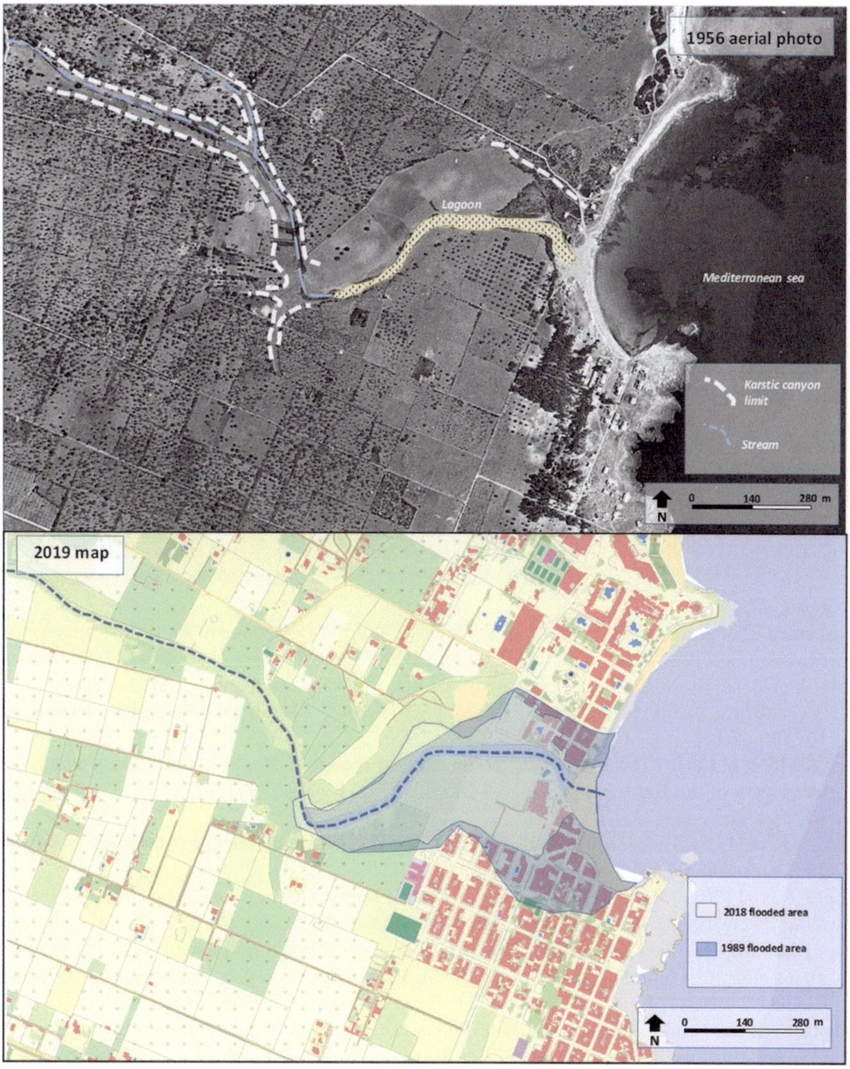

Fig. 3.5 Evolution of S'Illot's urbanization and floodable areas (*Source* Authors)

References

Alcántara Peña P (1891) Inundaciones de la ciudad de Mallorca (1403). Boletín de la Sociedad Arqueológica Luliana, IV, 151–154

Dorta Antequera P. et al (2020) Turismo y amenazas de origen natural en la Macaronesia Análisis Comparado. Cuadernos De Turismo 45:61–92. https://doi.org/10.6018/turismo.45.426041

Estrany J, Grimalt M (2014) Catchment controls and human disturbances on the geomorphology of small Mediterranean estuaries systems. Estuar Coast Shelf Sci 150:230–241

Fernández Muñoz S, Barrado DA (2011) El desarrollo turístico-inmobiliario de la España Mediterránea e insular frente a sus referentes internacionales (Florida y Costa Azul): un análisis comparado. Cuadernos De Turismo 27:373–402

Gallegos-Reina A (2023) Cambio en los patrones territoriales y análisis de inundabilidad y erodabilidad en cuencas de la provincia de Málaga, España (1956–2010). Agua y Territorio 21:69–83

Gallegos-Reina A, Perles-Roselló MJ (2022) Problemáticas de los patrones y dinámicas territoriales periurbanos del litoral mediterráneo español frente a los riesgos naturales: análisis aplicado en la provincia de Málaga. Ciudad y Territorio, estudios territoriales, LIV, 211:97–114 https://doi.org/10.37230/CyTET.2022.211.6

García Botella, E., Ramon Morte, A. (2023) Ephemeral Mediterranean watercourses strongly altered by growth in tourism: the case of Benidorm (Spain). Geosciences 13(8):247. https://doi.org/10.3390/geosciences13080247

Grimalt M et al (2006) Distribución especial y temporal de las precipitaciones intensas en Mallorca. In: Cuadrat JM, Saz MA, Vicente SM, Lanjeri S, de Luis M, González JC (eds) Sociedad y medio ambiente. Publicaciones de la AEC, Zaragoza, pp 411–420

Grimalt M et al (2021) The flood of October 9, 2018 in the city centre of Sant Llorenç des Cardassar (Mallorca). Cuadernos De Investigación Geográfica 47:265–286

Grimalt M, Rosselló J (2020) InunIB: analysis of a flood database for the Balearic Islands. Euro J Geogr 11(3):6–21. https://doi.org/10.48088/ejg.m.gri.11.3.6.21

IBESTAT (2023). Padrón de población 2022. Online at: https://ibestat.caib.es/ibestat/estadistiques/933ae75d-c922-494f-bc1a-04341d1f13a9/fe3be181-2d59-4205-a774-f703a3a671b9/es/pad_res01_22.px

Kaspersen PJ et al (2017) Comparison of the impacts of urban development and climate change on exposing European cities to pluvial flooding. Hydrol Earth Syst Sci 21:4131–4147

Lara A et al (2010) Social perceptions of floods and flood management in a Mediterranean area (Costa Brava, Spain). Nat Hazards Earth Syst Sci 10:2081–2091. https://doi.org/10.5194/nhess-10-2081-2010

Llasat MC et al (2013) Towards a database on societal impact of Mediterranean floods within the framework of the HyMeX project. Natl Hazards Earth Syst Sci 13:1337–1350. https://doi.org/10.5194/nhess-13-1337-2013

Llasat MC et al (2014) Flash flood evolution in North-Western Mediterranean. Atmos Res 149:230–243. https://doi.org/10.1016/j.atmosres.2014.05.024

López A et al (2019) Rainfall and flooding in coastal tourist areas of the Canary Islands (Spain). Atmosphere 10(12):809. https://doi.org/10.3390/atmos10120809

López Martínez F, Pérez Morales A (2017) Influencia del turismo residencial sobre el riesgo de inundación en el litoral de la Región de Murcia. Scripta Nova: revista electrónica de geografía y ciencias sociales, 21, https://raco.cat/index.php/ScriptaNova/article/view/329924

Máyer P (2002) Desarrollo urbano e inundaciones en la ciudad de Las Palmas de Gran Canaria (1869–2000). Investigaciones Geográficas 28:145–159

Máyer P et al (2006) Lluvias e inundaciones en los centros turísticos de Gran Canaria: el caso de San Bartolomé de Tirajana. Investigaciones Geográficas 41:155–173

Mínguez C (2012) The management of cultural resources in the creation of Spanish tourist destinations. Eur J Geogr 3(1):68–82

Munich Re (2023) Flood risk on the rise. Online at: munichre.com/en/risks/natural-disasters/floods.html

Olcina Cantos J et al (2010) Increased risk of flooding on the coast of Alicante (Region of Valencia, Spain). Nat Hazards Earth Syst Sci 10:2229–2234. https://doi.org/10.5194/nhess-10-2229-2010

Ortega-Mclear A (2019) Processos d'inundació als anys 1989 i 2018 al nucli de s'Illot. Reconstrucció a partir de testimonis, elements gràfics i indicis físics. Treball Final de Grau. Palma de Mallorca: Universitat de les Illes Balears. Departament de Geografia

Rentschler J et al (2023) Global evidence of rapid urban growth in flood zones since 1985. Nature 622:87–92

Ribas, A. et al (2020). More exposed but also more vulnerable? Climate change, high intensity precipitation events and flooding in Mediterranean Spain. Disaster Prevent Manage: An Int J 29

Rosselló-Geli J, Grimalt-Gelabert M (2021) Mapping of the flood distribution in an urban environment: the case of Palma (Mallorca, Spain) in the first two decades of the 21st century. Earth 2:960–971. https://doi.org/10.3390/earth2040056

Salvà P (1990) El turisme com element impulsor del procés d'urbanització a Balears (1960–1989). Estudis Baleàrics 38:63–70

Salvà P (2008) El turismo como transformador del territorio. In Alario M (coord). España y el Mediterráneo: una reflexión desde la geografía española. Túnez: XXXI Congreso de la IGU, 57–60

Sánchez Almodóvar E et al (2023) Floods and adaptation to climate change in tourist areas: management experiences on the coast of the province of Alicante (Spain). Water 15(4):807. https://doi.org/10.3390/w15040807

Stamataki I (2023) Greece's record rainfall and flash floods are part of a trend across the Mediterranean, the weather is becoming more dangerous. The Conversation. Online at: https://theconversation.com/greeces-record-rainfall-and-flash-floods-are-part-of-a-trend-across-the-mediterranean-the-weather-is-becoming-more-dangerous-213164

Toubes DR et al (2017) Vulnerability of coastal beach tourism to flooding: a case study of Galicia Spain. Environments 4:83. https://doi.org/10.3390/environments.4040083

Yang L et al. (2021) Urbanization exacerbated rainfall over European suburbs under a changing climate. Geophys Res Lett 48(21):e2021GLO95987

Chapter 4
Unveiling Fluvial Processes in Urban Rivers: The Case of the Congost River at Granollers, Catalonia, Spain

Joaquim Farguell Pérez, Lucero Ochoa, and Jhesibel Chavez

Abstract The Congost River, a tributary of the Besòs River, flows through the city of Granollers, a city in the metropolitan area of Barcelona with 62,950 inhabitants in 2023. The river was heavily fixed and channelized during the 1970s and 1980s of the last century to prevent the city urban expansion from flooding. However, with the aim of recovering a "fluvial aspect of the river" and encouraging a "social use of the fluvial space," a partial removal of the structures that kept the river fixed has been gradually undertaken since 2007 by the city council of Granollers. As a result, the fluvial morphology of the river has improved by recovering typical fluvial forms and recovering a "fluvial appearance." To evaluate whether this transformation represented a "real improvement" in the fluvial characteristics and morphodynamics, a hydro-geomorphological index (IHG, Ollero et al., Limnetica 27:171–188, 2008) was applied in two different sections of the Congost River. The first section conserves the fixing structures and no naturalization works have been done. The second one, however, is located downstream and represents the area where structures have been removed. The evaluation was undertaken by applying a hydro-morphological index, in which the naturality of the flow regime, the quality of the riparian forest structure and the morphology of the river channel are considered. Results from the assessment indicate that the naturalized section obtained a better evaluation than the urbanized section. Despite this, the naturalized section shows symptoms of geomorphological processes that have been hidden while the section was fixed and urbanized and now, after some time, appear as a consequence of the "freedom" provided to the river. Riverbed incision, lateral erosion, river migration and bed lowering are the major processes taking on in this river reach. The index used has been useful in the assessment of geomorphological changes, but to unveil current processes undergoing, additional evaluations must be applied to identify trends. The aim of this paper is to describe or unveil these processes that have been hidden for decades in fluvial

J. Farguell Pérez (✉) · L. Ochoa · J. Chavez
Department of Geography, GRAM (Grup de Rercerca Ambiental Mediterrània), University of Barcelona (UB), Barcelona, Spain
e-mail: jfarguell@ub.edu

J. Farguell Pérez
Institute of Water Research (IdRA), University of Barcelona, Barcelona, Spain

© The Author(s), under exclusive license to Springer Nature Switzerland AG 2024
J. Farguell Pérez and A. Santasusagna Riu (eds.), *Urban and Metropolitan Rivers*, The Urban Book Series, https://doi.org/10.1007/978-3-031-62641-8_4

urban areas, and that they should be considered by fluvial managers and authorities if a successful restoration project wants to be addressed.

Keywords Channelized rivers · Fluvial geomorphological processes · Urban rivers · Risk management · River restoration

4.1 Introduction and Aims of Study

Rivers transport water from higher altitude areas to the sea, but in this journey the water dissolves and drags rocks, and contains suspended particles of sand, silt and clay. In addition, according to the slope of the basin, its size, its land cover, the bedrock basin type, and the rainfall regime, determine the sediment and discharge regime of the river. The river processes involved are predominantly erosion in the upper part, transport in the middle reaches, and sedimentation and floodplain construction at the lower sections of the river (Schumm 1977). The river channel tends to widen as it flows downstream and its channel adapts to the river sediment and discharge regimes and to the frequency and magnitude of floods events (Leopold et al. 1964; Batalla et al. 2020). Floodplains are fertile areas where rivers deposit the fine sediment transported, but the urbanization processes have resulted in lowering the river bed and disconnect it from its own floodplains. Flooding is a river mechanism to decrease the amount of water carried in the main channel, decreasing the height and the velocity of water and the energy associated, and providing sediments to floodplains and deltas which are then refilled with nutrients and avoid its retreat.

However, the urban expansion and the occupation of floodplains and other fluvial areas have altered the natural processes of rivers, by channelization, river straightening and cutting off meanders (Gregory and Chin 2002; Gregory 2006). Moreover, the introduction of concrete structures in river reaches, designed and constructed to stabilize rivers and avoid the action of hydrological and geomorphological processes inherent to the nature of any river, has also been a widespread technique (Martin-Vide 1998). The effects of such structures on river reaches have resulted in the lowering of the riverbed and disconnecting it from its floodplains, to avoid floods and to assure human activities or urbanization in these areas.

Despite it, fluvial hydrological and geomorphological process may continue, although the river channel modifications and alterations introduced try to avoid them. Scour and incision processes may erode the base of the channel walls that, after several years of construction and lack of maintenance, may collapse and sink during high flood events. This is due to the river's trend to adapt its characteristics to the new slope resulting from channelization. Changes in water velocity, which increases due to the reduction of roughness, tend to increase the shear river stress, the capacity to erode, and finally, to harm the structures. Sediment transport may be altered, and fluvial forms may disappear, by decreasing the quality of the river ecosystem and functioning.

These processes, shaded or hidden during the river modification, arise when a process-based river restoration plans are executed, usually several years after the channelization projects. If no information on river processes is documented or detailed or considered when executing the renaturalization or restoration projects, these may fail or produce undesired effects on the river reaches affected, especially if no follow-up of the project is undertaken. In addition, “auto-healing” restoration processes may also occur since nothing is clearly done and the river is left to a self-adjustment after removing totally or partially the impact. The river then, adjusts the slope and the morphology according to the water flood regime and the sediment availability and characteristics (Kondolf 2011).

The case of the Congost River serves as an example. Channelization and riverbed fixing during the 1970s and 1980s aimed to disconnect of the river with its flooding area due to the need of urban and industrial expansion, and the recent catastrophic effects of the flood occurred in 1962.

The aim of this paper is to highlight and describe the river processes that have come to light or reappeared during a restoration river plan and the effects on the river section in the Congost River at Granollers, a 62,950 inhabitants city at 30 km north from Barcelona, Spain, within a Mediterranean fluvial regime highly altered due to urbanization processes and restoration processes.

4.1.1 The “Natural River” Fluvial Hydro-Geomorphological Processes

Although it should be considered that the concept “natural river” no longer exists, given that all fluvial systems have been modified or altered to a certain extent, and any modification in one specific river site may have an effect either upstream or downstream, rivers with relatively small direct human influence are mainly governed by natural hydro-geomorphological processes.

Such processes consist of complex interactions based on the transfer of mass and energy through the fluvial system, from the upper part of a basin to the lower, mobilized by gravity (Batalla 2021). Mass transfer is executed by fluvial sediment transport, which is moved depending on the frequency and magnitude of flood events (Leopold et al. 1964). According to the caliber and amount of sediment available and the slope of the river channel at a given point in a drainage basin, specific river morphology is developed, and several efforts of river classification according to the resultant morphology of these complex processes have been described (Church 2002).

4.1.2 Urban Rivers and the Hidden Processes

However, rivers circulating in urban areas, the freedom of space and movement and the associated sediment and hydrological processes have decreased or nearly forced to disappear. Channelization and riverbed lowering are modifications undertaken in urban rivers to disconnect the riverbed from its flooding area. These areas are usually used for urban and industrial expansion. In a second phase, transversal concrete structures (sleepers) are constructed along the riverbed, to fix it and avoid lateral migration (Martin-Vide 1998). Thus, river processes are minimized by reducing the slope and consequently, the speed of water and the sediment transport capacity are reduced and channel migration is avoided. When a process-based restoration is undertaken and removal of the fixing structures is the aim, the river tends to recover the hidden or shaded process to find an equilibrium between slope, fluvial regime and sediment characteristics of the riverbed.

4.2 The Congost River Stretch Case Study

4.2.1 The Congost River

The Congost River (221.94 km^2) flows from the western slopes of the Montseny Massif and the eastern slopes of Cingles de Bertí range, in the Pre-Litoral Catalan Range, with a maximum altitude 1319 m.a.s.l. The length of the river is 44 km and half of it flows through the mountain range while the other half flows in an open tectonic depression until the confluence with the Besòs River, and where cities and industrial areas have been constructed alongside (Fig. 4.1).

Rainfall at Granollers is about 530 mm yr^{-1} on average since 1950s (SMC 2022) although high intensity rain events have been recorded. As a typical Mediterranean river, flow can vary up to several orders of magnitude. The average discharge is 0.57 m^3s^{-1} (ACA 2022), but during flood events peak discharge can be several times higher. In 1994, a peak discharge of 1077 m^3s^{-1} was estimated in the nearest gauging station, at 9 km upstream. The last high event took place in January 2020 with a maximum flow of 220 m^3s^{-1}.

The river is highly affected by human intervention, especially in the lowland area, where population and economic activities take place. Granollers and the surrounding cities (Canovelles, les Franqueses del Vallès and Montornès) create an area of 135,000 inhabitants. The Congost River fluvial regime is altered due to the water dumped by water treatment plants, which increase the low flows and dilute the rainfall influence on the river regime. However, summer drought has a great influence on the river regime.

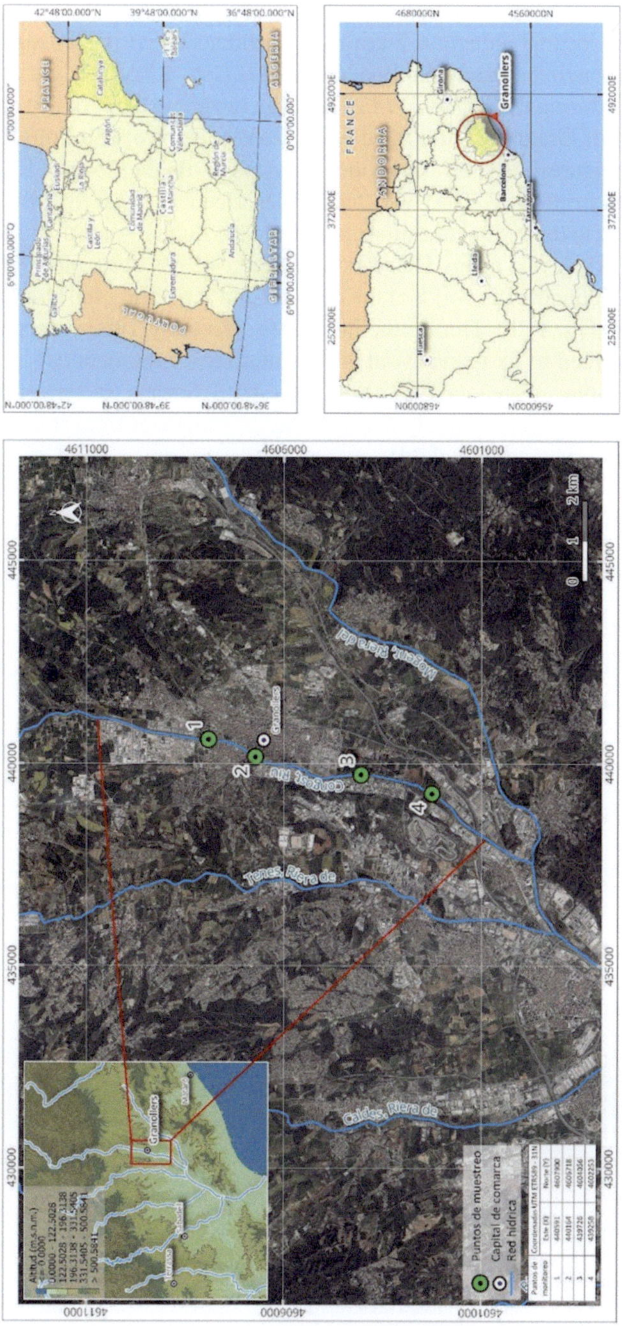

Fig. 4.1 Location of the study area and the sampling sites (*Source* The authors)

4.2.2 *The Study Area*

The study area is located along a 6.7-km river corridor, which is the length of the Congost River through the municipality of Granollers (Fig. 4.1). The river was channelized and fixed during the 1970s and 1980s of the twentieth century due to a rapid urban expansion of Granollers and there was a need to control and limit the flood area of the river. Figure 4.2a shows Granollers city and partly the Congost River in an aerial photograph from 1957. The river channel described a typical lowland braided channel according to the sediment caliber, slope and the flood regime at the period at which the photograph was taken. However, during the 1970s channelization works were undertaken reducing the width of the channel and separating the riverbed from its flooding area (Fig. 4.2b). During the late 1980s and early 1990s, sleepers were fixed into the riverbed to fix it and avoid lateral migration. Sediment bars and other morphological features were totally removed (Fig. 4.2c). Sewage and street drainage were connected to the river to rapidly drain the excess rainwater from streets and the increasing impervious area (Fig. 4.2d).

Fig. 4.2 **a** Aerial photograph of Granollers before channelisation (1957); **b** channelization works (Pere Crusellas 1978; **c**) inclusion of transversal structures to fix the river and avoid lateral migration (Esteve Gironella 1987); **d** sewage drainage into the river (Pere Espaulella 1992). Reproduced with the permission of Granollers Municipal Archives

Restoration works in the Congost River started back in 2007–2008 by enhancing bank stability by planting trees and creating an inexistent riparian community, and bioengineering techniques to restore biodiversity were also introduced (Romero 2023). Sleepers were removed during the period 2010–2012 in the lower part of the study area, with the aim of recovering longitudinal connectivity, where some agricultural land use persists and no danger in urban flooding could occur (Romero 2023). Since then, the river changed shape and looked like a river by self-healing hydro-morphological processes. That is, no rehabilitation nor restoration project was applied but removing sleepers along nearly 3 km south from Granollers urban area, while the sleepers remained in the upper area. Under these circumstances, the study assesses, by means of indicators, the degree of the auto-healing process and the improvement in the hydrogeomorphology process using a hydrogeomorphology index (IHG, Ollero et al. 2008). The study area was divided into two sections: The Upper Section is the unrestored river part that is closer to the north and central part of the city of Granollers, and Lower Section is the restored or auto-healed part due to the removal of sleepers, and it is located in the south and the peri-urban area of the city. The Upper Section is about 4 km long while the Lower Section is 2.7 km long (Fig. 4.1).

The IHG index is a method for the assessment of the functioning of the fluvial system and the riparian and river channel quality based on nine hydro-geomorphological parameters (Ollero et al. 2008). The index focuses on the human impact on fluvial systems and it is organized in three main sections: the degree or quality of the functionality of the fluvial system, the quality of the riverbed and the quality of the channel and the riparian section. Each section evaluates the degree of naturality of three subsections: water discharge, solid discharge, flooding space, morphology and shape of the section longitudinally, vertically and transversally, and the continuity of the river corridor, the width and the internal structure of the channel (Ollero et al. 2021). The index grades from 0 to 10 each subsection and the maximum possible score is 90 (30 points for each section). The score obtained by the index is classified in 5 different grades or ranges of quality depending on the total score obtained from the evaluation (Table 4.1). The lower score is given to the poorest geomorphological environments which are deeply transformed, while the highest score is given to natural rivers with little or no transformation.

Table 4.1 Scores obtained from the IHG index according to Ollero et al. (2008)

Evaluation	IHG score
Very good	75–90
Good	60–74
Moderate	42–59
Poor	21–41
Very poor	0–20

4.3 Results

4.3.1 Hydro-Geomorphological Index Scores

Scores obtained on the sampled sites in the study section were very low in the unrestored sampling sites in comparison with those on the restored area where the quality rating obtained is moderate (Table 4.2).

The river improves, geomorphologically speaking, from the upper to the lower sections given that the scores obtained for the IHG classify the Upper Section as a *very poor*, while the Lower Section is classified as a *moderate* rating. According to results, the index has detected an improvement in the Lower Section in the morphology of the river channel and a mobility and availability of sediment, resulting in the development of specific fluvial forms such as bars and a development of a braided channel form. In addition, it also gives high scores in terms of flow continuity and longitudinal connectivity given the absence of transversal structures. On the other site, the index evaluates negatively the items related to the structure and quality of the riparian forest, the naturality of the flow regime and the width limitations, given that the main channel walls have not been removed and the river has limitation on lateral migration and remains disconnected from its floodplains.The Upstream Section is totally fixed, preventing the river to move nor take any action that a river should do, as the resulting score from the IHG reveals.

However, the transition between both images (Fig. 4.3a, b) indicates that river processes are hidden, diminished or mitigated until the limiting factor disappears and then, new governing fluvial conditions are set. Although the IHG index measures the naturality of hydro-geomorphological processes in terms of mobility, it provides a static image of the state of a river reach in a specific time, and it cannot evaluate the magnitude of the processes hidden or undergoing that may be degrading or, on the contrary, improving the river reach. It is important to sequentially apply the index to identify such processes that may, sometimes, not be as positive as they could be, even if they provide an improvement in the scores on the IHG index.

Table 4.2 Results from the IHG index evaluation on the study section (Farguell et al. 2022)

Sampling site	River status	IHG score	Quality rating
Upper Section P1	Unrestored	10	Very poor
Upper Section P2	Unrestored	8	Very poor
Lower Section P3	Restored	46	Moderate
Lower Section P4	Restored	51	Moderate

Fig. 4.3 Images of the river Congost (Granollers, Barcelona). In **a** the river is channelized and immobilized by parallel walls and sleepers to avoid lateral migration, while in **b** the hydro-geomorphological processes and the channel's morphology have been recovered following the elimination of these roads (*Source* Images taken from Google Earth 2021)

4.3.2 Unleashing Fluvial Processes in the Restored Site

The improvement in the scores obtained by the IHG is mainly due to the recovery of river processes, basically sediment transport, and the longitudinal continuity of flow and sediment in the river channel, which is one of the main functions of a river system (Ollero et al. 2008; Church 2015). The recovery of processes and the transformation of the river morphology are due to the work that the river has been able to do according to the magnitude and frequency of flood events and the size and amount of sediment since the removal of the sleepers and other fixing structures. In other words, the river is healing itself by seeking to adjust to the new governing conditions of slope, flow frequency and magnitude and the size and quantity of sediment availability.

However, not all processes recovered lead to a good result. The restoration of the Congost River is incomplete because the main channelization walls and the disconnection between the riverbed and the flood area remain. That limits the fully recover of processes and the river auto-healing. In fact, this is consistent with results obtained from the index IHG applied. The scores obtained for the restored sections are higher than those obtained for the unrestored sections (Fig. 4.2 and Table 4.2), but they fell on the *moderate* quality range. Some of the consequences of the incomplete restoration can be summarized as follows:

(a) Limitation of the Lateral Migration of the River

Lateral migration has been possible by removing the sleepers and the riverbank control structures. The river has been moving along the available width to seek an adjustment with the slope. The river recovers the natural slope and thus the energy to entrain the sediment in motion and redistribute it and organize it according to the magnitude of the flood event. However, the width is still limited. In 1945 (Fig. 4.4a), the width of the river was three times greater than the current width (Fig. 4.4b), and the current morphology of the river, in a braided form, reminds that one from existing in 1945. This channel limitation has further consequences on the river as explained below.

(b) Channel Incision

According to the river processes previously described (Church 2002, 2015; Batalla et al. 2020; Batalla 2021), the stream power is dependent on the slope and the depth of water, and the greater both variables are the greater the energy of the river. In impervious channels, this implies a high velocity of water flowing through the channel, but if channels are natural, then there is energy to move sediment and reshape the morphology of channel, as it occurs in the Congost River. The removal of sleepers allowed the river to recover the natural slope (1% in the whole study area) (Farguell et al. 2022), gaining energy and capacity to transport sediment and reshape the riverbed. The river has created sediment bars and a system of step-pools structure. Given the limitation of the expansion space of the river, more water is concentrated on a small channel, with a higher stage and thus, greater energy to move sediment. The result is that the river is causing incision within the bed channel in the effort of expanding laterally and posing a threat for the integrity of the infrastructures (Fig. 4.5). Unless additional lateral space is given to the river to expand, walls will collapse and fall within the next flood events as the water flux is concentrating on the sides looking for a wider river ridge.

(c) Sediment supply and availability

Although the river has the capacity to mobilize, redistribute and resettle sediment and reshape the riverbed, sediment sources are limited. The sediment that is being mobilized in the river is the one that was previously evenly distributed during the sleeper's installation (Fig. 4.2c). In the Upper Section, sediment availability is limited, and floods, apparently, do not provide sediment upstream from the study area (Figs. 4.4a and 4.6), while in the Lower Section much more sediment is available. Considering that there are not important tributaries entering within the river along the study area, a reasonable hypothesis to establish is that the sediment transported or mobilized may be aiding to increase the intensity of the incision processes within specific reaches of the study area, and also contributing to aggradation in the lowermost reaches.

(d) "Gardening"

Given the urban constraints and encroachment of the river corridor and the limitations or closed opportunities to river expansion that these conditions create, some

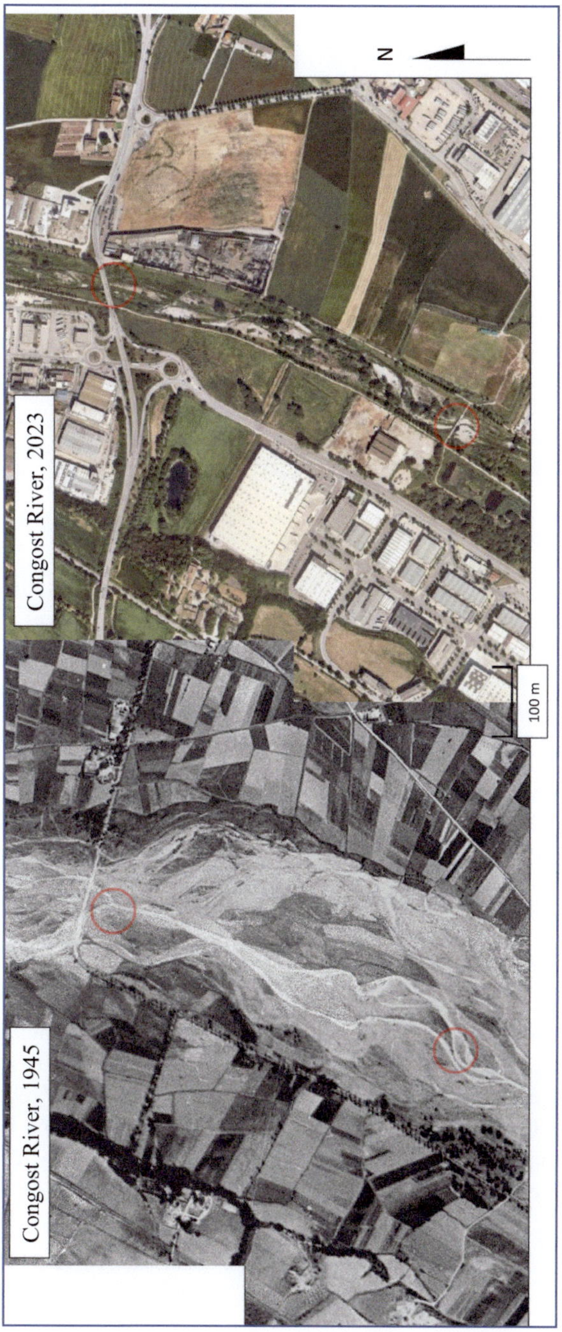

Fig. 4.4 Left image: Lower Section of the Congost River in 1945; Right image: Lower Section of the Congost River in 2023. Note the red circles in the left and right images to compare the same position in both images. River width in 2023 is three times smaller than in 1945, previous to engineering works (*Source* Cartographical and Geological Institute of Catalonia 2023)

Fig. 4.5 Examples of incision. The river tends to occupy a wider area than the currently available and erodes the river bed putting in danger the integrity of the infrastructures (*Source* Photos by the author 2022)

Fig. 4.6 Examples of sediment availability in **a** the Upper Section and **b** the Lower Section (*Source* Photos from the author 2022)

"gardening" projects may be appropriate to apply to find an intermediate balance between ecological goals and human issues (Kondolf 2011). These interventions are performed at a local scale and focused on a single item such as tree planting to faster create a riparian zone or create a bicycle lane or picnic area and must be maintained to ensure its purpose. Under these circumstances, the Congost River planners introduced areas to promote riparian vegetation and stabilize the river bank in a specific section to avoid lateral erosion once the river was able to migrate. They reduced the height of the artificial channel wall and smoothed the section to allow the growth of a riparian vegetation zone and increase biodiversity using bioengineering techniques and moved away the river flow to central part of the channel (Romero 2023; Fig. 4.7).

Fig. 4.7 Example of "gardening" when structures cannot be removed. **a** The original wall (2010); **b** wall covered, smoothed and seeded (2011); **c** current situation (2023) (*Source* Photos from the municipality of Granollers. Reproduced with permission)

4.4 Final Remarks and Conclusions

Although the river has recovered functionality, by reactivating hydrogeomorphological processes, and a characteristic river morphology has been developed, the auto-healing of the river is limited due to the physical barriers such as the levees or channel walls and the disconnection between the river and the floodplain (Kondolf 2011). The Congost River has a certain degree of recuperation or auto-healing because the riverbed is not impervious, and the stream power is relatively high to entrain bed material in motion. Flood events have reshaped the morphology of the river, and the river has created a step-pool sequence that has led to a certain ecological recovery. These comments agree with the scores recorded by the IHG index, which denote medium range values that are still far from a full geomorphological river recovery.

The IHG index has been applied in several studies and it is a good approach to provide information about the pressures and impacts of human interventions on fluvial processes (Ollero et al. 2008, 2021). The index cannot, however, evaluate or assess undergoing processes because it represents a static image of the river at a specific instant. Although sediment availability and mobility are scored as a good indicator, the index cannot decipher whether it is increasing incision or aggradation and at what stage. It is clear though, that it is needed a monitoring program either to detect the situation of the processes involved and the consequent evolution of forms, fluvial processes and also whether the river tends to incision or aggradation.

Thus, river process-based restoration and auto-healing can be achieved depending on the situation that a river has been transformed by human activities. The more the river is encroached by impervious surfaces and the smaller the river power the more difficult the restoration, and possibly it will be never achieved (Kondolf 2011). Moreover, it must be born in mind that fluvial systems that have been deeply altered or transformed will never recover a pre-transformation aspect and function (Dufour and Piégay 2009) and "gardening" might be the only possible solution (Kondolf 2011).

The Congost River restored section has a medium potential for self-healing. It is encroached by impervious surfaces, but the river has power given that the slope is still high. The restored section has shown that river processes have been reactivated but, at the same time, incision problems on the fluvial channel due to the limitation on the lateral migration and sediment dynamics and availability produce undesired effects. Despite this, planners and restorers should bear in mind that the effects of restoration practices or the auto-healing river processes will be seen at a medium or long term, depending on the magnitude and frequency of flood events occurring in the study section, and indicators or other follow-up measures should be clearly established to detect unexpected changes of river dynamics.

Due to the river limitations by urban constraints, it is difficult that the Congost River may improve unless the spatial planning in the Lower Section keeps the surrounding land as no urban and provide a greater space to the river to widen and develop a complete riverbank. To do so, planners, geomorphologists and inhabitants must agree to increase the space of liberty of the river. However, this proposal rises some questions: What is the limit in urban restoration proposals? Are we ready to allow an urban river to flood? and if so, what are the desired or hypothesized effects of such restoration plans? Research, planners and citizens will have to gather and discuss the future of the urban rivers.

Acknowledgements The authors acknowledge the financial support obtained from the Municipality of Granollers to carry out the study according to the resolution E-2371/2022 "Conservació llera riu Congost" and the research contained in the technical report and the Fundació Bosch i Gimpera to manage the project 3117013. This study has benefitted from the scientific and economic support of the Mediterranean Environmental Research Group (GRAM) of the University of Barcelona, (2021SGR00859), awarded by the Agency for the Management of University and Research Grants (AGAUR) of the Catalan Government (SGR 2021-2024).

References

Agència Catalana de l'Aigua (2022) Daily discharge of the Congost River at La Garriga gauging station. http://aca-web.gencat.cat/sdim21/filtre.do

Batalla RJ (2021) Reflexión sobre dinámica morfosedimentaria. Implicaciones para la gestion fluvial en un context de cambio global. Cuadernos De Geografía 107:175–190. https://doi.org/10.7203/CGUV.107.21372

Batalla RJ, Vericat D, Farguell J, Úbeda X, García C (2020) Processos hidrològics i geomorfològics als rius: context i exemples per a interpreter la seva resposta a episodis d'alta magnitude com el Glòria. Treballs De La Societat Catalana De Geografia 89:55–87. https://doi.org/10.2436/20.3002.01.191

Church M (2002) Geomorphic thresholds in riverine landscapes. Freshwater Ecol 47:541–557. https://doi.org/10.1046/j.1365-2427.2002.00919.x

Church M (2015) Channel stability: morphodynamics and morphology of rivers. In Rowinski P, Radecki A (eds) Rivers—physical, fluvial and environmental processes. Springer, Geoplanet: Earth and Planetary Sciences, p 613

Dufour S, Piégay H (2009) From the myth of a lost paradise to targeted river restoration: forget natural references and focus on human benefits. River Res Appl 25:568–581. https://doi.org/10.1002/rra.1239

Farguell J, Ochoa L, Chavez J (2022) Anàlisi geomorfològica i ambiental del riu Congost al seu pas per Granollers. Unpublished technical report, University of Barcelona

Gregory KJ (2006) The human role in changing river channels. Geomorphology 79:172–191. https://doi.org/10.1016/j.geomorph.2006.06.018

Gregory KJ, Chin A (2002) Urban stream channel hazards. Area 34(3):312–321. https://doi.org/10.1111/1475-4762.00085

Kondolf M (2011) Setting goals in river restoration: when and where can the river "Heal Itself"? In Stream restoration in dynamic fluvial systems: scientific approaches, analyses, and tools. Geophys Monograph Series 194:544, AGU. https://doi.org/10.1029/2010GM001020

Leopold LB, Wolman MG, Miller JP (1964) Fluvial processes in geomorphology. San Francisco, W.H. Freeman and Co

Martin-Vide JP (1998) Ingenriería fluvial. Politext, Àrea d'enginyeria civil. Edicions UPC, p 209

Ollero A, Ballarín D, Díaz B, Mora D, Sánchez M, Acín V, Echeverría MT, Granado D, Ibisate A, Sánchez L, Sánchez N (2008) IHG: Un índice para la valoración hidrogeomorfológica de sistemas fluviales. Limnetica 27(1):171–188. https://doi.org/10.23818/limn.27.14

Ollero A, Ballarín D, García H, Ibisate A, Mora D, Sánchez M (2021) Diagnístico fluvial, impactos en cauces y cambio global: aplicaciones del índice hidrogeomorfológico IHG. Geographicalia 73:295–316

Romero X (2023) Nature-based solutions in Granollers: from ecosystem restoration to the circular economy. Nature-Based Solutions 3. https://doi.org/10.1016/j.nbsj.2023.100072

Schumm SA (1977) The fluvial system, Ed. Wiley, Chichester, New York, 388

Servei Meteorològic de Catalunya (2022) Historic Rainfall data of Granollers (1950–2022). https://www.meteo.cat/wpweb/climatologia/dades-i-productes-climatics/series-climatiques-des-de-1950/

Chapter 5
Post-Project Appraisal in Hydromorphological River Restoration: Application to the Manzanares River at El Pardo (Madrid, Spain)

José Anastasio Fernández-Yuste and Carolina Martínez Santa-María

Abstract Post-project appraisal is a key component of river restoration scheme. It allows the success and failure of a project to be evaluated and provides essential information for proper adaptive management. This chapter presents a set of indicators whose objective is to evaluate the hydromorphological effectiveness of the actions included in the River Manzanares Restoration Project, in the area surrounding the Real Sitio de El Pardo (Madrid). All the proposed indicators evaluate variables indicative of processes, are based on the hydromorphological and ecological characteristics, have a precise and relatively simple estimation method, and their results are easily interpretable. The results obtained show the usefulness of the indicators (i) to evaluate the hydromorphological effect of the restoration actions carried out, (ii) as an adaptive management tool that will allow us to propose improvements and generate guidelines for other similar projects.

Keywords Peri-urban River · Boulder cluster · Hydraulic biotope · Benthic macroinvertebrates

J. A. Fernández-Yuste · C. Martínez Santa-María (✉)
Department of Forest and Environmental Engineering and Management, School of Forest Engineering and Natural Resources, Universidad Politécnica de Madrid, C/José Antonio Novais 10, 28040 Madrid, Spain
e-mail: carolina.martinez@upm.es

J. A. Fernández-Yuste
e-mail: tasio.fyuste@upm.es

© The Author(s), under exclusive license to Springer Nature Switzerland AG 2024
J. Farguell Pérez and A. Santasusagna Riu (eds.), *Urban and Metropolitan Rivers*, The Urban Book Series, https://doi.org/10.1007/978-3-031-62641-8_5

5.1 Introduction

River ecosystems are key elements for the environmental health of the planet because they are hubs of biodiversity, connect territories, build landscapes, and when they flow into the sea, they provide sediments and nutrients essential for life in transitional and coastal waters. They also provide essential ecosystem services.

The increase in human's capacity to modify the territory over the last 100 years has resulted in anthropogenic activities that have caused accelerated processes of degradation of river ecosystems, with losses of biodiversity and ecosystem services (Grill et al. 2015; Cantonati et al. 2020). The main causes of these alterations are linked to the construction of dams, channelling, occupation of river space (Schmutz and Sendzimir 2018) and discharges that affect water quality.

Urban streams are one of the most degraded aquatic ecosystems in the world (Francis and Hoggart 2008; Booth et al. 2016). These streams support the accumulation of anthropogenic pressures in their catchments, and direct alterations to their channels, banks and riparian zones by channelling, loss of space, urbanization and impermeabilization (Giller 2005). The effects of these pressures can generate significant alterations in the flow regime—affects the frequency, magnitude and duration—sediment delivery—affects the dynamic channel morphology and the variety and availability of habitats—(Zerega, Simões and Feio 2021), high concentrations of nutrients and pollutants—reduces biotic richness and increases dominance of tolerant species (Walsh et al. 2005; Smucker and Detenbeck 2014)—and reduces or removes river mobility space (Konrad and Booth 2005; Napieralski and Carvalhaes 2016).

Although the desire to recreate past conditions is tempting, science has shown that river systems follow complex trajectories that often make it impossible to return to an earlier state (Dufour and Piégay 2009; Yarnell et al. 2015). In urban areas, the restoration of streams—return to natural conditions—is most often not realistic due to the numerous unavoidable constraints brought by the urban environment (buildings, roads…) that cannot be removed. In addition, alterations to the environment may have started a long time ago and already caused dramatic changes in the structure and function of the stream (Booth 2005; Shoredits and Clayton 2013). Therefore, restoration objectives have been gradually moving away from the reference state due to the difficulty of achieving that state. The reference-based strategy (restoration) should be progressively replaced by an objective-based strategy (rehabilitation) (Batalla 2022), which contemplates the reclamation of as many of the stream's natural (predevelopment) components and functions as possible (Bradshaw 2002; Findlay and Taylor 2006).

Rehabilitation projects in urban streams should take into account (Lemm and Feld 2017; Rubin, Kondolf and Rios-Touma 2017; England et al. 2021; An et al. 2022; de Milleville et al. 2022): (i) historical information about the system to be restored, (ii) state and altered processes of the watershed and their influence both reach structural complexity and functional integrity, (iii) limitations for river space recovery, (iv) the expected future changes in land use, including climate change and (v) information from ecologically similar but less altered areas, or expert opinion based on empirical data and/or theoretical models.

Morphological dynamics of rivers are nowadays seen as a major part of the ecological river status (Grabowski and Gurnell 2016; Rinaldi et al. 2017). Any rehabilitation project should consider river morphology as a key element to be considered at all stages—characterization of the condition, diagnosis and proposal of actions—and their assessment is a crucial step in the development of management and restoration measures (Golfieri, Surian and Hardersen 2018).

These scientific considerations have also reached the legal framework. At the European level, the obligation to analyse and consider the hydromorphological component, in addition to the EU Water Framework Directive (WFD), is included in a wide spectrum of legislative frameworks and instruments (e.g., EU Birds and Habitats Directives, the European reference framework on Nature-Based Solutions, European Green Pact, the EU Biodiversity Strategy 2030, EU Green Infrastructure Strategy and Nature Restoration Act).

Therefore, projects for the rehabilitation of urban and peri-urban stream have also incorporated in their objectives the rehabilitation of geomorphological components and processes, as a necessary step to improve the composition and structure of the aquatic and riparian biota (An et al. 2022; de Milleville et al. 2022; Müller et al. 2022; Ghaforpur-Anbaran et al. 2023). However, most projects have not implemented effective monitoring to assess the outcome of hydromorphological restoration (Katz et al. 2007; Friberg et al. 2016).

The objective of this chapter is to (i) justify the importance of post-project appraisal hydromorphological response to river restoration, (ii) provide general guidelines on how to establish monitoring and post-project appraisal protocols and (iii) present a study case in a peri-urban stream in Madrid (Spain), one of the major European capitals.

5.2 Post-Project Appraisal (PPA)

Despite the large number of river restoration projects—mainly in USA, Europe, New Zealand, Australia and recently China—and the significant investments behind these projects, there is still no objective information on their successes and failures.

Monitoring and PPA are a key components of river restoration scheme (Fig. 5.1). It allows the success and failure of a project to be evaluated from the point of view of the initial design objectives and specific success criteria and provides essential information for proper adaptive management.

Including monitoring and post-project appraisal in a restoration scheme will (Kondolf and Micheli 1995; Skinner and Bruce-Burgess 2005; Environment Agency (Thames Region) 2007; Erwin et al. 2016; Belletti et al. 2018; England et al. 2021):

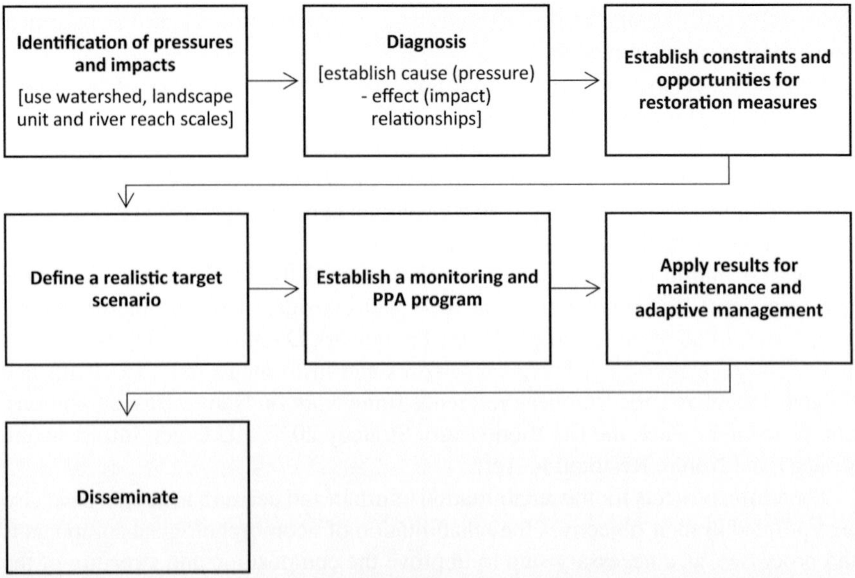

Fig. 5.1 Key components of river restoration scheme (adapted from Skinner and Bruce-Burges, 2005 and Environment Agency (Thames region), 2007)

- Demonstrate project effectiveness, providing further evidence on the benefits of restoration projects in achieving environmental objectives and recovery of ecosystem services.
- Identify the need for further restoration work and increase levels of confidence in the effects of restoration actions.
- Evaluate the changes associated with restoration in the short, medium and long term on components, processes and/or functions of the river ecosystem.
- Assess the relationships between physical changes and ecological responses.
- Identify which are the main factors for success or failure of the restoration: characteristics of the watershed and/or reach that limit or enhancing the effects of restoration actions.
- Promote advances in the science and practice of river restoration and understand which restoration measures provide the best results in different scenarios.
- Help establish PPA protocols and set timelines and costs.
- Ensure legislative compliance.

An adequate monitoring strategy and PPA requires (Toronto and Region Conservation Authority Geomorphic Solutions 2009):

1. Identification of Project Goals and Objectives.

   The complete and accurate definition of project goals and objectives is a critical component of the monitoring and PPA program design. It allows writing the questions that the monitoring program should answer and selecting the appropriate data collection activities.

2. Selection of Success Criteria.

The success criteria should be objective and allow any practitioner to assess whether the project achieves the objectives to an acceptable level.

The criteria should be established considering reference scenarios—not necessarily pristine or minimally altered—and the specific constraints of the reach and the project. The reference scenarios should be monitored at the same time and with the same methods as the project reach, thus generating comparable data to identify background variability.

The control scenarios—reaches with characteristics similar to those of the project in which no actions are carried out—should be monitored with the same methods and frequencies as the project reach. If no reference scenario is available, changes between the control and the project tranche can provide evidence of success or failure.

Monitoring during construction is also very important, since during this stage it is common for circumstances to arise that require modifications not included in the project. These modifications can significantly affect the objectives and should be considered when establishing the monitoring program.

3. Spatial Extent of Monitoring.

Ideally, monitoring should characterize and document the status of the entire project reach. However, in most projects, this is not economically feasible. In such cases, it is necessary to select portions of the reach that are representative of the project.

4. Duration and Frequency of Monitoring.

There is a general consensus that the minimum monitoring duration for stream restoration projects should be between three and five years, and that the ideal monitoring duration for a complex restoration initiative is at least ten years (Downs and Kondolf 2002; Toronto and Region Conservation Authority Geomorphic Solutions 2009) to be able to credit physical and biological changes and the synergies and dynamic processes between them.

Ten years of monitoring will increase expenditure, but this increase will be very small in proportion to the cost and effort associated with project design and construction and may make the difference between project success and failure over the long term.

The frequency of monitoring varies depending on the actions planned for the project. Each type of action will have expected effects with rates of change in monitoring variables that may be very different. Therefore, the frequency should be tailored to the characteristics of the effect being assessed.

5. Timing

The monitoring of both the reference scenario and the control scenario and those of the project sections should be carried out at the same time of the year to reduce the noise caused by the lack of simultaneity.

6. Selection of Variables and Monitoring Methods

   The monitoring parameters should be selected with efficiency criteria: to generate the best information to evaluate the achievement of the goals, objectives and success criteria with the minimum cost and effort. It is highly advisable to include variables linked to geomorphological process drivers (Columbia conservation district 2021). It should be ensured that the methods are standardized, or at least generally accepted by the scientific community.
7. Documentation and Data Management

   Monitoring should be documented in a report that includes objectives, variables, methodologies, success criteria and all the data obtained, together with any incidents that may have occurred during the process. It should allow professionals outside the project to adequately interpret all the information recorded.
8. Analysis of the Results

   In most projects it will be difficult to perform a rigorous statistical analysis of the quantitative monitoring data because the spatial and temporal extent of data collection is usually very limited by practical and economic constraints. For most parameters, the quality of the analysis will depend on it being carried out by skilled professionals using numerical or graphical comparison of data from different monitoring locations (i.e. project, reference, control), considering reach and/or project constraints and different monitoring events to assess the type and degree of change in the project area and progress towards success in the context of the established criteria.
9. Application and Adaptive Management.

   Partial results can be used to establish improvement actions in the reach or modify those already implemented. Considering the great variability of river types and their communities, their different degree of alteration and potential reversibility, the different scales of resilience of the physical and biological environments and the complex and dynamic interrelationships between hydromorphology and aquatic and riparian biota, it is very difficult to find two similar cases (O'Brien et al. 2017). Therefore, it is very important to take advantage of the results of each specific case to establish adaptive management guidelines.
10. Dissemination

   Considering the scarce knowledge currently available on PPA, it is very important to disseminate the results in order to contribute to improve the knowledge of the effects of river restoration projects (Erwin et al. 2016).

In the specific case of the hydromorphological status assessment most of the methods were developed in Europe, USA or Australia, e.g. MQI (Rinaldi et al. 2013), PHM (Dirección general del agua 2019.), CEMs (Schumm, Harvey and Watson 1984), RSF (Brierley and Fryirs 2008).

Numerous papers have been published in recent years that can serve as a guide for monitoring and evaluating the effects of river restoration on hydromorphology

(Environment Agency (Thames Region) 2007; Erwin et al. 2016; Klösch and Habersack 2017; Rinaldi et al. 2017; Belletti et al. 2018; Columbia conservation district 2021; González del Tánago et al. 2021; Hinshaw et al. 2022; Mondal and Patel 2022; Müller et al. 2022). The reach characteristics, the project objectives, the time and money, are variables that, among many others, will determine the selection of one of these methods or the use of recommendations of several of them.

5.3 Study Case

5.3.1 Study Area

The Manzanares river flows ~90 km from its headwaters in the Guadarrama Mountains to its outlet into Jarama river. Since the 1970s, the Manzanares river has been subject to extensive flow regulation: Santillana reservoir was constructed (1969) at the headwaters to provide water to Madrid, and El Pardo reservoir was constructed (1975) to provide flood protection. It has also been straightened, channelized and diked within the municipality of Madrid.

The study reach is located on the middle Manzanares River, downstream of the "El Pardo" dam, within the municipality of Madrid and through the village of "El Pardo" (Fig. 5.2). In 2016, the *Confederación Hidrográfica del Tajo* (official water management agency) approved the restoration project for this reach, with the purpose of mitigating the main impacts on the river and its riparian area (Fig. 5.2 and Table 5.1). Project designers translated the objectives into techniques and works suitable for accomplishing the goals (Table 5.2). In 2018, the actions included in the project were carried out, and in 2019 the post-project appraisal program was initiated. Table 5.3 summarizes the highlights of this program. This paper presents only the results most directly related to morphological components and processes.

5.3.2 Project Actions, PPA Data Collections and Methods

To provide diversity in hydraulic biotope and increase refuges for fish, nature-based solutions for restoring the rivers (Addy et al. 2016) are considered in the restoration project.

Boulder clusters (Fig. 5.3) were installed to improve diversity in bed particle size, depth and velocity profiles and macroinvertebrate community. These in-stream structures are groups of boulders placed in the stream channel to provide, under a variety of flow conditions, a diversity of water depth, substrate and velocity, thereby increasing habitat diversity of an otherwise homogeneous streambed conditions (Oregon Department of Forestry 2010; Cramer 2012; Iowa Department of Natural Resources 2018; Park, Lee and Kim 2018). In the project, the boulder clusters are a

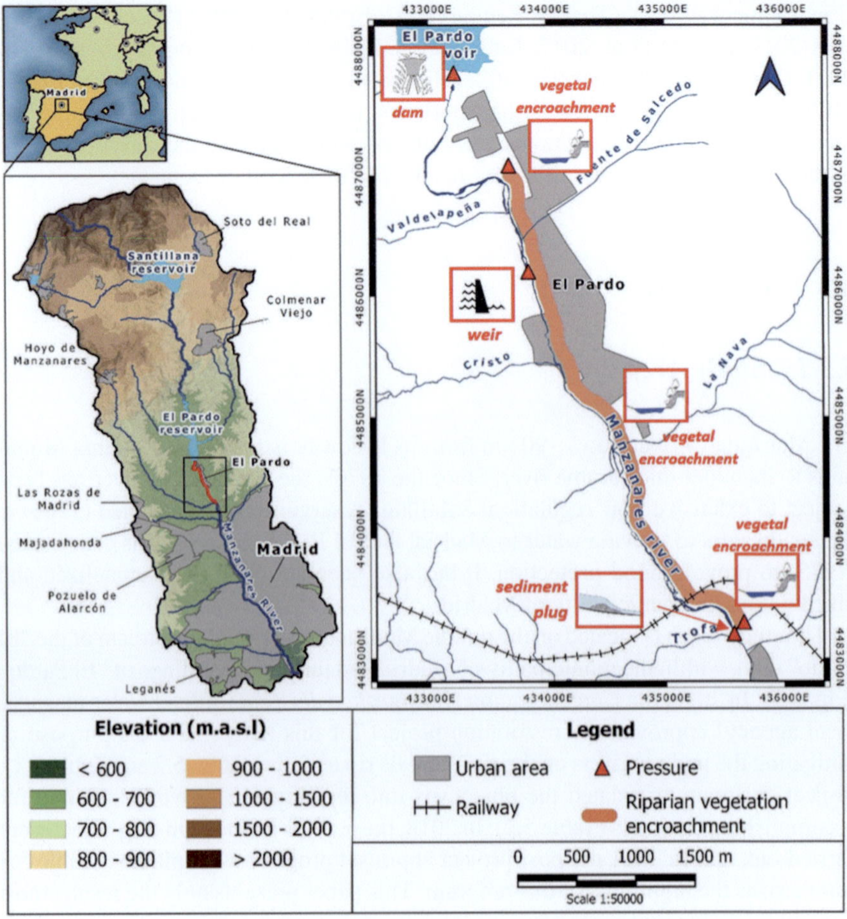

Fig. 5.2 Study area with indication of the main existing pressures

pilot test to analyse their effectiveness; with the results of the appraisal program, it is expected to assess their use in more points of the river.

For the appraisal program, four river cross sections were established: three of them were located in the zone of influence of the boulder cluster and the fourth (control section) on a section not affected by these actions. In 2017—pre-project situation—the four sections were marked in the field and bed particle size, depth, velocity and macroinvertebrate were estimated. Post-project data were taken in those same sections in 2019, 2020, 2021 and 2022. All measurements were taken at the same time of the year and at similar streamflow (range: 1.3–1.5 m^3/s).

To increase the available refuges for fish, root wads and large wood complexes structures (Fig. 5.4) were installed. Root wads and large woody provide a variety of cover habitats for fish, aquatic insects and wildlife (Hicks and Reeves 1994; Sylte

Table 5.1 Pressures and impacts in the study reach

Pressures	Impacts	Effects		
		Components	Processes	Functions
"El Pardo" dam	Alteration of the hydro-sedimentological regime: –Reduction of flows –Loss of inter–and intranual variability –Loss of habitual, geomorphological and connectivity with floodplain floods	Ichthyofauna: –Proliferation of exotic species due to homogenization of the flow regime –Riverbank vegetation: loss of succession processes due to homogenization of the flow regime	Hydromorphological dynamic: –Stabilization of the riverbed –Homogenization of mesohabitats –Homogenization of the aquatic biotope	Corridor: –Loss of longitudinal connectivity Habitat: –Loss of aquatic and riparian biotope diversity
Riparian vegetation encroachment	Alteration of the river mobility space	Riparian space: –Loss of riparian space due to channelling	Morphological dynamic: –Loss of capacity for lateral mobility	Corridor: –Loss of lateral connectivity Social use: –Limitations on access to the river

(continued)

Table 5.1 (continued)

Pressures	Impacts	Effects		
		Components	Processes	Functions
Weir	Alteration of longitudinal continuity: –Impassable to fish Creation of backwater: –Permanent increase in water level –Change from lotic to lentic conditions	Ichthyofauna: –Prevents upward migration for spawning –The backwater provides an aquatic biotope suitable for exotic species Riverbank vegetation: –Loss of vegetation due to permanent waterlogging –Changes in composition and density of bank margin Helophytes: –The backwater provides an aquatic biotope suitable for helophytes proliferation	Morphological dynamic: –Loss of fluvial dynamism –Sedimentation processes and interruption of sediment flow	Corridor: –Loss of longitudinal connectivity Habitat: –Loss of aquatic and riparian biotope diversity
Sedimentation at the confluence of the Trofa creek with the Manzanares river: development of a “weir or plug” that blocks the river and grows progressively with each episode of sediment entrainment	Creation of backwater: –Change from lotic to lentic conditions	Ichthyofauna: –The backwater provides an aquatic biotope suitable for exotic species Riverbank vegetation: –Loss of vegetation due to permanent waterlogging –Changes in composition and density of bank margin Helophytes: –The backwater provides an aquatic biotope suitable for helophytes proliferation	Morphological dynamic: –Loss of fluvial dynamism –Sedimentation processes and interruption of sediment flow	Hábitat: –Loss of aquatic and riparian biotope diversity

Table 5.2 Pressures, project actions to correct/mitigate and goals

Pressures	Limitations	Opportunities	Actions	Goals
"El Pardo" dam	It must maintain the capacity for flood protection in Madri	The legal obligation to establish a regime of ecological flows will make it possible to recover floods that clean the bed and activate the geomorphological dynamics	Proposal for prescribed floods—for bed clearing and geomorphological dynamics—in the ecological flow regime Create in-stream structures for aquatic biotope diversification	–Increase the diversity of mesohabitats –To guarantee the suitability of the bed for the spawning of autochthonous species –To mobilize bed materials by increasing the heterogeneity of the substrate –Create refuges for ichthyofauna –Diversify depth, velocities and substrate to improve physical and hydraulic habitats –Increase benthic macroinvertebrate diversity
Riparian vegetation encroachment	Houses and streets in the fluvial space in approximately half of the section	Possibility of laying artificial slopes and bank restoration on 1/3 of the reach	Recover riparian space and plantation Creation of public access areas	–To recover the composition and diversity of the vegetation in the recovered riparian space –Favouring access to the river for citizens

(continued)

Table 5.2 (continued)

Pressures	Limitations	Opportunities	Actions	Goals
Weir	It must be maintained so that it can continue to provide water supply services for forest fire defence facilities	Possibility to reduce the height of the weir Possibility to build a fish passage ramp	Reducing the crest height Construction of a natural fish ramp	–Reduce backwater, reduce surface occupied by helophytes, restore lotic environment –Recovering previously flooded river space –Restore continuity of the upstream and downstream movement of fish and other organisms
Sedimentation at the confluence of the Trofa creek with the Manzanares river: development of a "weir or plug" that blocks the river and grows progressively with each episode of sediment entrainment	Nones	Proactive attitude of municipalities and administrations with land in the basin to solve erosion processes	Initiate an action project in the catchment area of the stream to control erosion processes Remove sedimenta "plug"	–To develop the action project in the Trofa stream basin –Reduce backwater, reduce surface occupied by helophytes, restore lotic environment –Recovering previously flooded river space

Table 5. 3 Project actions and indicators for appraisal. Only the indicators presented in this case study are included

Actions	Purpose	Indicators of change	Field methods	Spatial scale	Monitoring frequency and duration
Create engineering structures for aquatic biotope diversification	Increase aquatic biotope diversity	Depth, velocity, substrate granulometry	Bathymetry, flow measurement and characterization of the substrate granulometry	Representative sections of project (3) and control (1)	Pre-construction monitoring Post-construction once per monitoring year
		IBMWP index (Benthic macroinvertebrates index)	IBMWP protocol		
	Increase refuge for fish	Persistence and functionality	Field verification of persistence and functionality with expert judgment	Representative structures of project (6 structures)	Post-construction once per monitoring year
Remove sediment plug	Reduce backwater Restore lotic environment	Surface occupied by backwater Length occupied by backwater	Unmanned aircraft system: Aerial images georeferenced and orthorectified	Total reach length	Pre-remove monitoring Post-remove once per monitoring year
	Reduce surface occupied by helophytes	Surface occupied by helophytes			

Fig. 5.3 Boulder cluster. Streamlines are drawn: In blue those causing increased velocity, entrainment of silt and sand and thus increased sediment size and depth. In yellow those causing backflow, with low velocities, deposition of fine sand and reduction of sediment diameter and depth

and Fischenich 2000; Bureau of Reclamation and US Army Corps of Engineers 2015; Iowa Department of Natural Resources 2018; Heaton et al. 2022). Over the four years of assessment, these in-stream structures were only evaluated visually, checking only their integrity by experts.

Boulder cluster, root wads and large wood were placed in the bed as a module that was repeated in the selected reaches (Fig. 5.5).

To reduce the backwater area, the Trofa plug was removed. The helophytes were removed mechanically (amphibious backhoe). The effects of these actions were assessed with orthophotographs taken from unmanned aircraft systems (UAS) (Erwin et al. 2016; Layzell et al. 2022). In 2017, the flight was made to establish the pre-project condition and was repeated in 2019, 2020, 2021 and 2022. All flights were done at the same time of the year and at similar streamflow.

Fig. 5.4 Root wads and large woody to increase habitat variety for fish, aquatic insects and wildlife

Fig. 5.5 Boulder cluster, root wads and large wood module placed on the river

5.3.3 Bed Particle Size: Proposed Methodology for Assessing Changein Substrate Granulometric Diversity.

In the aquatic (i.e., submerged) cross sections samples of the substrate were taken every 0.5 m and classified according to Table 5.4. The Shannon–Wiener-Index—SWI—(Spellerberg and Fedor, 2003), defined to assess species diversity, has also been used to assess substrate diversity (Jähnig and Lorenz 2008; Verdonschot et al. 2016; Belletti et al. 2018):

$$\text{fi} = \text{relative frequency of class } i \text{ in the section} = \frac{\text{length of substrate class } i \text{ in section}}{\text{section length}}$$

i = number of substrate classes in the section

As an indicator of the diversity of particle size in the substrate, we use the Shannon equitability index (SEI): Shannon diversity index divided by the maximum diversity:

$$\text{SEI} = \frac{\text{SWI}}{\ln(k)}$$

$\ln(k)$ = maximum SWI value. Occurs when all substrate classes have the same relative frequency.

This normalizes SWI index to a value between 0 and 1. Less values of SEI indicate lower diversity while higher values indicate more diversity.

Table 5.4 Values of the indicators IQR -depth, velocity-, SWI, SEI, IBMWP and granulometry classification

Section		Year	RIC		No. sub.	Types of substrate (USC, Unified soil classification) %						SWI	SEI	IBMWP
			Velocity	Depth		Fines (<0.075 mm)	Fine sand (>0.075 mm)	Medium sand	Coarse sand	Gravel	Cobbles			
RS 1	Pre	2017	0.37	0.39	3	24.0	0.0	48.0	0.0	0.0	28.0	1.05	0.33	57
		2019	0.35	0.26	3	15.0	56.0	0 0	0.0	0.0	28.0	0.97	0.30	36
	Post	2020	0.29	0.31	4	24.0	3.0	24.0	0.0	0.0	44.0	1.25	0.33	57
		2021	0.30	0.36	3	8.0	0.0	16.0	0.0	0.0	76.0	0.70	0.22	47
		2022	0.41	0.47	3	24.0	0.0	0.0	24.0	0.0	52.0	1.03	0.32	36
RS 2	Pre	2017	0.40	0.31	3	14.3	28.6	0.0	57.1	0.0	0.0	0,96	0.31	54
		2019	0.30	0.20	4	4.8	23.3	61.9	9.5	0.0	0.0	101	0.33	65
	Post	2020	0.24	0.20	3	0.0	33.3	52.4	14.3	0.0	0.0	0.98	0.32	39
		2021	0.20	0.30	3	0.0	33.3	47.6	10.0	0.0	0.0	1.04	0.34	37
		2022	0.13	0.30	3	0.0	19.0	28.6	52.4	0.0	0.0	101	0.33	30
	Pre	2017	0.24	0.68	3	48.1	22.2	22.2	0.0	0.0	7.4	1.21	0.37	33
		2019	0.13	0.34	4	25.9	29.6	22.2	0.0	0.0	22.2	1.38	0.42	51
RS 3	Post	2020	0.16	0.24	3	0.0	66.7	7.4	0.0	0.0	25.3	0.81	0.25	41
		2021	0.13	0.34	2	0.0	66.7	0.0	0.0	0.0	33.3	0.64	0.19	19
		2022	0.16	0.30	5	3.7	37.0	37.0	7.4	0.0	14.8	1.33	0.40	27
	Pre	2017	0.48	0.50	3	4.8	19.0	76.2	0.0	0.0	0.0	0.67	0.22	43
		2018	0.27	0.37	2	33.3	0.0	66.7	0.0	0.0	0.0	0.64	0.21	46
RS control	Post	2020	0.29	0.35	2	19.0	0.0	81.0	0.0	0.0	0.0	0.43	0.16	33
		2021	0.11	0.40	2	0.0	33.7	0.0	14.3	0.0	0.0	0.41	0.13	40
		2022	0.12	0.15	2	0.0	90.5	3.5	0.0	0.0	0.0	0.31	0.10	37

5.3.4 Water Depth and Velocity Profiles: Proposed Methodology for Assessing Change in Hydraulic Diversity

Depth and velocity were measured every 0.5 m. Depth measuring accuracy was 0.5 cm. Current velocity was measured at 0.6 of the water depth from the water surface with an accuracy of ± 0.015 m/s (for the range 0–3 m/s).

To evaluate the evolution over time of depth and velocity diversity, in each section and for each measurement campaign (one per year), the interquartile range (IQR = difference between the first quartile—the 25th percentile—and the third quartile—the 75th percentile) was calculated. The benefit of using IQR to measure the spread of values in a dataset instead of range (difference between the minimum and maximum value in a dataset) or coefficient of variation (Jähnig and Lorenz 2008) is that IQR not affected by extreme values. Increase in the IQR means an increase in the diversity of the variable.

5.3.5 Benthic Macroinvertebrates

Sampling of the sections has been performed according to the official Spanish norm (Ministerio de medio ambiente 2013). As an indicator of the quality of the benthic fauna, the official Spanish index (IBMWP) was used (Alba-Tercedor et al. 2002), and computed according to "protocol for calculating the IBMWP index" (Ministerio de medio ambiente 2013). This index combines the number of taxa at the family level with a tolerance value of the different taxa to water quality loss. The final value is obtained as the sum of the tolerance values of each of the identified families, ranging from 0 (maximum tolerance to pollution) to 10 (minimum tolerance to pollution). A higher IBMWP value involves a greater diversity of families with greater intolerance to contamination.

5.3.6 *Refuge for Fish*

Due to budget limitations, fish use monitoring of root wads and large wood debris could not be carried out. It was limited to checking their persistence, integrity and that, in the opinion of the experts, they were functional. Consequently, only post-project data were taken, years 2019–2022.

5.3.7 *Backwater, Lotic Environment and Helophytes*

With the orthophotographs taken from UAS, the surface and length occupied by the backwaters were measured. The change in surface is used as indicator of the effect of the action on the pressure. The change in backwater length is used as an indicator of the recovery of the lotic condition. The change in the area occupied by helophytes is the indicator of the effect of the actions carried out to control helophytes spread.

5.4 Results

Table 5.4 shows the values of the indicators—interquartile range (IQR) for velocity and depth, Shannon equitability index (SEI) for substrate and IBMWP for macroinvertebrates. To allow comparisons, for each section and for each indicator, the percentage change with respect to the pre-project value has been calculated (Table 5.5).

$$\%\text{change of indicator}\, p_k^j = \frac{\text{value of the indicator}\, \mathrm{p}_\mathrm{k}^\mathrm{j}}{\text{value of the indicator}\, p_{2017}^j}$$

where:

$p =$ indicator (IQR_{vel}, IQR_{depth}, SEI, IBMWP)
$j =$ section (1,2,3, control)
$k =$ year (2019, 2020, 2021, 2022)

Table 5.5 Values of the indicators of backwaters: length, surface and surface occupied by helophytes

			Length (m)	Surface (m^2)	Surface occupied by helophytes (m^2)
Backwater: weir	Pre	2017	491	13,360	5661
	Post	2019	471	11,366	3776
		2020	469	15,363	4377
		2021	468	15,651	4147
		2022	468	15,846	4738
Backwater: confluence with Trofa stream	Pre	2017	1783	73,626	14,610
	Post	2019	1057	28,728	9047
		2020	1075	28,978	7149
		2021	1104	30,068	5881
		2022	1144	31,163	7964

Figure 5.6 shows the temporal evolution of the % change. The % changes have been calculated for the four sampling campaigns compared to the 2017 sampling. Figure 5.6A show that the % change in IQR depth increases from 2017 to 2020 in RS1 and RS2 and decreases in RS Control and RS3, although this decrease in RS3 has been smaller than that estimated in the control section. With respect to the % change in IQR velocity (Fig. 5.6B) only RS2 increases. In the remaining sections the indicator decreases although this decrease has been smaller in RS1 and RS3 than that estimated in the control section. In the sample period, the SEI indicator (Fig. 5.6C) has practically not changed in the three sections and has decreased significantly in the control section. In IBMWP (Fig. 5.6D) there is a reduction in the % change in all sections, being lower in the control section.

Figure 5.7 presents the averages of the % change, grouping the sections that reflect the effects of the actions. In general, the values of the four indicators decrease with respect to pre-project, although this reduction is always greater in the control section, except in the case of IBMWP.

These results highlight the importance of establishing a control section: the actions carried out have improved the values of the indicators—except in IBMWP—with respect to the pre-project values.

Table 5.5 shows the results obtained in the monitoring of the two backwaters in the stretch. The indicators measured were length and surface of the backwater and surface occupied by helophytes. As with the previous indicators, the % change with respect to the reference campaign (2017) is also calculated. The results are shown in Fig. 5.8.

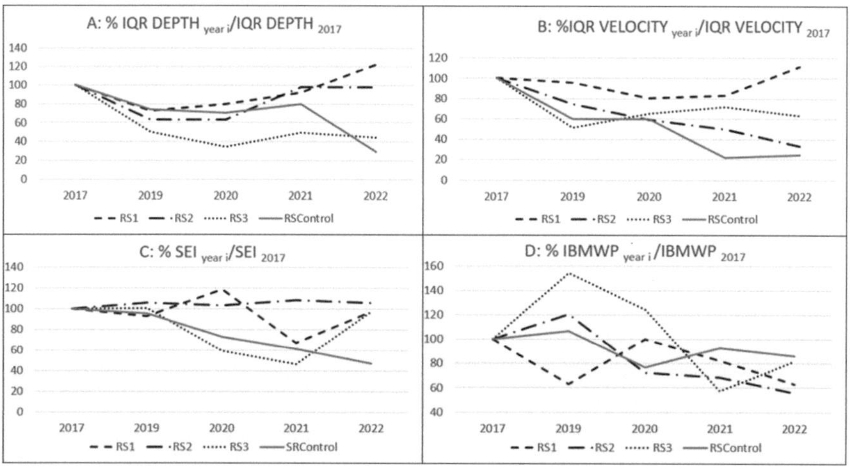

Fig. 5.6 Temporal evolution of the % change: **A** indicator IQR vel; **B** IQR depth; **C** SEI; **D** IBMWP. The % changes have been calculated for the four sampling campaigns compared to the 2017 sampling.

These results show the favourable effects of the actions with respect to the initial situation in all the variables measured, except for the backwater surface of the weir, which increases slightly.

5.5 Discussion and Conclusion

5.5.1 Boulder Clusters

Compared to the pre-project state, the boulder clusters have not improved diversity in terms of depth, velocity, substrate or macroinvertebrates. However, the reduction compared to the control section has been smaller, except for macroinvertebrates (Fig. 5.5). These poor results of the boulder clusters can be explained by considering the drivers of the bed mobilization processes. The arrangement of the boulders generates streamlines that in some areas produce an increase in velocity and shear stresses, with a greater capacity to move materials, both in size and quantity. In other areas velocities are reduced and smaller particles are not entrained. These processes are expected to occur around boulders. But for this to occur and to generate notable changes in the diversity of sediment size and in depths and velocities, the presence of boulders is not enough. It needs to act as a driver of the process: flows with the capacity to induce effective bed mobilization, generating diversity in substrate, draughts and velocities (Kasvi et al. 2017).

The regulation generated by the "El Pardo" dam is very intense, with a practically constant flow regime and very low values, a regulation which has intensified in recent years (Fig. 5.9). In the restoration project, it was foreseen to establish prescribed flows to guarantee the opportunity to mobilize the bed materials. However, these flows were not released by those responsible for the dam (see Table 5.6 with percentiles of exceedance of 1, 5 and 10% of natural and post-dam regime). The management of the dam has left the river without the opportunity to mobilize bed particles: the absence of the driver explains the limited effect of the boulders.

In the case of macroinvertebrates, the assumption was that the improvement in substrate diversity, depth and velocity would increase habitat diversity. This increase should lead to an increase in macroinvertebrate diversity. The results have shown that this hypothesis has not been fulfilled.

Increasing habitat heterogeneity to increase macroinvertebrate biodiversity does not always have significant effects (Friberg et al. 2014). If not removed or mitigated, environmental stressors, such as water quality, catchment land use and flow alterations, source populations of species are lacking, the effects of increased habitat diversity may be irrelevant (Kitto et al. 2015; Verdonschot et al. 2016; Schneider and Petrin 2017).

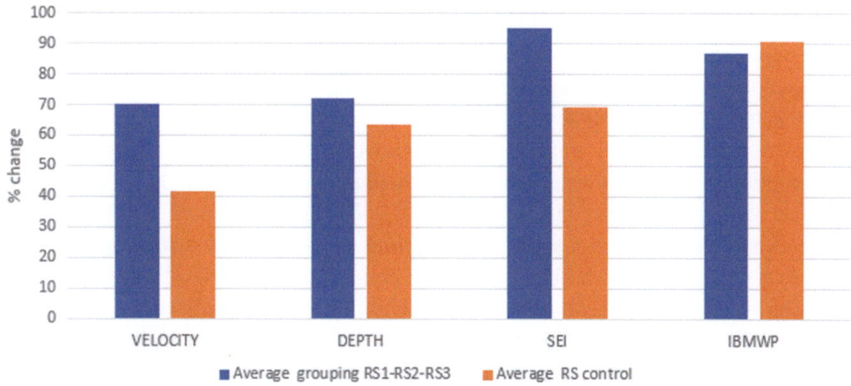

Fig. 5.7 Average of the % change grouping the sections with project actions and in the control sections

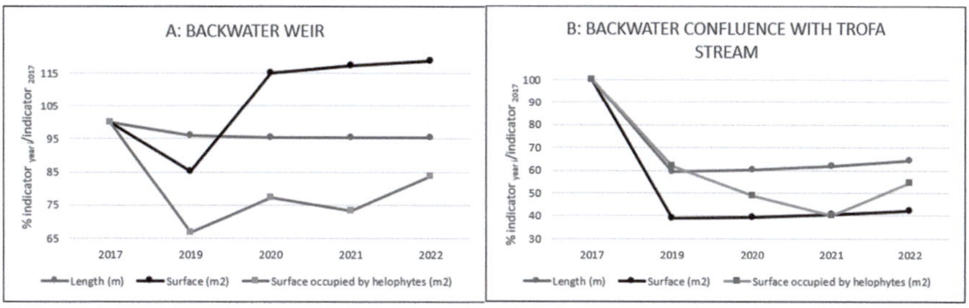

Fig. 5.8 **Temporal evolution of the % change in backwaters indicators**. A: weir; B: confluence with Trofa stream. The % changes have been calculated for the four sampling campaigns compared to the 2017 sampling

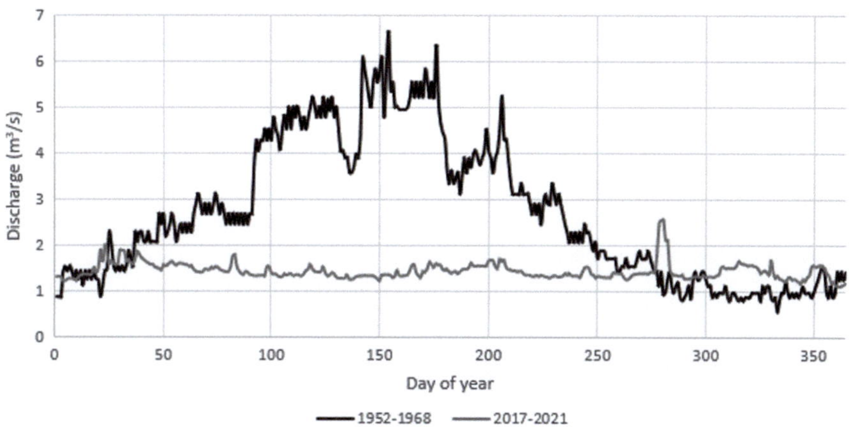

Fig. 5.9 Median annual hydrograph developed from mean daily discharge data at 3070 GS for the pre-project period (1952–1968) and post-project period (2017–2021)

In general, in the sections with boulder clusters, there was no increase in diversity in the parameters studied (depth, velocity and substrate), but rather, with very localized exceptions, diversity decreased. However, this decrease was much lower than that detected in the control section.

However, macroinvertebrates have not followed this behaviour. In the reach there is a sampling point of the water quality monitoring network. The PO_4^{-3} concentration data recorded presents a mean value (2018–2022) of 0.35 mg/l. This concentration does not exceed the threshold for assigning the water body a status worse than good (0.5 mg/l), but it is a high value. The concentration of NH_4^+ has an average value during the same period of 0.45 mg/l. This concentration also does not exceed the threshold for a worse than good status (0.6 mg/l), but it is also a high value. These high values of PO_4^{-3} and NH_4^+ may be responsible for the practically null response of the macroinvertebrate community, either by themselves or by influencing other aspects of the habitat—e.g. development of filamentous algae covering the substrate (Cattaneo et al. 2013; Oberholster et al. 2017).

In conclusion, the use of boulder clusters in future actions in the reach to increase substrate diversity, depths, velocities and macroinvertebrate community should only be considered if (i) it is ensured that the dam managers are willing to provide prescribed floods that activate sediment dynamics processes—the minimum flow value to ensure this activation must be estimated using river morphology equations that ensure the removal of bed materials in the vicinity of the boulder—and (ii) the quality thresholds of authorized releases are revised to ensure a reduction in PO_4^{-3} and NH_4^+ concentrations.

5.5.2 Refuge for Fish

Monitoring of fish refugia limited to assessing their persistence and functionality is not sufficient. It is necessary to establish a sampling protocol (Bureau of Reclamation and US Army Corps of Engineers 2015) to make an objective assessment of the use of these refuges before proposing that they be applied to other reaches of the Manzanares River. Being in a peri-urban environment and upstream of an urban reach with a high population density, it is also necessary to establish a risk assessment to evaluate the effect of the mobilization of root wads and large wood (Bureau of Reclamation and US Army Corps of Engineers 2015; Thorne et al. 2015).

5.5.3 Removed Trofa Stream Plug

The effect in reducing the backwater has been noticeable, but monitoring shows that the backwater tends to grow again. Considering that the causes of the sediment plug caused by the Trofa stream, it is necessary (i) after each major storm, to check the confluence in case it is necessary to dredge the sediments and thus prevent the plug from developing again and (ii) to address an action project in the Trofa stream catchment to reduce the processes of erosion and sediment transport.

5.5.4 *Reducing the Weir Crest Height*

Due to the low slope of the thalweg, the effect of this action (reduction of 0.5 m) has not been very noticeable in the indicators monitored. However, there is evidence of a considerable improvement in the availability of aquatic and riparian habitat, with the recovery of bank areas and the emergence of previously flooded islands that have been rapidly colonized by vegetation.

5.5.5 *Mechanical Removal of Helophytes*

The reduction in the area occupied by helophytes has also been notable. It is important to bear in mind that the diverse and efficient reproductive strategies of these species mean that their mechanized removal will only have a short-term effect if they are not accompanied by actions that increase shading and create depth and velocity conditions that are detrimental to them (Fortier et al. 2010; Deltoro Torró et al. 2012; Carley et al. 2016; Sender, 2016). In this case, willow staking, poplar planting and the emerged surface have allowed this reduction to be maintained in the areas where these works have been carried out.

As a final conclusion, the most important one, this study case shows how a PPA has made it possible to: (i) know the effects of the actions carried out, (ii) evaluate their efficiency with respect to the objectives, (iii) identify pressures that have conditioned the results, (iv) provide guidance to establish actions in reaches that present a similar situation and (v) propose adaptation measures that complement the actions of the initial project and enhance its effects.

References

Alba-Tercedor J et al (2002) 'Caracterización del estado ecológico de ríos mediterráneos ibéricos mediante el índice IBMWP (antes BMWP').', Limnetica, 21(3–4). https://doi.org/10.23818/limn.21.24

An JH et al (2022) Evaluation on the restoration effects in the river restoration projects practiced in South Korea. Water 14(17):2739. https://doi.org/10.3390/w14172739

Batalla RJ (2022) 'Reflexión sobre dinámica morfosedimentaria. Implicaciones para la gestión fluvial en un contexto de cambio global', Cuadernos de Geografía de la Universitat de València, (107). https://doi.org/10.7203/cguv.107.21372

Belletti B et al (2018) Assessing restoration effects on river hydromorphology using the process-based morphological quality index in eight European river reaches. Environ Manage 61(1). https://doi.org/10.1007/s00267-017-0961-x

Booth DB (2005) Challenges and prospects for restoring urban streams: a perspective from the Pacific Northwest of North America. J North Am Benthological Soc. 10.18990887-3593(2005)024\2.0.CO;2

Booth DB et al (2016) Global perspectives on the urban stream syndrome. J Freshwater Sci 35(1):412–420. https://doi.org/10.1086/684940

Bradshaw AD (2002) Introduction and philosophy. In Handbook of ecological restoration. Cambridge: Cambridge University Press, pp 3–9

Brierley GJ, Fryirs KA (2008) Geomorphology and river management: applications of the river styles framework, geomorphology and river management: applications of the river styles framework. https://doi.org/10.1002/9780470751367

Bureau of Reclamation and US Army Corps of Engineers (2015) National large wood manual: assessment, planning, design, and maintenance of large wood in fluvial ecosystems: restoring process, function, and structure. Seattle

Cantonati M et al (2020) Characteristics, main impacts, and stewardship of natural and artificial freshwater environments: consequences for biodiversity conservation. Water (Switzerland). https://doi.org/10.3390/w12010260

Carley J et al (2016) Response of macrophyte communities to flow regulation in mountain streams. Environmental

Cattaneo A et al (2013) Hydrological control of filamentous green algae in a large fluvial lake (Lake Saint-Pierre, St. Lawrence River, Canada). J Great Lakes Res 39(3):409–419. https://doi.org/10.1016/j.jglr.2013.06.005

Columbia conservation district (2021) Geomorphic assessment and restoration prioritization. Dayton

Cramer ML (ed) (2012) Stream habitat restoration guidelines. Olympia, Washington: Co-published by the Washington Departments of Fish and Wildlife, Natural Resources, Transportation and Ecology, Washington State Recreation and Conservation Office, Puget Sound Partnership, and the U.S. Fish and Wildlife Service. Available at: https://wdfw.wa.gov/sites/default/files/publications/01374/wdfw01374.pdf (Accessed: 12 March 2023)

Deltoro Torró V, Jiménez Ruiz J, Vilán Fragueiro X (2012) Bases para el manejo y control de Arundo donax L. Available at: https://www.miteco.gob.es/va/ceneam/grupos-de-trabajo-y-seminarios/red-parques-nacionales/BasesparaelmanejoycontroldeArundodonax_tcm39–169319.pdf

Dirección general del agua (2019) Guía para la evaluación del estado de las aguas superficiales y subterráneas. Ministerio para la transición ecológica. https://www.miteco.gob.es/es/agua/publicaciones/guia-para-evaluacion-del-estado-aguas-superficiales-y-subterraneas_tcm30-514230.pdf

Downs PW, Kondolf GM (2002) Post-project appraisals in adaptive management of river channel restoration. Environ Manage. https://doi.org/10.1007/s00267-001-0035-X

Dufour S, Piégay H (2009) From the myth of a lost paradise to targeted river restoration: forget natural references and focus on human benefits. River Res Appl 25(5). https://doi.org/10.1002/rra.1239

England J et al (2021) Best practices for monitoring and assessing the ecological response to river restoration. Water (Switzerland). https://doi.org/10.3390/w13233352

Environment Agency (Thames Region) (2007) Geomorphological monitoring guidelines for river restoration schemes. Bristol

Erwin SO, Schmidt JC, Allred TM (2016) Post-project geomorphic assessment of a large process-based river restoration project. Geomorphology, 270. https://doi.org/10.1016/j.geomorph.2016.07.018

Findlay SJ, Taylor MP (2006) Why rehabilitate urban river systems? Area, 38(3). https://doi.org/10.1111/j.1475-4762.2006.00696.x

Fortier J et al (2010) Nutrient accumulation and carbon sequestration in 6-year-old hybrid poplars in multiclonal agricultural riparian buffer strips. Agricult Ecosyst Environ. Elsevier B.V., 137(3–4):276–287. https://doi.org/10.1016/j.agee.2010.02.013

Francis RA, Hoggart SPG (2008) Waste not, want not: the need to utilize existing artificial structures for habitat improvement along urban rivers. Restorat Ecol 16(3). https://doi.org/10.1111/j.1526-100X.2008.00434.x

Friberg N et al (2014) The River Gelså restoration revisited: habitat specific assemblages and persistence of the macroinvertebrate community over an 11-year period. Ecol Eng. Elsevier B.V., 66:150–157. https://doi.org/10.1016/j.ecoleng.2013.09.069

Friberg N et al (2016) Effective river restoration in the 21st century: from trial and error to novel evidence-based approaches. In Advances in Ecological Research. https://doi.org/10.1016/bs.aecr.2016.08.010

Ghaforpur-Anbaran P, Ahmadabadi A, Ghanavati E (2023) Hydro-Morphological Analysis of Karaj River in the Urban Area from Beylqan to the Railway Bridge. Geography Environ Sustain 13(1):21–39

Giller PS (2005) River restoration: seeking ecological standards. Editor's introduction. J Appl Ecol. https://doi.org/10.1111/j.1365-2664.2005.01020.x

Golfieri B, Surian N, Hardersen S (2018) Towards a more comprehensive assessment of river corridor conditions: a comparison between the morphological quality index and three biotic indices. Ecol Indicators, 84. https://doi.org/10.1016/j.ecolind.2017.09.011

González del Tánago M et al (2021) Improving river hydromorphological assessment through better integration of riparian vegetation: scientific evidence and guidelines. J Environ Manage. https://doi.org/10.1016/j.jenvman.2021.112730

Grabowski RC, Gurnell AM (2016) Hydrogeomorphology-ecology interactions in river systems. River Res Appl 32(2). https://doi.org/10.1002/rra.2974

Grill G et al (2015) An index-based framework for assessing patterns and trends in river fragmentation and flow regulation by global dams at multiple scales. Environ Res Lett 10(1). https://doi.org/10.1088/1748-9326/10/1/015001

Heaton MG, Grillmayer R, Imhof JG (2022) Ontario's stream rehabilitation manual. Streams O (ed). Belfountain, Ontario

Hicks BJ, Reeves GH (1994) Restoration of stream habitat for fish using in-stream structures. Restoration of Aquatic Habitats, (Kaufmann 1987), pp 67–91

Hinshaw S et al (2022) Development of a geomorphic monitoring strategy for stage 0 restoration in the South Fork McKenzie River, Oregon, USA. Earth Surface Processes and Landforms, 47(8). https://doi.org/10.1002/esp.5356

Iowa Department of Natural Resources (2018) River Restoration Toolbox Practice Guide

Jähnig SC, Lorenz AW (2008) Substrate-specific macroinvertebrate diversity patterns following stream restoration. Aquatic Sci 70(3). https://doi.org/10.1007/s00027-008-8042-0

Kasvi E et al (2017) Flow patterns and morphological changes in a sandy meander bend during a flood-spatially and temporally intensive ADCP measurement approach. Water (switzerland) 9(2):6–11. https://doi.org/10.3390/w9020106

Katz SL et al (2007) Freshwater habitat restoration actions in the Pacific Northwest: a decade's investment in habitat improvement. Restoration Ecol 15(3). https://doi.org/10.1111/j.1526-100X.2007.00245.x

Kitto JAJ et al (2015) Meta-community theory and stream restoration: evidence that spatial position constrains stream invertebrate communities in a mine impacted landscape. Restor Ecol 23(3):284–291. https://doi.org/10.1111/rec.12179

Klösch M, Habersack H (2017) The hydromorphological evaluation tool (HYMET). Geomorphology, 291. https://doi.org/10.1016/j.geomorph.2016.06.005

Kondolf G, Micheli E (1995) Evaluating stream restoration projects. Environ Manage 19:1–15

Konrad CP, Booth DB (2005) Hydrologic changes in urban streams and their ecological significance. Am Fisheries Soc Sympos 2005(47)

Layzell AL et al (2022) UAS-based assessment of streambank stabilization effectiveness in an incised river system. Geomorphology 408:108240. https://doi.org/10.1016/j.geomorph.2022.108240

Lemm JU, Feld CK (2017) Identification and interaction of multiple stressors in central European lowland rivers. Sci Total Environ 603–604. https://doi.org/10.1016/j.scitotenv.2017.06.092

Milleville de L et al (2022) The heterogeneity of the hydromorphological responses of a stream to the urbanization of its basin. Earth Surf Process Landforms. https://doi.org/10.1002/esp.5514

Ministerio de medio ambiente (2013) Protocolo de muestreo y laboratorio de fauna bentónica de invertebrados en ríos vadeables. Madrid. Available at: https://www.miteco.gob.es/es/agua/temas/estado-y-calidad-de-las-aguas/ML-Rv-I-2013_Muestreoylaboratorio_Faunabentónicadeinvertebrados_Ríosvadeables_24_05_2013_tcm30-175284.pdf (Accessed: 12 March 2023)

Mondal S, Patel PP (2022) Incorporating hydromorphological assessments in the fluvial geomorphology domain for transitioning towards restorative river science—context, concepts and criteria. In Fluvial Systems in the Anthropocene. Cham: Springer International Publishing, pp 43–75. https://doi.org/10.1007/978-3-031-11181-5_4

Müller H et al (2022) Hydromorphological assessment as the basis for ecosystem restoration in the Nanxi River Basin (China). Land, 11(2). https://doi.org/10.3390/land11020193

Napieralski JA, Carvalhaes T (2016) Urban stream deserts: Mapping a legacy of urbanization in the United States. Appl Geogr. https://doi.org/10.1016/j.apgeog.2015.12.008

O'Brien GR et al (2017) A geomorphic assessment to inform strategic stream restoration planning in the Middle Fork John Day Watershed, Oregon, USA. J Maps, 13(2). https://doi.org/10.1080/17445647.2017.1313787

Oberholster PJ et al (2017) The interplay between environmental conditions and filamentous algae mat formation in two agricultural influenced South African rivers. River Res Appl 33(3):388–402. https://doi.org/10.1002/rra.3081

Oregon Department of Forestry (2010) Guide to Placement of Wood, Boulders and Gravel for Habitat Restoration. Available at: https://www.roguenativeplants.org/guide-to-placement-of-wood-boulders-and-gravel-for-habitat-restoration/ (Accessed: 12 March 2023)

Park K, Lee KS, Kim YO (2018) Use of instream structure technique for aquatic habitat formation in ecological stream restoration. Sustainability (Switzerland), 10(11). https://doi.org/10.3390/su10114032

Rinaldi M et al (2013) A method for the assessment and analysis of the hydromorphological condition of Italian streams: the morphological quality index (MQI). Geomorphology, 180–181(null):96–108. https://doi.org/10.1016/j.geomorph.2012.09.009

Rinaldi M et al (2017) 'New tools for the hydromorphological assessment and monitoring of European streams. J Environ Manage Acad Press 202:363–378. https://doi.org/10.1016/J.JENVMAN.2016.11.036

Rubin Z, Kondolf GM, Rios-Touma B (2017) Evaluating stream restoration projects: what do we learn from monitoring? Water (Switzerland). https://doi.org/10.3390/w9030174

Schmutz, Sendzimir (2018) Riverine ecosystem management: science for governing towards a sustainable future. Aquatic Ecology Series

Schneider SC, Petrin Z (2017) Effects of flow regime on benthic algae and macroinvertebrates—a comparison between regulated and unregulated rivers. Sci Total Environ. Elsevier B.V., 579:1059–1072. https://doi.org/10.1016/j.scitotenv.2016.11.060

Sender J (2016) The effect of riparian forest shade on the structural characteristics of macrophytes in a mid-forest lake. Appl Ecol Environ Res 14(3):249–261. https://doi.org/10.15666/aeer/1403_249261

Shoredits AS, Clayton JA (2013) Assessing the practice and challenges of stream restoration in urbanized environments of the USA. Geography Compass, 7(5). https://doi.org/10.1111/gec3.12039

Skinner KS, Bruce-Burgess L (2005) Strategic and project level river restoration protocols—key components for meeting the requirements of the water framework directive (WFD). Water Environ J 19(2). https://doi.org/10.1111/j.1747-6593.2005.tb00561.x

Smucker NJ, Detenbeck NE (2014) Meta-analysis of lost ecosystem attributes in urban streams and the effectiveness of out-of-channel management practices. Restorat Ecol 22(6). https://doi.org/10.1111/rec.12134

Spellerberg IF, Fedor PJ (2003) A tribute to Claude-Shannon (1916–2001) and a plea for more rigorous use of species richness, species diversity and the "Shannon-Wiener" Index'. Glob Ecol Biogeogr 12(3). https://doi.org/10.1046/j.1466-822X.2003.00015.x

Sylte T, Fischenich C (2000) Rootwad composites for streambank erosion control and fish habitat enhancement. Habitat, 10

Thorne C et al (2015) Project risk screening matrix for river management and restoration. River Res Appl 31(5):611–626. https://doi.org/10.1002/rra.2753

Verdonschot RCM et al (2016) The role of benthic microhabitats in determining the effects of hydromorphological river restoration on macroinvertebrates. Hydrobiologia, 769(1). https://doi.org/10.1007/s10750-015-2575-8

Walsh CJ et al (2005) The urban stream syndrome: current knowledge and the search for a cure. J N Am Benthol Soc. https://doi.org/10.1899/04-028.1

Yarnell SM et al (2015) Functional flows in modified riverscapes: hydrographs, habitats and opportunities. BioScience. Oxford University Press, pp 963–972. https://doi.org/10.1093/biosci/biv102

Zerega A, Simões NE, Feio MJ (2021) How to improve the biological quality of urban streams? Reviewing the effect of hydromorphological alterations and rehabilitation measures on benthic invertebrates. Water (Switzerland). https://doi.org/10.3390/w13152087

Part II
Planning

Chapter 6
From De-Urbanisation to the Renaturation of Riverbanks in France: The Emergence of an Ecology of Reconciliation in Urban Design?

Sylvain Rode

Abstract Many of France's urban rivers have been the focus of redevelopment projects, with the main aim of turning their banks into attractive public spaces. In addition to emblematic projects in the central areas of metropolises this chapter focuses on more modest reclamation projects currently underway in small and medium-sized towns and/or on small urban watercourses. The aim is to decipher the multiple development issues simultaneously addressed by projects that take the flood risk issue as their starting point and respond to it by deciding to de-urbanise areas close to the river. How are these riverside areas redeveloped? Do these projects reflect a renewed vision of urban rivers and their development? At a time when the ecological and climate crisis requires us to reinvent our relationship with the living world and the environment, is the redevelopment of these de-urbanised areas helping to make more room for ecological and climate issues?

Keywords De-urbanisation · Ecological and climatic issues · Flood risk · Public spaces · Renaturation

6.1 Introduction

Urban waterfronts have been the focus of public attention for several decades. In North America, the issue of reclaiming riverbanks emerged in the 1960s, when port cities began to redevelop river and coastal areas abandoned by industry and port activities (Gravari-Barbas 1998). In France, Lyon was the first urban area to take up the challenge of reclaiming its rivers, the Rhône and the Saône and redeveloping their banks from the 1980s onwards (Gérardot 2004). Other major cities have followed suit: Bordeaux, Lille, Montpellier, Nantes, Paris and Toulouse. But medium-sized

S. Rode (✉)
Department of Geography and Planning, UMR 5281 ART-Dev, University of Perpignan Via Domitia (UPVD), Perpignan, France
e-mail: sylvain.rode@univ-perp.fr

© The Author(s), under exclusive license to Springer Nature Switzerland AG 2024
J. Farguell Pérez and A. Santasusagna Riu (eds.), *Urban and Metropolitan Rivers*, The Urban Book Series, https://doi.org/10.1007/978-3-031-62641-8_6

cities have also followed suit: Angers, Le Mans, Limoges, Nancy, Orléans, Perpignan, Poitiers, Rouen, Tours, etc., (Carcaud et al. 2019). These projects represent major challenges in terms of urban development. They should contribute to the revitalisation of local neighbourhoods and the renewal of the city's image. Big names in landscape and architecture have been called in (Michel Corajoud in Bordeaux, the TER agency in Toulouse, Jean Nouvel and Marc Mimram in Perpignan, etc.). More recently, small towns have also been concerned about their rivers and have set up development projects. From metropolises to small towns, from major rivers to small urban rivers, from rivers cut off from the city by dykes or expressways that we want to reclaim to buried rivers that we want to reopen (the Vilaine in Rennes, the Bièvre in the Paris region, etc.), the situations vary widely. However, the objective underlying all these projects is similar: to turn rivers and their banks back into genuine urban amenities, contributing to the quality of life in the city. To achieve this, the riverbanks are being redeveloped and transformed into attractive public spaces.

But while watercourses are mobilised in the urban narrative as amenities that convey positive values, their presence within the city is also synonymous with the risk of flooding. As the French geographer Paul Fénelon pointed out in relation to the Loire, a river is a sort of Janus with two faces (Fénelon 1978), a source of amenities and risks, both ally and enemy (Carré and Deutsch 2015). We know that, historically, rivers have helped to establish settlements and that many towns have been built on their banks. France is no exception, with the valleys of the country's main rivers representing major concentrations of population. Since the mid-twentieth century, urbanisation has developed in flood-prone areas in a largely uncontrolled manner, contributing to a sharp increase in the stakes. In France today, it is estimated that no fewer than 17.1 million permanent residents and 9 million jobs are exposed to the risk of flooding from overflowing rivers. Given these trends, it's easy to understand why the floods that regularly hit the country are causing ever-increasing damage. To tackle the risk of flooding, the public authorities have implemented several different strategies. After having favoured protection strategies through so-called structural measures (protective dykes, dams), since the 1990s the public authorities have favoured prevention strategies, by controlling urban development in flood-prone areas through the Natural Risk Prevention Plans (PPRN) created by the Barnier law of 1995. However, when certain forms of land use appear to be unsuited to the risk and highly vulnerable, exposing their inhabitants to a high level of danger, the public authorities sometimes resort to a radical solution: de-urbanising certain at-risk areas by relocating the issues at stake. This involves acquiring and then destroying properties close to a watercourse in order to free these areas from all construction and re-establish flood expansion zones. These correspond to "*areas that are not or only slightly urbanised and little developed, and where the flood can store a large volume of water, such as natural areas, farmland, urban and peri-urban green spaces, sports grounds, car parks,* etc."[1]. They allow floods to "*dissipate their energy at the cost*

[1] French circular of 24 January 1994 on the definition of floods and the management of flood-prone areas.

of limited risks to human life and property[2]". In terms of flood risk management, favouring the development of flood expansion zones over the erection of protective structures such as dykes and dams is seen by some as a "*shift from a 'vertical' strategy to a more 'horizontal' approach to hazard management*" (Fournier et al. 2021).

This chapter analyses four projects currently underway in four small French towns. What these projects have in common is that they are concerned with the river from the point of view of flood risk. They deal with this risk by de-urbanising areas close to the river. How are these riverside areas being redeveloped following their de-urbanisation? Do these projects reflect a renewed vision of urban rivers and their development? At a time when the ecological and climate crisis requires us to reinvent our relationship with the living world and the environment, is the redevelopment of these de-urbanised areas helping to make more room for ecological and climate issues?

6.2 Flood-Prone Areas to be De-Urbanised

As the French architect and urban planner Frédéric Bonnet points out, thinking about resilience leads to "*defining different strategic axes in the face of the hazard*" (Bonnet 2016, p. 19): while in some cases it is possible to adapt urban development to flooding by designing resilient neighbourhoods, in other cases it is necessary to act on the withdrawal by restoring certain sites to the natural environment, restoring "*margins of fluctuation*" (Rossano 2021, p. 252) for watercourses. This is why public authorities sometimes choose to de-urbanise certain areas deemed to be too risky by acquiring and then demolishing properties close to watercourses.

While such a choice comes up against many obstacles and resistance—economic and social acceptability, in particular (Rode 2008; Mineo-Kleiner 2017)—property buybacks/demolitions are nevertheless being implemented in France in several areas when the conditions are right. By de-urbanising, the aim is to give space back to the river and restore its floodplain. Such a choice echoes, for example, the "Room for the river" programme implemented in the Netherlands since 2006 (Ribas Palom et al. 2017). Land use is adapted to the natural functioning of the environment and to flooding. This means taking note of the unsuitability of the land use developed over time by human societies in relation to the natural functioning of the environment and trying to put things right.

Some relocations are implemented following a disaster. They can be described as curative, insofar as the aim is to resolve a situation of high vulnerability of an area and its inhabitants, revealed by the disaster and rendered socially unacceptable. This is the most common scenario. Following the flooding that hit the municipalities of Trèbes and Villegailhenc in the south of France in October 2018, it was decided to de-urbanise whole sections of the urban fabric close to watercourses. The aim is to reduce the vulnerability of these areas by eliminating the issues most exposed

[2] *Idem.*

to flooding, to avoid a repeat of the 2018 disaster. In Trèbes, no fewer than fifty buildings will be bought out and demolished, most of them on the right bank of the Aude in the Faubourg and Aiguille neighbourhoods (Fig. 6.1), where water levels were the highest (more than 2 m in habitable areas for two-storey houses, or more than 80 cm in the absence of a refuge area).

In Villegailhenc, 38 houses and 4 barns are to be bought out and demolished, concentrated mainly along the Trapel, in the old town centre at the confluence with the Merdeau (Fig. 6.2).

The aim is to make more room for water, with the removal of the urban fabric making it possible to widen the bed of the Trapel as it flows through the old town centre.

Other relocations can be described as preventive. They are decided on and implemented outside a situation of emergency and post-disaster emotion. Such a decision to de-urbanise neighbourhoods that have not recently been hit by flooding as a preventive measure is more difficult to take and then to accept in the face of incomprehension and discontent among the residents concerned. Nevertheless, there are a few examples of this situation in France. In Blois, for example, the local authority decided in 2003 to restore the Bouillie spillway (Fig. 6.3).

The aim was to restore this area, which had become urbanised over time, to a non-urbanised floodplain. This meant relocating all the buildings, people and activities of the two districts that had developed there over time. Not only were the 400 residents of the Glacis and Fouleraie districts, located in the axis of the spillway, directly exposed to violent flooding from the Loire, but the safety of the residents of the neighbouring Blois-Vienne dyke district was also threatened, as the spillway's flow capacity was no longer guaranteed. The aim of the public authorities was to acquire all the properties located in the spillway to de-urbanise it and restore it to working order. By the beginning of 2023, the de-urbanisation of the site was almost complete, with 135 of the 143 properties acquired and 131 destroyed.

In Villeneuve-Saint-Georges, a town on the outskirts of Paris that was severely flooded in June 2016 and again in January 2018, the town council decided in 2010 to buy back several houses in the Belleplace-Blandin district on the banks of the Yerres river out of court to demolish them (Fig. 6.4).

Following the flooding in January 2018, the local mayor wanted to extend the project to the entire highly flood-prone area, believing that this was the only solution that would provide lasting protection for the residents. In total, no fewer than 174 plots of land will be acquired and around a hundred buildings demolished in this area on the banks of the Yerres (a tributary of the Seine) (Table 6.1).

6.3 Imagining New Functions for De-Urbanised Spaces

The de-urbanisation of these flood-prone areas close to rivers raises major issues for their redevelopment. The void created in the urban space by the disappearance of built-up areas needs to be thought through and qualified (Sowa 2022). What new

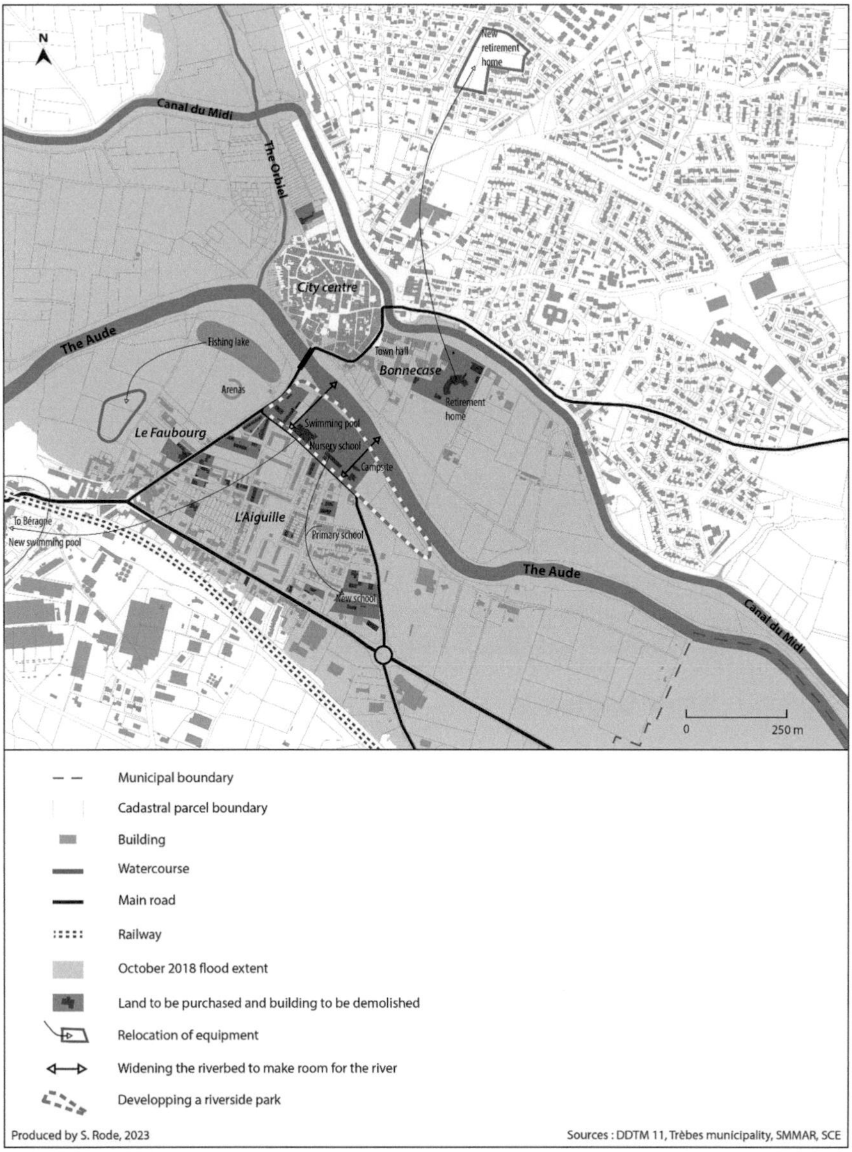

Fig. 6.1 De-urbanisation and gentle redevelopment of the banks of the Aude in Trèbes (*Source* Sylvain Rode 2023)

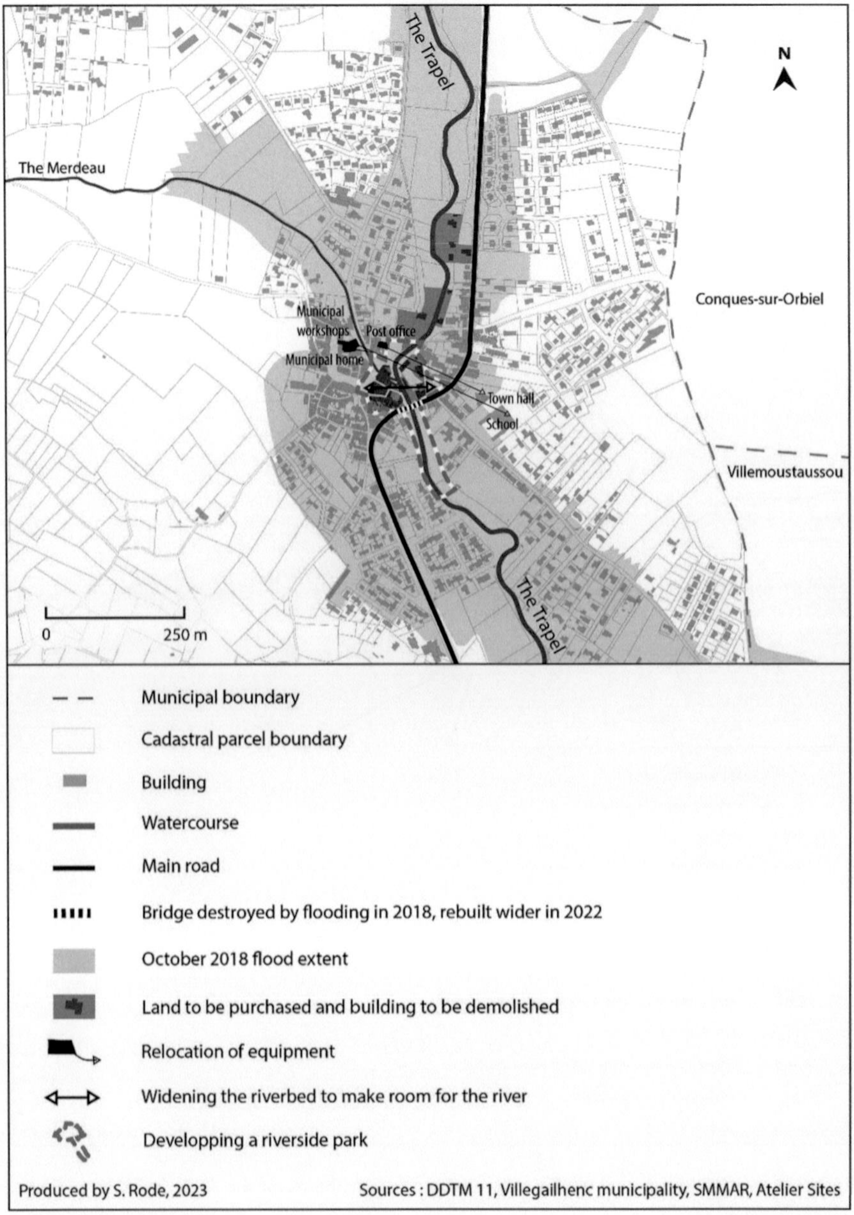

Fig. 6.2 De-urbanisation and gentle redevelopment of the banks of the Trapel in Villegailhen (*Source* Sylvain Rode 2023)

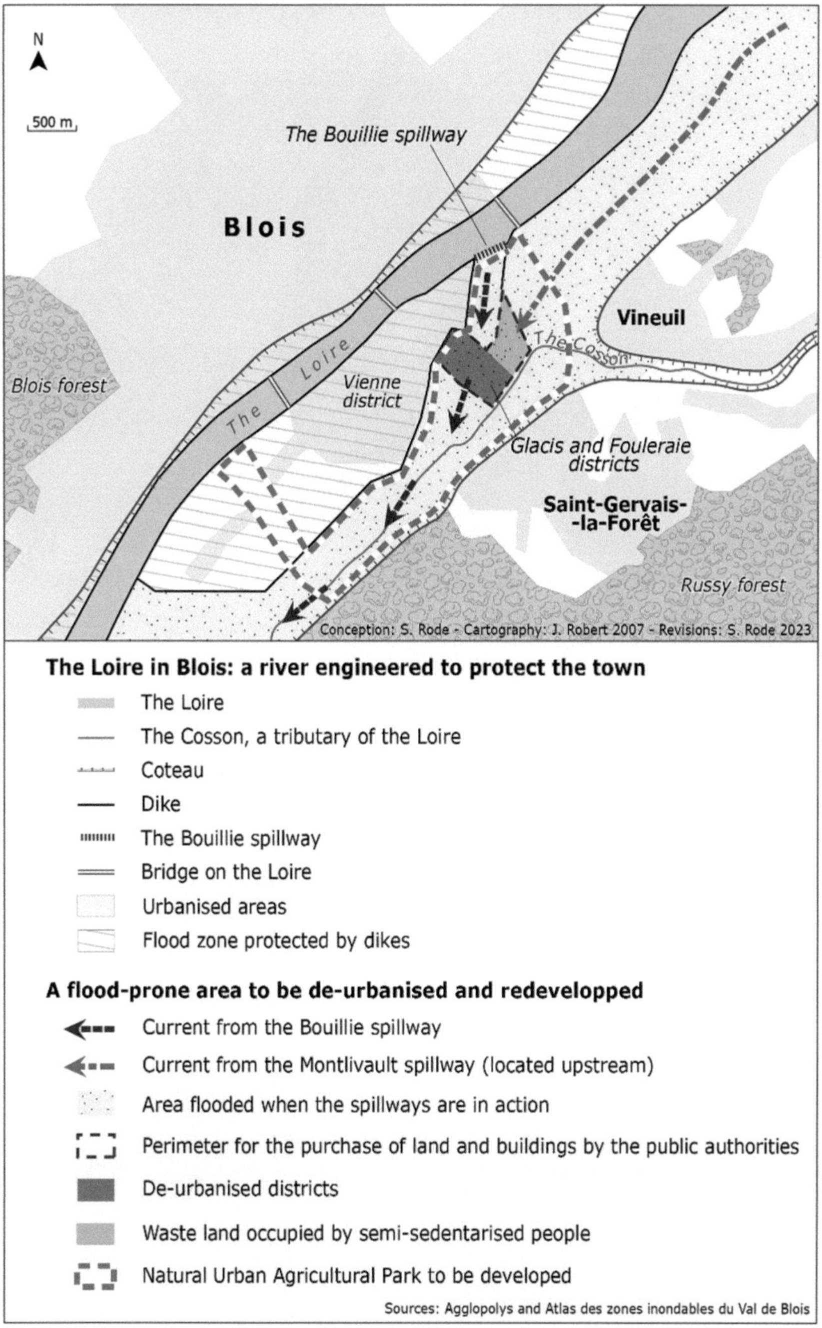

Fig. 6.3 De-urbanisation of a Loire spillway in Blois (*Source* Sylvain Rode 2023)

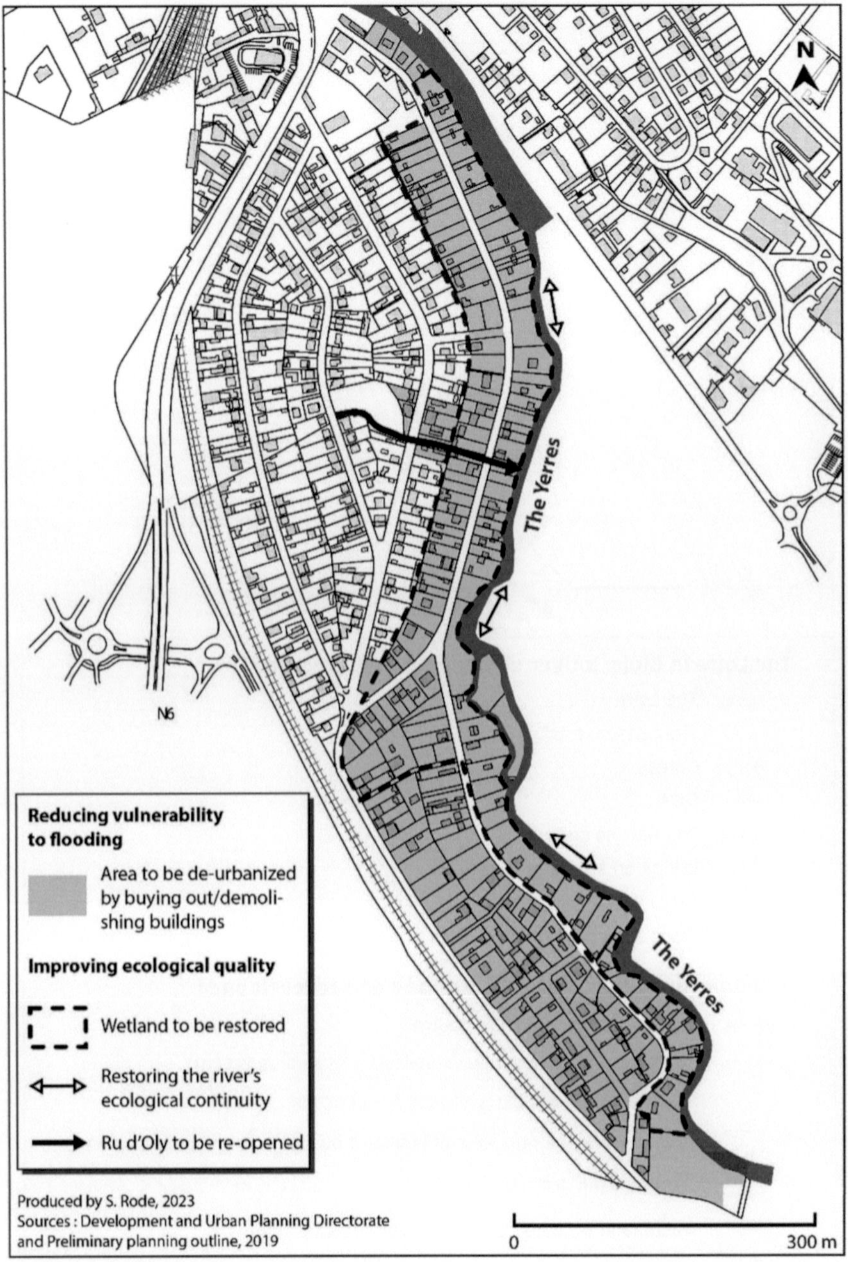

Fig. 6.4 The Belleplace-Blandin district in Villeneuve-Saint-Georges: from an urbanised flood zone to a restored wetland (*Source* Sylvain Rode 2023)

Table 6.1 Four riverside areas undergoing de-urbanisation

Town	Villegailhenc	Trèbes	Villeneuve-Saint-Georges	Blois
Population (INSEE 2020)	1658	5438	34 845	46 660
Watercourse	Trapel et Merdeau	Aude	Yerres	Loire et Cosson
Location of de-urbanised area	Riverbanks	Riverbanks	Riverbanks	Spillway
Start of demolition	2020	2020	2014	2006
Number of buildings bought and demolished	46	52	100	143
Type of relocation	Curative	Curative	Preventive/curative	Preventive

Source Sylvain Rode (2023)

developments and new uses can be imagined for these de-urbanised spaces, given that they are now strictly unbuildable? How can these areas be redeveloped, considering the regulatory constraints and natural characteristics of these flood-prone sites? To understand this, we need to analyse the redevelopment choices made. Let's take a look at the four case studies developed in the previous section.

In the four examples analysed, fairly soon after taking the decision to de-urbanise the areas along the river, the public authorities launched studies aimed at proposing new ways of developing these areas.

In three of the four towns, the idea put forward is to develop a park on the site of these de-urbanised areas. Turning these de-urbanised flood plains into parks seems logical, almost a "natural" solution. However, the term "park" can cover a wide range of issues and functions, with contrasting uses and forms of development.

In Blois, the guiding idea adopted by the elected representatives is that of a Natural Urban Agricultural Park (PANU), which meets several challenges. First of all, there is the issue of the memory of water and flooding, which will be physically reflected in the landscape via a string of ponds whose water level will vary with the seasons. Through these changing environments and living landscapes, the aim is to make people see, feel and understand the hydraulic function of this spillway, which has been restored to working order following its de-urbanisation. Then there is the issue of agricultural production, by developing pastures and hay meadows, agricultural activities that are compatible with the regulations governing this flood-prone area. Lastly, the recreational aspect: this natural area at the gateway to the city is not intended to be set aside as a nature reserve for no other use, but rather to become a public space open to city dwellers for recreation, relaxation and walks. However, the aim is not to create a classic urban park on this site, a "*twentieth-century park […] in a dense urban fabric […]. For us, talking about these places as a natural space and less as an urban park is what would seem most appropriate*[3]". The Urban Natural Agricultural Park is in fact a large-scale space on the edge of the city, a hybrid space that can support productive activities, recreational uses and ecological wealth, as we will come back to later.

In Trèbes, the reorganisation of the municipal territory following the 2018 flood should provide an opportunity to reclaim the de-urbanised banks of the Aude. It has been proposed to develop a "riverside park" and a "sports plain" on the left bank of the Aude. The aim is to "*live with the river, to make the bet that the river is an asset*[4]" for the area and its inhabitants.

In Villegailhenc, a "Trapel park" has been proposed on the site of the demolished dwellings in the town centre. As the landscape note produced by the firm of architects and landscape architects selected to design the detailed redevelopment project for the municipality points out: "*The stream and its surroundings become the showcase of the village, its breathing space and its future centre of life for the residents*".

In Villeneuve-Saint-Georges, on the other hand, the term "park" is not used to describe the future of this de-urbanised area. From the outset, the main objective was to renaturalise the banks of the Yerres and restore a wetland. Three different scenarios were then proposed in December 2020 by the project management team responsible for outlining the future of the district in its overall ecological and landscape study: preserved nature (a vast, protected natural area, with uses limited to contemplation and wandering within a defined area); urban agriculture (the creation of pastures, allotments, an urban farm and orchards); sport and nature (a canoeing and kayaking base, a natural children's play area, sports fields and equipment, a fitness trail). Consideration of the redevelopment of the banks of the Yerres is therefore marked by a tension between the desire to develop uses that will enhance the value of this renaturated area and the temptation to make it a sanctuary (or at least to drastically restrict visitor numbers and uses) in order to preserve its natural environments.

Whatever the various redevelopment choices, these de-urbanised areas are designed to be multifunctional. Their primary function is to ensure the free flow of water in the event of flooding, thereby reducing the risk of flooding. But it is also a question of turning them into public spaces that support new uses for city dwellers. Is de-urbanising these areas not also an opportunity to renaturalise them and restore natural environments? An opportunity to take better account of ecological and climate issues in urban planning?

---

[3] Interview on 8 April 2021 with the landscape designer commissioned to carry out the pre-operational study for the redevelopment of the Bouillie sector, which began in 2020.

[4] Interview of 18 April 2023 with the Director General of Services of the town of Trèbes.

6.4 Reinventing the Links Between City and River to Face Ecological and Climatic Issues?

While the ecological and climate crisis we are experiencing is due to profound disturbances in the functioning of environments and living organisms, it is also, according to the French philosopher Batiste Morizot, "a crisis in our relationship with the Living" (Morizot 2020, p. 16). Reinventing our relationships with living beings and environments is therefore necessary if we are to be able to work towards making the Earth habitable once again. It is about "thinking and acting with nature" (Larrère and Larrère 2018) rather than against it. Urban planning must contribute to these new relationships with nature, to this new paradigm, by designing cities that integrate into their environment in a more harmonious and less destructive way (Rode 2023). Are these de-urbanisation operations part of the paradigm of more ecological urban planning? From an urban planning that takes care of the environment? Are the redevelopment projects for de-urbanised areas succeeding in developing more symbiotic links between the city and the natural environment?

Flooding and the proximity of water are both constraints that justify de-urbanisation and assets that should enable these areas to be redeveloped in a more environmentally friendly way. In Blois, the vision for redevelopment of the de-urbanised area put forward by the team in charge of the pre-operational study is strongly underpinned by ecological considerations. This is due not only to the sensitivity of the landscape architect commissioned to carry out the study, but also to the presence of an ecologist, integrated into the group in order to "*have a fairly global vision of the issues of management, environmental coherence and ecological continuity*[5]".

The approach adopted by the landscape designer is that of a gentle, minimalist development, seeking to allow the natural dynamics to express themselves. The aim is for the project to preserve and enhance the ecological richness and diversity of the environments present on the site (the Loire and its tributary the Cosson, the riparian forests of these two watercourses, ponds, semi-humid grasslands, dry grasslands, areas of fructicaceae, etc.), rather than turning them upside down. It's even a laissez-faire attitude that's claimed: "*The idea is not to be in a big* tabula rasa*, we're going to plant, bring back plants, no! The environment has all its potential, so that's what we're going to capitalise on with a lot of patience, observation and monitoring, and it's the environment that will transform itself*[6]". It's a question of respecting the dynamics of the environment and the living world, by simply activating and extending them. Trust in the dynamics of living things (Morizot 2020) is at the heart of the landscaping choices made for this area. Such a choice is entirely in line with the desire expressed by French landscape architect Gilles Clément to "do as much as possible with, as little as possible against" (Clément 1999, p. 256), and which he put into practice in his Jardin "en mouvement". The elected representatives share

---

[5] Interview on 8 April 2021 with the landscape designer commissioned to carry out the pre-operational study for the redevelopment of the Bouillie sector, which began in 2020.

[6] *Idem.*

this desire to encourage the naturalness of this de-urbanised site, even if it must also become a place for gentle social uses.

In both Villegailhenc and Villeneuve-Saint-Georges, the redevelopment of de-urbanised areas along the river is seen as an opportunity to promote biodiversity and combat the urban heat island.

In Villegailhenc, the Trapel park, laid out in the heart of the urban fabric along the river and on the site of deconstructed buildings, will "*bring about a dynamic return to nature, making way for a biodiversity that will need to be instilled through the work of a rustic local plant palette with added value for the local fauna*[7]". In Villeneuve-Saint-Georges, the de-urbanisation of the banks of the Yerres is an opportunity to restore ecological continuity and a whole mosaic of habitats: a large wet meadow and woodland, orchards and small meadows, as well as a peaty depression. The aim here is "*to restore a previous ecological state, specific to the bocage environments formed by the meadows and woodlands naturally present in the alluvial plain of the Yerres*[8]".

The interplay of players involved seems to explain this choice of gentle development, favourable to the functioning of natural environments. The contracting authority for this project is the Yerres river management association,[9] which is responsible for restoring and maintaining the riverbanks and wetlands, as well as implementing ecological continuity on and along the river (green and blue framework). His vision is supported by the Agence de l'Eau (Water Agency), a key funder of the project: "*The Agence de l'Eau's funding is really geared towards natural areas and biodiversity, so it's clear that today we're really embarking on this approach of wetland requalification, sanctuary, biodiversity, with perhaps a few small developments on the fringes, such as access by footpaths, raised wooden decking, observation pontoons, educational trails,* etc. *But not beyond that in terms of the quality of the environment. But no more than that in terms of facilities*[10]". The idea would be to open up this area to the public on a very limited basis via footpaths and small observation promontories and to keep most of it in a protected area to avoid trampling and degradation of the environment by the public. The aim is to "*channel the public and preserve as much of the natural area as possible*[11]". The project will also provide an opportunity to restore the ecological continuity of the Yerres (by removing the weir at the old Villeneuve-Saint-Georges mill) and to open up a small tributary of the Yerres called the ru d'Oly—which is currently buried—in its downstream section, just before it flows into the Yerres. This will contribute to the reconstitution of a natural environment within an urbanised area. A transitional ecological management

[7] Atelier Sites, landscape note.

[8] SyAGE, *Dossier d'autorisation au titre du classement de la vallée de l'Yerres*, 2019, p. 6.

[9] SyAGE, which brings together 85 municipalities and 20 groupings of municipalities in the Val-de-Marne, Essonne and Seine-et-Marne départements.

[10] Interview on 19 May 2021 with the head of the SyAGE water and aquatic environment management department.

[11] SyAGE, *Dossier d'autorisation au titre du classement de la vallée de l'Yerres*, 2019, p. 7.

plan (prior to and during the forthcoming works) has been drawn up to minimise the impact on existing environments.

In these two municipalities, the reconstitution of green natural spaces at the very heart of the urban fabric is explicitly designed to help cool the local atmosphere, in conjunction with the proximity of the river, since it is known that the presence of a water surface amplifies the cooling effect of green spaces (ADEME 2021). Renaturing the banks of the Yerres at Villeneuve-Saint-Georges will help to reduce the urban heat island. The project's impact study emphasises that "*the sector will be much less sensitive to episodes of extreme heat and should make it possible to significantly reduce the urban heat island effect, including for neighbouring residential areas, thanks to the shade provided by the tree vegetation and the circulation of cooler air from the renaturated area*[12]". Similarly, the landscape designer in charge of the "Parc du Trapel" project in Villegailhenc stresses the importance of "*reintegrating plants as an element [...] that refreshes the atmosphere*[13]".

The projects analysed therefore illustrate, each in their own way, a marked concern for the care of these de-urbanised and then renatured floodplains.

6.5 Conclusions—The Redevelopment of De-Urbanised Riverbanks: from the Ecology of Conservation to the Ecology of Reconciliation?

Ecology is at the heart of these projects to redevelop riverside areas. While the starting point for this choice of de-urbanisation is indeed the desire to reduce vulnerability to the risk of flooding and to restore a flood expansion zone, these projects simultaneously address a number of issues: restoring natural environments and ecological continuities to enable biodiversity to develop, adapting to climate change, improving the quality of use of the areas and the quality of life of the people living there.

Reclassifying wetlands with diverse environments within urbanised areas and then managing them to encourage the dynamics of the environment and the development of non-human life could be seen as an illustration of a logic of separation between town and nature, considered as two separate entities. Rather, is it not an illustration of an urban territory now seen as "a place where humans and nature live together" (Radu 2018, p. 42)? As a place where nature must regain its rightful place so that humans can experience a reconnection with nature (Salomon-Cavin and Granjou 2021)? We see this as an operationalisation, in terms of urban planning, of the ecology of reconciliation proposed by the American ecologist Michael Rosenzweig, who refers to a desire to "reconcile human uses of the planet with the uses of other species" (Rosenzweig 2003). The aim is to go beyond conservation ecology by emphasising the challenge of integrating nature into highly anthropised areas, particularly cities. And for this, urban planning has a role to play, by providing "space and a diversity of habitats for

---

[12] Notice from the Environmental Authority, 2022, p. 21.

[13] Interview on 4 January 2023.

other species when we design our cities" (Maris 2018, p. 190). For M. Rosenzweig, this is the condition for a win–win ecology, i.e. one that benefits both human societies and non-human life forms. Today's urban designers are increasingly sensitive to ecological issues and are gradually giving them a greater role in the projects to which they contribute (Rode 2023). Projects developed in areas where buildings have been demolished provide a good opportunity to "renew the practice of architectural, urban and landscape design" (Sowa 2022, p. 194), insofar as they involve imagining new functions for these undeveloped areas, which can be opportunities for the restoration of environments and soft uses.

References

ADEME (2021) Rafraîchir les villes, des solutions variées, 79, https://librairie.ademe.fr/cadic/5604/recueil-rafraiçhissement-urbain-011441.pdf

Bonnet F (dir.) (2016) Atout risques. Des territoires exposés se réinventent, Marseille, Éditions Parenthèses, 173

Carcaud N, Arnaud-Fasseta G, Évain C (dir.) (2019) Villes et rivières de France, Paris, CNRS Éditions, 293

Carré C, Deutsch J-C (2015) L'eau dans la ville. Une amie qui nous fait la guerre, La Tour d'Aigues, Éditions de l'Aube, 320

Clément G (1999) Le jardin pour la maison de l'homme. In Younès C (dir.), Ville contre-nature. Philosophie et architecture, Paris, Éditions La Découverte, pp 254–273

Fénelon P (1978) Les pays de la Loire. Flammarion, Paris, p 501

Fournier M, Bonnefond M, Debray A (2021) La servitude de sur-inondation: un mécanisme capable de penser les solidarités entre espaces ruraux de fonds de vallées et espaces urbains inondables ? Le cas du bassin-versant de l'Oudon. Géographie, Économie, Société 23:489–506. https://doi.org/10.3166/ges.2021.0018

Gérardot C (2004) Les élus lyonnais et leurs fleuves : une reconquête en question. Géocarrefour 79(1):7–84

Gravari-Barbas M (1998) La "festival market place" ou le tourisme sur le front d'eau. Un modèle urbain américain à exporter. Norois, n°178, pp 261–278. https://doi.org/10.3406/noroi.1998.6868; http://www.persee.fr/doc/noroi_0029-182x_1998_num_178_1_6868

Larrère C, Larrère R (2018) Penser et agir avec la nature. Une enquête philosophique, Paris, Éditions La Découverte/Poche, 406

Maris V (2018) La part sauvage du monde. Penser la nature dans l'Anthropocène, Paris, Éditions du Seuil, 259

Mineo-Kleiner L (2017) L'option de la relocalisation des activités et des biens face aux risques côtiers: stratégies et enjeux territoriaux en France et au Québec, thèse de doctorat en géographie, Université de Bretagne occidentale

Morizot B (2020) Manières d'être vivant. Enquêtes sur la vie à travers nous, Arles, Actes Sud, 256

Radu F (2018) Typologie de formes de cohabitation humain-nature. Urbia. Les cahiers du développement urbain durable, n° 21, pp 39–56, https://www.unil.ch/ouvdd/files/live/sites/ouvdd/files/shared/URBIA/urbia_21/04_Urbia%20n21_Florinel%20Radu.pdf

Ribas PA, Sauri PD, Olcina CJ (2017) Sustainable land use planning in areas exposed to flooding: some international experiences. In: Vinet F (ed) Floods, vol 2. Risk Management. ISTE Press, Elsevier, pp 103–117

Rode S (2008) La prévention du risque d'inondation, facteur de recomposition urbaine ? L'agglomération de Blois et le déversoir de la Bouillie. L'information Géographique 72(4):6–26

Rode S. (2023), Écologiser l'urbanisme. Pour un ménagement de nos milieux de vie partagés, Lormont, Le Bord de L'eau éditions, p 216

Rosenzweig ML (2003) Win-win ecology: how the earth's species can survive in the midst of human enterprise. Oxford University Press, New York, p 224

Rossano F (2021) La part de l'eau. Vivre avec les crues en temps de changement climatique, Paris, Éditions de la Villette, 270

Salomon-Cavin J. et Granjou C. (dir.) (2021) Quand l'écologie s'urbanise, Grenoble, UGA Éditions, 385

Sowa C (2022) L'émergence de la déconstruction comme théorie et comme pratique en architecture. In Chavassieux P, Gay G, Kaddour R, Morel Journel C, Sala Pala V (dir.), (Dé)construire la ville. Les villes en décroissance, laboratoires d'une production urbaine alternative, Presses de Saint-Étienne, 192–214

Chapter 7
One Garden to Rule Them All? Exploring Recent Changes in Spain's Urban Riverscapes

Albert Santasusagna Riu

Abstract Recent transformations of urban rivers have opened the door to the implementation of urban policies that seek to prioritize uses of sociability for a city's residents. Large river parks, recreational areas, sports fields and tracks and cycling routes have come to epitomize this paradigm shift in the river-citizen relationship. However, there are certain aspects of this transformation of uses that need to be critically examined in the broader debate as to whether this "gardenscape model"—to use the terms in which it is currently being presented—is the one best suited to urban life in the coming decades. Based on the study of several urban rivers in Catalonia (Spain), a set of recommendations are made regarding the need to reformulate certain objectives of this model to address the series of problems identified, most notably the uniformity and homogenization that the implementation of a common solution to environments characterized by their rich diversity and complexity represents.

Keywords Riverfronts · Urban rivers · Green urban spaces · Urban centrality · Spain

7.1 Introduction

Since the late 1960s, urban development, revitalization and renewal projects have proliferated throughout Europe focused specifically on the zones of contact between bodies of water and urban fringes, be it waterfronts or riverfronts (Benson 2002; Schubert 2010; Brownill 2013; Sepe 2013; Porfyriou and Sepe 2016). Many such projects have come about as a result of a loss of functionality, triggered primarily by economic factors, including stages in the industrial delocalization of coastal strips and river ports (Hoyle 2000; Bruttomesso 2004). The transformation of these spaces has converted zones without any use (or with highly dispersed uses, often held in

---

A. Santasusagna Riu (✉)
Department of Geography. GRAM (Grup de Rercerca Ambiental Mediterrània) and Institute of Water Research (IdRA), University of Barcelona (UB), Barcelona, Spain
e-mail: asantasusagna@ub.edu

© The Author(s), under exclusive license to Springer Nature Switzerland AG 2024

J. Farguell Pérez and A. Santasusagna Riu (eds.), *Urban and Metropolitan Rivers*, The Urban Book Series, https://doi.org/10.1007/978-3-031-62641-8_7

low esteem by society) into new axes of urban centrality, becoming in some cases success stories even on the international stage (Jauhiainen 1995; Jones 2017; Samant and Brears 2017). The analysis of the prior state in which such zones found themselves, the progressive or drastic abandonment of uses, the specific urban designation afforded them, earlier projects backed by one or more municipal authorities and the support offered by suprastate institutions have been key determinants in their global transformation or, conversely, a change centred solely on a very specific tract of land.

The literature addressing this aspect of urban renewal traces the transformation of these spaces—especially the ports and waterfronts of the cities of North America—to a number of specific historical events (Sieber 1991; Lehrer and Laidley 2008; Castonguay and Evenden 2012; Kostopoulou 2013). The delocalization of the port industries is widely considered the first factor of functional change, but it is neither the sole nor the most frequent factor when we focus strictly on riverfronts. The diversity of cases here is much wider, and probably the best way to structure this vast body of differing circumstances is to focus on the situation prevailing immediately prior to a recent transformation. What specific use (or uses) had been given to this urban river space in recent decades? And if we go back a little further in time, are any notable changes apparent? What model of economic use has been prioritized in this space? Or have there, perhaps, been no specific uses, its recent history being marked by a process of abandonment? Is the space a reflection of the rural–urban duality, in which the pressures of urbanization have had a clearly detrimental impact on the agricultural spaces? And what role have catastrophic events associated with such hazards as flooding played in its design? Recent uses combine with historical uses, sometimes as a result of a see-sawing process of river-city rapprochement and estrangement, in what is never a static relationship. Undoubtedly, the best way to address this phenomenon of change is to treat each case separately, without this preventing us from observing recurring patterns in the way such spaces have been treated in recent decades. Nor does it exclude our being able to anticipate future risks associated with their current state.

The objective of the present study is to reflect critically on different aspects of the present-day management of urban river spaces but focusing attention specifically on one of them: the increasing uniformity of solutions for the transformation of urban rivers that seek to prioritize aspects of the sociability of their use. What, for many years, has been seen as an eminently positive element (one that has consigned abandoned, polluted spaces of next to no use for urban living to the past) is inevitably associated with other elements that need to be examined critically. Here, the focus is specifically on the geographical setting of Catalonia (Spain) and involves a series of case studies that allow this discussion to emerge. In what follows, I briefly make mention of a conceptual debate concerning urban rivers, namely: Is river restoration synonymous with urban regeneration? (7.2). I then seek to highlight the particular facets of what I choose to call the "gardenscape model" (7.3), before presenting a series of case studies with the aim of highlighting their most critical aspects (7.4). Finally, I provide some conclusions in what constitutes an open reflection (7.5).

7.2 Restoration Versus Regeneration. An Open Debate

From an examination of the literature in fields as diverse as biology, geomorphology and river ecology, it can be inferred that the treatment of river ecosystems pursues different strategies depending on the technical capacity to restore the structure and functions of the river to its pre-altered state (Gurnell et al. 2007; Beechie et al. 2008; Francis 2012; Sabbion 2016; Morandi et al. 2022). Yet, this sum of strategies responds to a common objective: the improvement—to varying degrees—of the river ecosystem and the prioritization of aspects that favour the river's environmental quality. At present, various concepts typifying the management of the river environment coexist and these concepts can be applied in different ways depending on the river in question (Dufour and Piégay 2009; Smith et al. 2014; Klaus and Kiehl 2021). However, the goals of river restoration (or rehabilitation) differ, conceptually, from those of the urban regeneration (or revitalization) of a riverfront. Indeed, the two processes pursue different strategies for what are primarily two reasons.

First, the context in which these transformative processes take place is critical. River restoration, in seeking to optimize any process of intervention, can be undertaken essentially in those spaces that, although degraded, have not been fully urbanized (Findlay and Taylor 2006; Guimarães et al. 2021). In contrast, regeneration acquires greater sense in those stretches of river typical of the urban fabric, previously canalized, diverted or exploited for a specific anthropic activity (Eden and Tunstall 2006; May 2006; Kondolf and Pinto 2017). Second, the end user of both processes may be different and, for this reason, the interventions to achieve the proposed objectives may also differ (Table 7.1). The demolition of weirs and dams (to recover a river's longitudinal continuity), the elimination of fords (to re-establish its transversal continuity), the regreening of riverbanks, the recovery of old river features and a river's freedom spaces are some of the main actions of a process of river restoration. In contrast, a process of urban regeneration sets itself another goal: the sociability of the river space. A greater public presence is pursued by means of the creation of river parks and other green spaces, and via the remodelling of the riverbanks for leisure activities or sport, the creation of a new communication infrastructure and the promotion of elements of educational, social and pedagogical potential.

7.3 The New "Gardenscape Model" for Urban Rivers

The "domestication" of rivers by municipal councils has involved, in most cases, attempts at regulating and controlling their flow, by means of large public infrastructure such as reservoirs. It has also sought to establish new river courses (by diverting the riverbed) and to limit the space occupied by the river, thanks to canalizations, retaining walls and other related works (Petts et al. 2002). However, domestication does not terminate with the creation of this new infrastructure; rather, the change

Table 7.1 Primary goals and actions of river restoration/rehabilitation and the regeneration/revitalization of urban rivers

River restoration/rehabilitation	Urban river regeneration /revitalization
Primary goal Recover the ecosystem values of river spaces: restore the river to its pre-altered state *Principal actions* Demolition of weirs and dams (longitudinal continuity) Elimination of crossings and fords (transversal continuity) Regreening of riverbanks Recovery of old river features Recovery of river's freedom spaces Elimination of canalizations and deurbanization	*Primary goal* Promote the social and economic values of river spaces, changing uses on the riverbanks *Principal actions* Gardening and creation of river parks and green spaces Creation of crossings and the design of paths running parallel to the river course Creation of new urban spaces (residential, commercial and for sports activities, among others) Creation of a new road and railway infrastructure to achieve social connectivity Promotion of elements of social, heritage, educational and pedagogical potential

Source Author, based in Ollero (2015)

in uses promoted by urban renewal, fostering the recreational anthropization of the riverbanks, can be interpreted as one more step in this global process. For many years, the political discourse on rivers has treated them as a space that should be hidden away, and for this reason cities have tended to "turn their backs on the river". Thanks to the promotion of new uses, based on generous endowments of green spaces, urban rivers have recovered a leading role thanks to their capacity to develop new centralities (Everard and Moggridge 2012; Francis 2014; Tort-Donada et al. 2020; Zingraff-Hamed et al. 2021). The "gardenscape model" has consisted of exploiting the urban riverbanks for a series of uses and activities that can be linked to four main functions: social, aesthetic, economic and environmental (Fig. 7.1).

The social function of the model seeks to develop leisure and recreational activities and to serve as a place where citizens can meet, by creating play areas, sports fields, rest places, observation points and walking and cycling paths, routes and circuits that take advantage of the continuity of the riverbanks to connect neighbourhoods, cities and even larger-scale territories. By focusing on this function, planners have been able to reflect on the quality and accessibility of the river's longitudinal and transversal connections with the rest of the urban space and to develop broad global projects, when funding has so allowed. The promotion of the model's social function has logically implied a change in the riverfront's users, which has also led to an increase in the number of people visiting the area. In summary, the model has succeeded in making the urban river a key element in the planning of a city's green space and in adapting it to new discourses of sustainability (e.g., ecosystem services, nature-based solutions and green infrastructure).

Aspects of an aesthetic nature have, likewise, been of some weight in the implementation of the model, specifically in consolidating the new social function of urban rivers. The emphasis has been placed on greening as the basis of the social imaginary

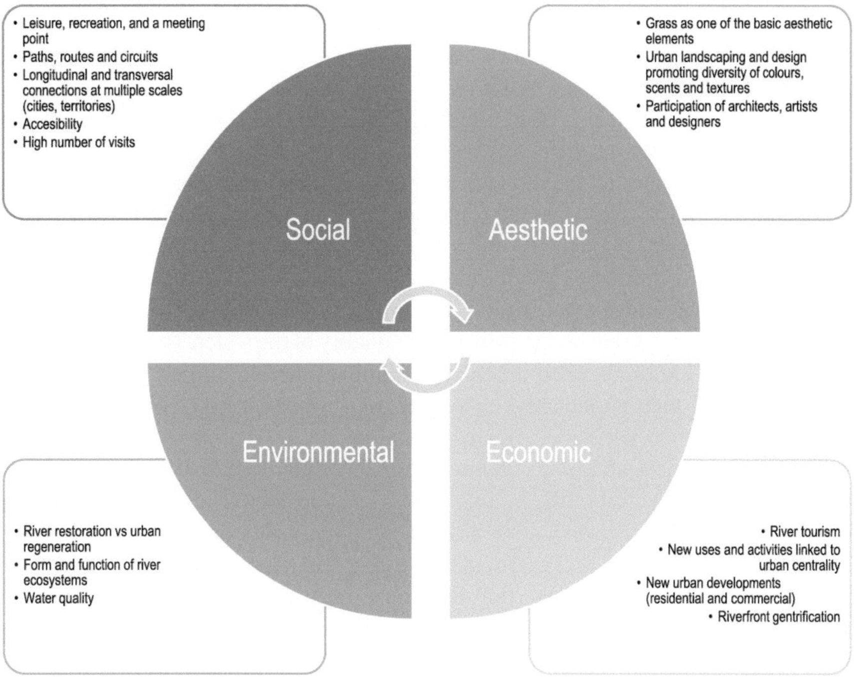

Fig. 7.1 Conceptual diagram of the "gardenscape model" adopted by most of the world's urban rivers today (*Source* Author)

of urban nature, replacing areas of farming or urbanized strips with expanses of grass and the planting of shrubs and bushes. By so doing, gardenscapes have been created, which play with the combination of colours, scents and textures and, in certain cases, these green areas have been able to count on the expertise of prestigious architects, artists and designers who have contributed to exalting the symbolic values of the river and generating new interpretations and images. Moreover, in what once were scenarios of serious environmental degradation, grassy banks have come to constitute a new aesthetic that has corrected clumps of damaged vegetation, inconvenient slopes, untidy pathways and decaying historical heritage. It has been a uniform solution, which has provided urban rivers with a new aesthetic, endowing them with all those qualities of the landscape that are perceived as recreational nature, made suitable and available for human use.

Economic uses and activities have also been the object of this new model, especially when opportunities to generate new spaces of tourist and commercial attraction have been identified. In the first of these cases, river tourism has played a prominent role as a promoter of tourist activities both on the river (river trips and different types of navigation as well as sports activities on the water) and on its banks (exploiting activities coherent with its newfound social functions: fairs, markets, special events, etc.). But the forces of urban regeneration have also been used to attract new urban developments, both residential and commercial. This has led to a change of uses that in some cases have boosted the local economy, but it has also opened the door to processes of riverfront gentrification.

Finally, the environmental function has been limited to those opportunities that each urban regeneration project has had to incorporate aspects of river restoration. In only a few cases around the world has it been possible to achieve the primary goal of restoration, that is, to recover the form and function of the river's ecosystem. But, in most urban rivers, this has not been possible, and therefore, the environmental function has been limited to much more modest goals, often focused on a specific stretch of the river under regeneration. Moreover, one of the problems that remains unaddressed in many urban rivers is that of the quality of the water. Thus, it is possible to find examples of landscaped urban rivers, transformed into linear parks of great aesthetic quality that, lamentably, have very poor water quality. This constitutes a further limit for their socioeconomic exploitation.

To end this section, an initial conclusion can be offered: the "gardenscape model" comprises many pieces, but they do not always fit together. Thus, we find rivers in which, in some of their urban stretches—depending on the prevailing fashion, the budget available to them and political interests—this new model proliferates. Yet, the possibility also exists of starting the house from the roof, that is, developing parks of great public interest but with rivers of very poor water quality. Each urban river is a case in its own right, with all its historical baggage, its geomorphological idiosyncrasies and its concrete plans of management being operated by different levels of government. The rediscovery of these spaces is based on an approach that prioritizes uses, designs and projects that are remarkably similar to each other and which transcend borders.

7.4 A Critical Appraisal of the Model Based on Two Case Studies

Having described the main characteristics of this new urban model, we next consider two case studies that contribute to the debate regarding the needs and shortcomings of the "gardenscape model" being implemented in a large number of river cities. First, we turn our attention to matters of an aesthetic nature, as highlighted by two settlements in the Segre river basin which forms part of the western watershed of Catalonia. And, second, we focus on Terrassa, Catalonia's third-largest city, with a geography marked by small ephemeral water courses or *rieras*. In the first case, the model was implemented during the 1980s and 1990s, while in the case of Terrassa, the city's municipal council is planning to regenerate a large torrential river by applying this model.

7.4.1 Aesthetic Replication

The existence of an innovative, pioneering, high-profile project attracts attention not only from civil society and urban and regional agents, but also (and, more particularly) from planners. It is from this perspective that we can begin to understand how, at the global scale, the landscapes inherent to urban rivers have become increasingly more and more similar. Firms, consultants and studios of architects, because of the projects commissioned by municipal governments around the world, have become specialized in this field, and opt time and again for a very similar aesthetic model and series of uses. Today, such actions as the landscaping of a square or an intervention of tactical urbanism in the framework of a process of urban renaturation can be considered specific niches of such projects of regeneration.

In Spain, two urban settlements in the Segre river basin, Lleida (140,000 inhabitants) and Balaguer (17,000 inhabitants), are paradigmatic examples of this process. At the beginning of the 1980s, the basin suffered a flood that was to take a huge economic toll and which highlighted the need to reconfigure the urban river spaces in the main settlements of the basin, but above all in Lleida and Balaguer. The flood triggered negotiations between the municipal councils and the public water agency (Ebro River Basin Confederation) to determine the terms under which this transformation should take place. Although the initial plan of the river basin agency was to canalize the Segre river using a rectangular cross-section, the municipal councils favoured a trapezoidal section that would, at least, ensure the river banks remained in direct contact with the water. This modification was an opportunity to create green spaces along the urban stretches of the Segre in both Lleida and Balaguer, with the same model being adopted on both banks: that is, the surface of the water, a sparse covering (where present) of shrubby riparian vegetation, an area of grass planted with trees and a retaining wall (serving to separate the recreational river space from the parallel road and residential space). The project also involved stabilizing the riverbed and winning back space from the river for the benefit of the new green area, particularly in the case of Lleida—with the development of a larger linear park (2.5 km vs. 1.5 km in the case of Balaguer)—which extends unbroken on both rivers banks (but limited to just the left bank in Balaguer).

The precarity of the situation and the need to intervene strategically so as to be able to address future flood episodes meant the adoption of a uniform model, based on an almost identical intervention in terms of the landscaping of the river spaces (Fig. 7.2): first, a similar deployment of landscaping elements—a grassy area, lines of trees and the ornamental plants—that achieve a very similar effect in both cases; and, second, the exposition of the facades of the urban residential fabric running adjacent to the river, of similar construction and height, which contribute to creating this sense of a shared landscape. Here, the linear engineering conducted on this stretch of river is a key factor as is the towering presence in both instances of religious heritage of immense historical value (the Seu Vella in Lleida—the city's Romanesque cathedral—and the church of Santa Maria in Balaguer) perched on promontories overlooking the river. Before the intervention, the respective riverscapes had many features in common, but these have been bolstered and consolidated thanks to the creation of the two linear riverparks.

A shared riverscape can have certain limitations, especially in these two cases in the Segre basin. First, it should be stressed that the city of Lleida starts from a position of some advantage: it is the capital of the province, with a population eight times that of Balaguer's, and is more likely to become a benchmark for urban regeneration, given that the area occupied by its river park makes it one of the largest in Catalonia. Indeed, Lleida is frequently cited as an example of urban river integration, while the project completed in Balaguer is often overlooked and has been the subject of little interest from planners. Additionally, Lleida organizes a considerably greater number

Fig. 7.2 The linear parks of Lleida (top) and Balaguer (bottom) (*Source* Lleida City Hall (2018) (top) and a photograph (bottom) taken by Pepita Sotelo Paradela 2021)

of social activities of a traditional nature and as part of the city's calendar of *fiestas* in the river park (including, most notably, the *Aplec del Cargol*—a popular gastronomic festival at which snails are the protagonists—held in the landscaped area on the left bank of the Segre and the adjacent park, known as the Champs Elysees). However, Balaguer has taken a number of steps to differentiate itself in recent decades. For example, the town organizes the *Transsegre*, a summer event first held in the 1980s that targets young people and which seeks to promote the river park as a place for a range of activities (including the descent of the Segre in original homemade boats) and games on the water. Similarly, particular attention has been paid to the colours of the facades of the buildings facing the river (reminiscent of those that stand on the banks of the Onyar river in the Catalan city of Girona). Fighting this uniformity of uses by promoting differentiated social and economic activities in favour of a more specialized local development is, perhaps, one of the best ways to avoid this constant aesthetic replication and which benefits the original, more ambitious intervention.

7.4.2 Without Form or Function: River Parks Without Water

Terrassa, located in the Barcelona metropolitan region, is Catalonia's third-largest city, with more than 220,000 inhabitants. It lies in an alluvial cone, formed by the accumulation of sediments transported and deposited by small ephemeral water courses or torrents. The city has grown up and expanded in the heart of this network of torrents, with calamitous consequences for the city's history. The response of Terrassa's municipal council to successive historical flood episodes differed, but it was to prove especially fateful in the 1960s. In September 1962, a sudden change in synoptic conditions led to a cold drop that caused the Llobregat and Besòs river basins to overflow, affecting various *comarcas*, but above all that of the Vallès Occidental. The consequences were devastating; in what is considered one of the worst floods in the recent history of Spain, hundreds of people died, thousands of homes were destroyed and much of the area's infrastructure disabled. One of the worst-hit cities was, precisely, that of Terrassa. Although the flooding impacted the entire city, it was to have a particularly disastrous effect in the areas of informal housing, built by those who had arrived over the previous decades as part of Spain's internal migration. Cut off from any access to affordable rents, and with few possibilities of owning their own homes, much of the migrant population settled on the broad floodplain of these torrential river courses or *rieras*, including that of the *riera de les Arenes*, in the east of the city. The migrant population were quite unaware of the vulnerability of the site, just as the council, immersed in the benighted days of Francoism, had demonstrated no concern in its urban planning for the grave flood hazard to which they were exposed.

The response to this tragic event was swift, and in the years following the catastrophe the city's urban structure was completely reconfigured. Large urban operations were set in motion, the primary goal of which was to tame the torrential river courses, identified now as a potentially dangerous element—a time bomb that could go off at any moment—which had to be excised from the urban life of Terrassa. In the case of the *riera del Palau*, to the west of the city, it was decided to divert the river course by means of a peripheral canalization and, then, to urbanize the stretch that runs through the city centre. In so doing, one of the most important urban arteries of Terrassa was born: the *Rambla d'Ègara*, a commercial avenue that runs from the north of the city to the south. The *torrent de Vallparadís*, in the city centre, an area given over to a series of allotments, was gradually absorbed as a green space and the council recovered former proposals from its planning archives to convert it into a park, which would eventually be opened at the end of the last century. Meanwhile, on the eastern outskirts of the city, the *riera de les Arenes* was the object of a radical project of river engineering, its river bed canalized to limit it as far as possible and its banks transformed into both physical and mental boundaries. This project merits a more detailed explanation.

The *riera de les Arenes* in Terrassa is, today, an entirely peripheral space. Lying to the east of the city, it is hemmed in on both of its banks by the traffic that circulates along the *Avinguda del Vallès,* and which connects the city with the rest of the metropolitan region. The space is given over almost entirely to industrial and commercial uses that occupy several large estates. The neighbourhoods adjacent to the *riera de les Arenes* present a complexity of socioeconomic challenges, most notably Ca n'Anglada. The aesthetics of the neighbourhood are fairly deplorable, an element that is further aggravated by the existence of massive electricity pylons occupying the pavement along which, presumably, local residents should be able to stroll to get a view of the river course. To this we should add the deficient design of the communication infrastructure, which, in certain stretches, finds itself below the level of the riverbed (specifically, the intersection with the road to the municipality of Rubí), and which is prone to flooding when there are episodes of heavy rain. The situation is further exacerbated by an inefficient drainage system at various points, which becomes blocked due to the build-up of rubbish, leaves and other plant debris.

In recent years, the Terrassa municipal council has taken the decision to initiate a comprehensive project for the renaturation—the ecological rewilding—of the city's river courses, with the aim of converting these spaces into areas of attraction for the city's residents based on their new landscape value. The project is an ambitious one and has involved changes in municipal governance (with the creation of a specific council department to oversee it), as well as the setting up of a commission made up of politicians, experts and neighbourhood representatives. At the time of writing, work is underway to draft a strategic plan which, based on an exhaustive diagnosis of the area, is to include a series of proposals that have the backing of experts and neighbours alike. From a strictly political point of view, it is very good news that the problem of the *rieras* is to be addressed and, thanks to a process of public participation, that the future challenge of planning these spaces is to be tackled. Some of the general lines of action that the municipal council wants to implement in the *riera de les Arenes* have already been made public and are worth examining.

The plan, as things currently stand, for the *riera* in the eastern outskirts of the city involves the application of the "gardenscape model" along most of the almost four kilometres of channel that runs through the urban area. It is hoped that the funding for this plan and its subsequent execution will be made available, in part, through the NextGeneration EU budget. Yet, the immediate advantages of implementing this model are not necessarily functional in the specific case we are dealing with. Not only because the space is peripheral, highly urbanized and with what is a primarily industrial use, but also because it does not have a basic element of attraction: a constant flow of water. The project proposes maintaining the roads on either bank, but regreening the pavements and burying the high voltage lines. Another important debate centres on how to deal with the riverbed: an intervention is proposed that favours accessibility to the local population with the creation of stairways that connect the two levels (the banks and the bed), as well as the construction of paths and small walkways within the riverbed itself. In practice, what is wanted is to convert the bed of the *riera* into what we might call a "dry river park".

The planned intervention does not involve burying the *riera* underground (as the municipal council had proposed, many decades ago) nor does it contemplate the continuity of the floodplain's current reality. Rather, it is a very clear reflection of the will to convert something that is neither in the main green, suitable for human transit, or stable over time into a transitable, stable green space. The *riera* today is essentially dry, its bed is deeply uneven and heavily cracked in parts with gully incisions. The render with which the proposal is presented imagines a reality radically different from its current state (Fig. 7.3), and it is relatively easy to connect with this image in idealistic terms, albeit that its application in practical terms will be extremely challenging. As more specific details of the proposal become known, it will be critical to know if, at a technical level, the artificialization of the riverbed is advisable, and what the consequences of this action are likely to be during episodes of intense rainfall and the flooding of the torrential river course. Something that might be highly desirable at a social level may be neither possible nor recommendable at a technical level. Similarly, very careful thought will need to be given to the expected effects of the project at this social level, as this is something that will require a change of mentality of the city's residents. To this, we need to add the great counterweight currently presented by the Vallparadís park in the centre of the city. This situation, however, might be alleviated by the creation of green corridors connecting the two parks, taking advantage of some of the city's main or secondary streets, which could, in some cases, be fully pedestrianized. Maintaining "green" spaces in a climate scenario of increasingly intense and prolonged droughts is another matter to consider, as is the absence of shady spaces along the riverbed—indeed, an endless number of questions—some more general in scope, others of detail—that will have to be brought to the table once the decision has been taken to go ahead with the project.

Fig. 7.3 The *riera de les Arenes* as it is today (top) and the municipal council's proposal (bottom) (*Source* Photograph taken by Albert Santasusagna in 2021 (top) and Terrassa City Hall (bottom))

7.5 Closing Remarks: Does One Functional Urban River Model Fit All?

What kind of urban rivers do we want? Should our priority be to promote limitless processes of renaturation? Can such interventions actually be implemented, bearing in mind the high degree of urbanization that has typically gone before? Are they compatible with the human presence and sociability required to convert these spaces into new elements of urban centrality? Is a *river-garden* with a *river-habitat* a feasible proposition? These are the concerns that currently assail planners, geographers, engineers, architects and ecologists as they sit down to discuss the future of our urban rivers. All of the above issues can be addressed by what might be considered largely generic responses, actions characterized by varying degrees of adaptability and coherence depending on the particular urban river under debate. However, a number of universal trends have been detected that point to transformations that are shared by an increasing number of urban rivers. Today, the "gardenscape model" very much typifies many of those rivers that have experienced tragic episodes of flooding or which have been stigmatized as peripheral, pestilential elements of the city environment.

This universal model, while offering obvious advantages in terms of sociability and the social use of urban river spaces, faces a series of risks that we cannot ignore. The first is that this model has been designed to transform rivers into elements at the service of society, which means ecological and environmental concerns are pushed into the background. The recovery of the river ecosystem's function and form is not the priority; rather, the primary goals are ensuring its accessibility, aesthetics and the creation of spaces that promote human relations and uses. Moreover, what this model does is to homogenize uses and landscapes. In the absence of any clear vision of the social, physical and patrimonial identity of the local environment, all interventions are reduced to a process of *greenwashing* and the eradication of what makes each space, each city or each region unique. But is it really so easy to implement this model across the board? In the course of this chapter we have seen a case study that addresses this very question, typified by the enormous complexity inherent in such spaces given the absence of a constant river discharge for most of the year. Has there been a global shift from city councils that sought to eliminate their urban rivers by burying them underground to authorities that seek to intervene in these river environments to the point of considering them an element for limitless exploitation, redefinition and redesign? In many cases, the answer is probably yes. The environmental and social implications may well differ, but the incessant will to intervene is the same.

Can this model be deemed useful for the future? In answering this question, two issues must be taken into account. The first concerns the future climate scenarios we can expect, prone, in the case of the Mediterranean basin, to intense, prolonged drought. As such, the model needs to be recast in favour of plant species whose water consumption is low, and here it is reasonable to assume that a large expanse of grass—an element that has become normalized in our collective urban way of thinking—is not the most appropriate solution. Second, the interaction between society and nature may change in the future. How will we use our green spaces? How will we move around our cities? Will we continue to see rivers as spaces for recreational activities? The cycleways in the river parks of the 1990s are today paths shared by e-scooters and segways. And what about other forms of mobility and leisure that might emerge over the coming years? The kites of yesteryear are the drones of today. What are likely to be the main traits of the future users of river parks? In some countries, populations are ageing as birth rates dwindle. Assuredly, we will have to readjust activities and uses to meet the needs of new users.

Further risks might be identified that have not been specifically addressed in this chapter. For some years now, various authors have warned of the risk of green gentrification resulting from the regeneration of river spaces. Just as projects of urbanism designed to create new green spaces in a city's streets, avenues and squares can increase land values and result in the social exclusion of the local population, riverside spaces can have a similar impact when considered as new green spaces. Gentrification never comes unaccompanied; it usually comes hand in hand with control. In cases where it is necessary to establish flood risk control, it may also be or become a control of uses and activities and, ultimately, some form of social control. Surveillance systems, active in various green spaces around the world, can, in this sense, represent a step backwards in the free and democratic use of public space, as well as a step towards homogenization.

The best way to address the homogeneity in the planning of urban river spaces is, probably, by introducing and promoting collaborative mechanisms of control, participation and decision-making, incorporating both local citizens and expert professionals who can advise on basic ecological principles when intervening in these ecosystems. Additionally, efforts need to be made to introduce environmental education, with a specific focus on increasing awareness of the ecological functioning of rivers, in schools, neighbourhood associations and local government. Various urban river renewal projects have been undertaken around the world in which citizens have been given the opportunity to express their ideas, and in many cases this has strengthened community decision-making and promoted the consideration of environmental questions, as well as ensuring that questions of local history and heritage are taken into account. Undoubtedly, many different cases might be cited and it is equally difficult to offer an alternative, global, sustainable model to that of "gardenscape". However, some basic ecological principles must be taken into account when drawing up a river plan: intervene as little as possible, allow more intense human uses only in certain stretches, prioritize nature-based solutions, improve water quality, promote citizen science activities to control the quality of the river in urban environments and, above all, understand the river as a living ecosystem that must be preserved even in the most conflictive scenarios. In the face of pollution and urban chaos, a river-garden may be a good solution. Faced by an urban river that largely retains its form and function, opting to prioritize greater access for visitors and to plant the riverbanks with grass is probably not the best solution. Moderation, restraint and common sense when intervening in urban rivers should be interpreted as the beacons of environmentally sensitive urban and spatial planning.

Acknowledgements This study has benefitted from the scientific and economic support of grant 2021SGR00859 awarded by the Agency for the Management of University and Research Grants (AGAUR) of the Catalan Government of the *Generalitat* (SGR 2021-2024).

References

Beechie T, Pess G, Roni P, Giannico G (2008) Setting river restoration priorities: a review of approaches and a general protocol for identifying and prioritizing actions. North Am J Fish Manag 28(3):891–905

Benson E (2002) Rivers as urban landscapes: renaissance of the waterfront. Water Sci Technol 45(11):65–70

Brownill S (2013) Just add water. waterfront regeneration as a global phenomenon. In: Leary ME, McCarthy J (eds) The Routledge companion to urban regeneration. Routledge, New York, pp 45–55

Bruttomesso R (2004) Complexity on the urban waterfront. In: Marshall R (ed) Waterfronts in post-industrial cities. Spon Press, London, pp 47–58

Castonguay S, Evenden M (2012) Urban rivers: remaking rivers, cities, and space in Europe and North America. University of Pittsburgh, Pittsburgh

Dufour S, Piégay H (2009) From the myth of a lost paradise to targeted river restoration: forget natural references and focus on human benefits. River Res Appl 25(5):568–581

Eden S, Tunstall S (2006) Ecological versus social restoration? how urban river restoration challenges but also fails to challenge the science—policy nexus in the United Kingdom. Environ Plan c: Polit Space 24(5):661–680

Everard M, Moggridge HL (2012) Rediscovering the value of urban rivers. Urban Ecosyst 15:293–314

Findlay SJ, Taylor MP (2006) Why rehabilitate urban river systems? Area 38(3):312–325

Francis RA (2012) Positioning urban rivers within urban ecology. Urban Ecosystems 15:285–291

Francis RA (2014) Urban rivers: novel ecosystems, new challenges. Wires Water 1(1):19–29

Guimarães LF, Teixeira FC, Pereira JN, Becker BR, Oliveira AKB, Lima AF, Veról AP, Miguez MG (2021) The challenges of urban river restoration and the proposition of a framework towards river restoration goals. J Clean Prod 316:128330

Gurnell A, Lee M, Souch C (2007) Urban rivers: hydrology, geomorphology, ecology and opportunities for change. Geogr Compass 1(5):1118–1137

Hoyle B (2000) Global and local change on the Port-city waterfront. Geogr Rev 90(3):395–417

Jauhiainen JS (1995) Waterfront redevelopment and urban policy: the case of Barcelona Cardiff and Genoa. Europ Plan Stud 3(1):3–23

Jones AL (2017) Regenerating urban waterfronts—creating better futures—from commercial and leisure market places to cultural quarters and innovation districts. Plan Pract Res 32(3):333–344

Klaus VH, Kiehl K (2021) A conceptual framework for urban ecological restoration and rehabilitation. Basic Appl Ecol 52:82–94

Kondolf GM, Pinto PJ (2017) The social connectivity of urban rivers. Geomorphology 277:182–196

Kostopoulou S (2013) On the revitalized waterfront: creative milieu for creative tourism. Sustainability 5(11):4578–4593

Lehrer U, Laidley J (2008) Old mega-projects newly packaged? waterfront redevelopment in Toronto. Int J Urban Reg Res 32(4):786–803

May R (2006) "Connectivity" in urban rivers: conflict and convergence between ecology and design. Technol Soc 28(4):477–488

Morandi B, Cottet M, Piégay H (2022) River restoration. Political, Social, and Economic Perspectives. Oxford: Wiley

Ollero A (2015) Guía metodológica sobre buenas prácticas en restauración fluvial. Manual para gestores. Zaragoza: Universidad de Zaragoza, Centro Ibérico de Restauración Fluvial

Petts G, Heathcote J, Martin D (2002) Urban rivers. Our inheritance and future. IWA Publishing, London

Porfyriou H, Sepe M (2016) Waterfronts revisited. European Ports in a historic and global perspective. Routledge, New York

Sabbion P (2016) Urban river restoration. In Perini K, Sabbion P (eds) Urban sustainability and river restoration. Green and blue infrastructure. Chichester: Wiley Blackwell, 76–92

Samant S, Brears R (2017) Urban waterfront revivals of the future. In Yok Tan P, Yung Jim C (eds) Greening Cities. Forms and functions. Springer Singapore: Singapore, 331–356

Schubert D (2010) Waterfront revitalizations. From a local to a regional perspective in London, Barcelona, Rotterdam, and Hamburg. In Desfor G, Laidley J, Stevens Q, Schubert D (eds) Transforming urban waterfronts. Fixity and Flow. New York: Routledge, 90–114

Sepe M (2013) Urban history and cultural resources in urban regeneration: a case of creative waterfront renewal. Plan Perspect 28(4):595–613

Sieber RT (1991) Waterfront revitalization in postindustrial port cities of North America. City Society 5(2):120–136

Smith B, Clifford NJ, Mant J (2014) The changing nature of river restoration. Wires Water 1(3):249–261

Tort-Donada J, Santasusagna A, Rode S, Vadrí MT (2020) Bridging the gap between city and water: a review of urban-river regeneration projects in France and Spain. Sci Total Environ 700:134460

Zingraff-Hamed A, Bonnefond M, Bonthoux S, Legay N, Greulich S, Robert A, Rotgé V, Serrano J, Cao Y, Bala R, Vazha A, Tharme RE, Wantzen KM (2021) Human-river encounter sites: looking for harmony between humans and nature in cities. Sustainability 13(5):2864

Chapter 8
City and River in the Garonne Valley: The Geohistory of a Renewed Space (Toulouse, Agen, Bordeaux)

Philippe Valette and Matthew Hatvany

Abstract Sixty-five towns exist in the Garonne Valley, and their origins and locations are the result of multiple historical factors. Today's urban river landscape is strongly marked by different flood control policies as well as different actions aimed at enhancing river resources. It is possible to identify five stages in the geohistory of Garonne urban river landscapes since the Roman period. The last began about twenty years ago, following a period of rejection of the river, with the cities of Toulouse, Agen, and Bordeaux developing policies to reconnect with the Garonne through the requalification of the river front. The Garonne becomes a source of landscape amenities centered around leisure and cultural activities. This reconquest was initially centered essentially in the old city centers, but urban regeneration has today broadened to a larger metropolitan scale (metropolis) that includes many other initiatives.

Keywords Garonne · Urban river landscapes · Revitalization · Renewal · Evolution

8.1 Introduction

In today post-industrial context, urban development policies, elected officials, developers, and urban planning services are returning waterways to the "heart of urban development strategies and ensuring redevelopment in the name of making rivers accessible to urban inhabitants" (Carré and Deutsch 2015). In France, many rivers, large and small, are the subject of an aesthetic and landscape revaluation, carried

P. Valette (✉)
Department of Geography, Planning and EnvironmentGeode UMR 5602 CNRS, University Toulouse Jean Jaurès, Toulouse, France
e-mail: philippe.valette@univ-tlse2.fr

M. Hatvany
Department of Geography, Laval University, Quebec, Canada
e-mail: matthew.hatvany@ggr-ulaval.ca

© The Author(s), under exclusive license to Springer Nature Switzerland AG 2024
J. Farguell Pérez and A. Santasusagna Riu (eds.), *Urban and Metropolitan Rivers*, The Urban Book Series, https://doi.org/10.1007/978-3-031-62641-8_8

out as part of landscape architectural studies. As one such study succinctly notes, "It is a time of reunion between cities and their rivers; it is a time for staging festivities to make everyone aware of the blessed alliance of urban heritage and nature rediscovered" (Rossiaud 2016: p. 7).[1]

In this time of urban renewal with rivers, it is worthwhile considering if and how this change is being manifested at the scale of the river valley? Our work is interested in the whole of the Garonne Valley, focusing on the object of urban river landscapes. The Garonne is located in South–West France and is the smallest of the country's river systems (530 km long with a flow rate of 650 m^3/s). It is a river system with a volatile character, its history a succession of floods among the highest in France (8.32 m above flood level at the Pont Neuf in Toulouse in 1875; 12.50 m in Agen in 1435). This volatility has marked the urban river landscapes of the Garonne with different flood control policies according to location, as well as different actions aimed at enhancing river resources.

The objective of this study is to create an inventory by which to show the evolution of the city/river relationship in the Garonne Valley. The first part focuses on characterizing the diversity of the early urban settlements sites in the valley. The second part, through a geohistorical approach, exposes the different stages of the elaboration of urban river landscapes in Toulouse, Bordeaux, and Agen, the three largest cities of the valley. Finally, the last part deals with the shift of river landscapes, where after a long period of distancing from the river, the cities of Toulouse, Bordeaux, and Agen have developed policies aimed at reconnecting their populace with the Garonne through the restoration of river fronts.

8.2 The "Origins" of Urban Settlement on the Banks of the Garonne

One of the age-old questions in geography is explaining the reasons for the growth of an urban core at one site rather than another. It has given rise to numerous reflections and criticisms on the existence or otherwise of geographical determinism (Dolfus 1985; Bergevin 1992; Friedberg 1992). The question asked here is: is the presence of the Garonne a fundamental element for the development of cities?

8.2.1 Typology of Garonne Towns

According to INSEE (National Institute of Statistics and Economic Studies), a city is "a municipality or a group of municipalities with a continuous building area (no gap of more than 200 m between two constructions) which has at least 2000 inhabitants". Applying this criteria to the Garonne Valley, there are 65 urban sites with more than

[1] Author's translation.

2000 inhabitants where it is possible to observe diverse orders of grandeur ranging from towns to cities (Fig. 8.1). The largest category is that of towns (2000–5000 inhabitants), accounting for 51% of the urban sites. Places with populations between 20,000 and 200,000, like Agen, are considered big cities and account for 12% of the urban sites. Finally, cities of more than 200,000 inhabitants, like Toulouse and Bordeaux (3% of the whole), are considered metropolises.

It is possible to distinguish several types of urban establishment in the Garonne Valley (Fig. 8.2). The best-represented category (with 51%) is linked to locations on alluvial terraces sheltered from flooding by distance from the river itself. Here, urbanization has resulted in 39% of all development being exclusively on the alluvial terrace and out of reach of flood waters. Some 12% of this development, however, has spilled over into the flood zone. The second category corresponds to establishments on alluvial terraces, but in the immediate vicinity of the Garonne (26%). The goal of such a location was to be protected from flooding, while benefiting from the proximity of river resources. Here, some 12% of the cities have continued their development on alluvial terraces but out of reach of the river, while others (14%) have seen strong urban development within the floodplain. The distinction between near and distant alluvial terraces is in part dependent on the Garonne, which has displayed a certain historical mobility by changing its route. Consequently, some towns far from the river at their origins now find themselves much closer. Only 23% of towns in the Garonne were originally established in a flood zone, either near the banks (17%) or on local bulges (6%). In the end, few cities were established directly on the banks of the river or on a terrace edge near the Garonne (29%). This aspect testifies to a strong fear of floods on the part of society.

The analysis of the "original" sites of towns in the Garonne Valley offers relatively few actual riverfront cities. In this category, it is possible to mention: Cazères, Muret, Toulouse, Agen, Tonneins, La Réole, Langon, Cadillac, Langoiran, and Bordeaux.

8.2.2 The Garonne as a Limit and Border

At these different sites, the banks of the Garonne undoubtedly served as a natural limit, which is explained by the technical obstacles imposed by the crossing of the Garonne. As Valette et al. (2011: p. 77) note, "Like any border, the river invented its places of obligatory passage, first materialized by a ford or by a simple ferry, then quite quickly by a bridge". The Garonne was for a very long time a river without bridges and the fords of Toulouse and Agen acted to link one bank with the other. The width of the river bed, the dangers of the floodplain, and the frequency of flooding discouraged the construction of bridges. This has long resulted in an asymmetry in the urban expansion of Garonne towns. Indeed, most cities spanning the river are not equally balanced on both sides, the earlier developed side often being considerably larger. Toulouse, Castelsarrasin, Agen, Tonneins, and Marmande all developed on the right bank of the river, with smaller developments on the opposite shore. But this is not an absolute rule as Carbonne, Portet-sur-Garonne, Muret, Langon, and

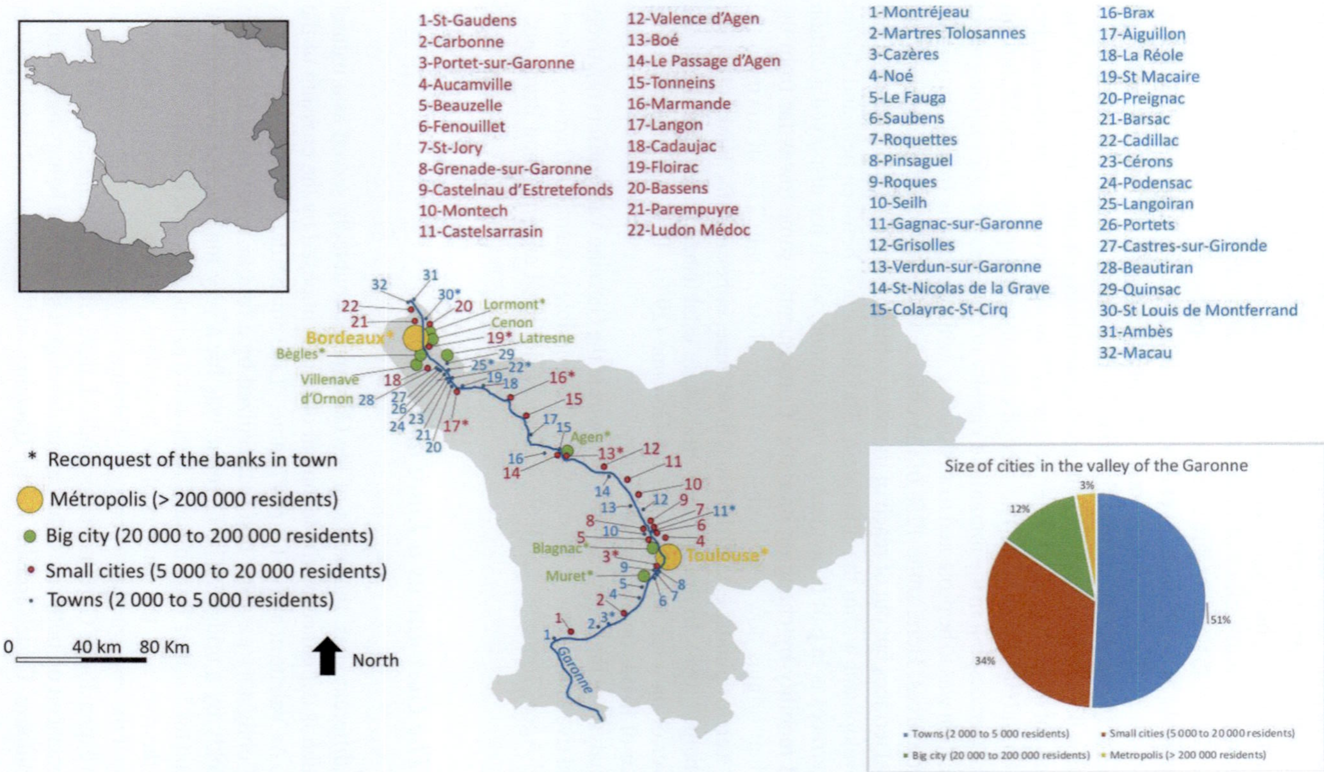

Fig. 8.1 Location of urban centers in the Garonne Valley according to population (*Source* Valette 2023)

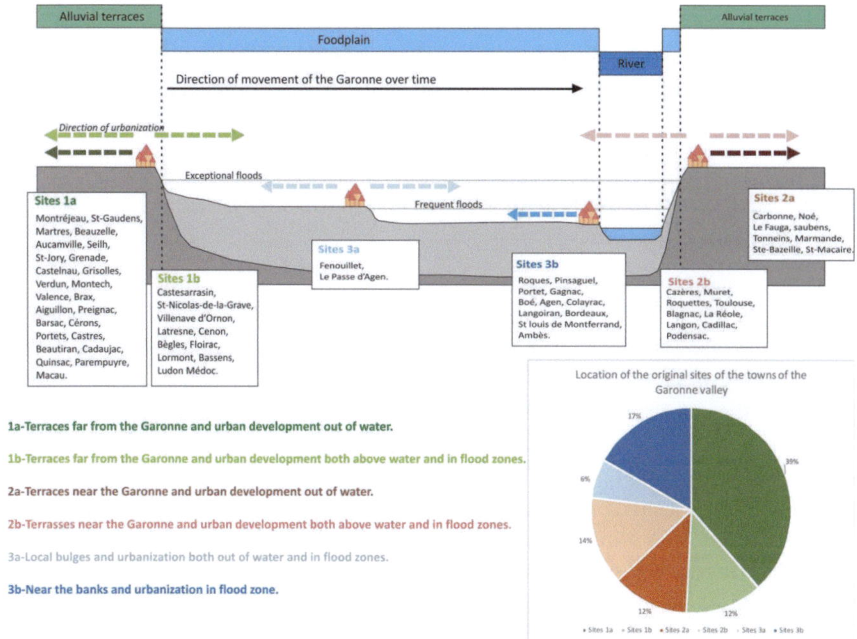

Fig. 8.2 City locations in the Garonne Valley (*Source* Valette 2023)

Bordeaux developed on the left bank, with smaller developments on the other side. The possibility of crossing the river, whether by means of a ford or a bridge, quickly allowed for the development of a town on the opposite shore, where the roads of a region came together. Thus, at specific points in its course, the Garonne stops acting as a barrier and, on the contrary, becomes a place of exchange and commerce at the origin of urban cores.

8.2.3 *The Socioeconomic Advantages of the Establishment of Towns Near the Garonne*

The initial advantages of all urban sites (fords, topography) have been accentuated by the transformation of the sites by society. The waters constituting the river are the basis for the birth of many cities—its use taking many forms. It is, as Pelletier (1990: p. 233) argues, a "natural resource used for food, hygiene, recreation, industry and transport". While the city of Bordeaux was established on a site sheltered from flooding, it was above all chosen because of the good drinking water available thanks to streams descending from the Landes plateau. In addition, Bordeaux's development owes much to its tidal port, giving access to the Atlantic Ocean (Poussou 2011). The city of Toulouse, for its part, owes part of its wealth to the presence of numerous mills that used the hydraulic power of the river (Mercié 2007) as well as its strategic position between the Pyrenees, the Atlantic, and the Mediterranean.

8.2.4 *Adapting to Raw and Flooding of the Garonne in the City*

It is notable that 77% of Garonne towns are located away from the flood zone of the Garonne (see Fig. 8.2). Over the centuries, floods and flooding left many traces and were part of the daily life of the inhabitants of riverside towns (Lambert 1982). Among all the catastrophic events in the Garonne, the flood of June 24, 1875, which left 500 victims in its wake continues to be remembered for its destruction of life and property. And yet, despite such lessons, continual urbanization has led many cities, in need of land, to expand their territories into flood zones. Some urban sites in the valley are located entirely or almost entirely in a flood zone: Pinsaguel, Portet-sur-Garonne, Fenouillet, Gagnac-sur-Garonne, Boé, Le Passage, Langoiran, Ambès.

If the site of installation of an urban core is the result of several geographical factors, they are insufficient to fully explain the blooming and development of a city. Without falling into determinism, we can speak at most of geographical aptitudes. Consequently, the urban river landscapes of the Garonne are the result of complex interactions between natural processes (raw, floods) and socioeconomic processes linked to urbanization. Above all, the river-city relationship in the Garonne Valley is marked by dynamism, changing over the course of history. Using a geohistory approach, it is possible to reconstruct this dynamism using the examples of Toulouse, Agen, and Bordeaux as examples.

8.3 Geohistory of the Urban River Landscapes of Toulouse, Agen, and Bordeaux

Geohistory is an approach in geography which allows, through the analysis of historical sources, to reconstruct the evolution over time of the relationship between cities and the Garonne. This involves identifying major phases of evolution over time based on old texts and maps. Within the framework of the Garonne Valley, it is possible to determine five periods, during which the city/Garonne is different (Fig. 8.3).

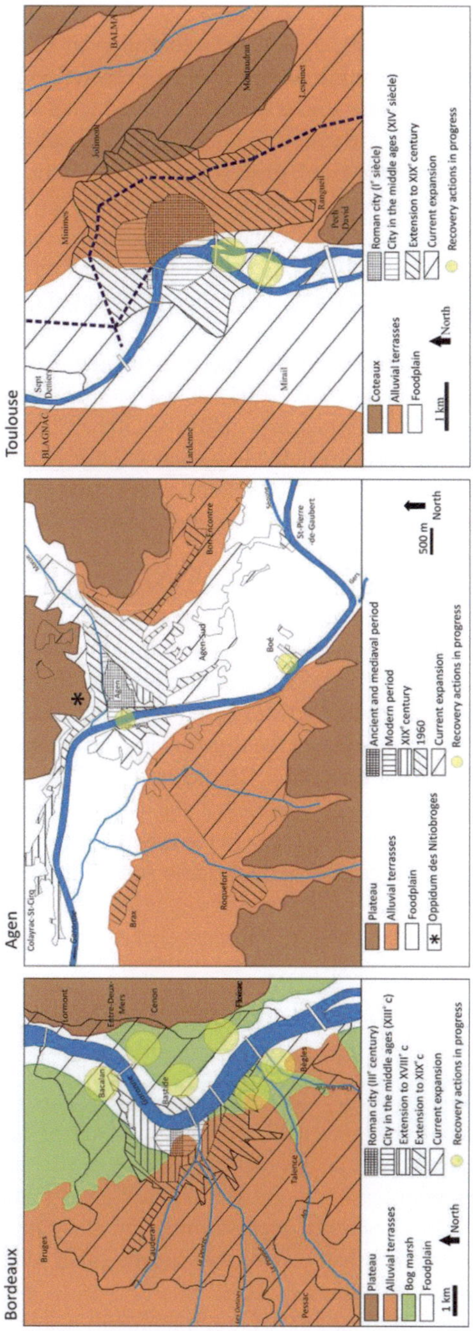

Fig. 8.3 Sites of urban evolution at Toulouse, Agen, and Bordeau (*Source* Valette 2023)

8.3.1 The Development of Cities Under Roman Influence

According to Barraud et al. (2009: p. 21), the excavation of the Grand Hôtel de Bordeaux in 2004 showed a "genuine proto-urban agglomeration" on the left bank terrace "with well-drained soils and whose area is estimated at between three and four hectares". It "overlooks the Devèze valley on one side and the Palu des Chartrons on the other…". After the conquest of Gaul, the Romans increased the size of Burdigala (their name for Bordeaux) to give it an area of 80 hectares in the first decades AD. In this space on the terrace looking toward the Garonne, the Bituriges Vivisques, the Celts of the region, gradually established themselves. But from the outset, Bordeaux was a small sheltered Roman port located at the meeting point of the Devèze and the Garonne.

The Romans were also responsible for the establishment of Tolosa, today's Toulouse (Pailler 2002; Aramond et al. 2004). It was a strategic location for the Romans at the crossroads of trade routes, especially that of the Garonne which served as a navigation artery. Tolosa was surrounded by several Celtic nuclei of the Volques Tectosages. The oppidum of Vieille (Old) Toulouse was 150 ha in area, overlooking the Garonne near its confluence with the Ariège and St Roch (80 ha) play a major role here.

A similar situation prevailed in Agen. There the early settlement of the Nitiobroges de l'Hermitage oppidum (164 m above sea level) was abandoned in the first-century BC (under the reign of Augustus) in favor of a site within the floodplain (Aginnum, located 45 m above sea level), bordering the Garonne and the Masse marshes. Moreover, an ancient port was located not far from the confluence between the Masse and the Garonne (Rue du Fond Raché) inside the Roman city, the configuration of which recalls on a smaller scale that of the port of Bordeaux.

Under several centuries of Roman influence the Garonne became an axis of communication of primary importance for both the towns of the valley and between the Mediterranean and Atlantic coasts. In the Middle Ages this role would continue and in many ways be enhanced.

8.3.2 The Medieval Period: Towns Subject to the Nature of the River?

In medieval times, Bordeaux remained confined to the left bank of the Garonne while Agen and Toulouse remained confined to its right bank. Nevertheless, development on the opposite banks emerged with the growth of urban areas featuring less desirable infrastructure including hospitals, leprosarium, and slaughterhouses (Valette and Caroza 2013). In Agen, the urban core of the Passage solidified on the left bank around marine activities.

The Middle Ages towns of the Garonne were commercially centered on the river—the Garonne acting as an economic engine (Valette 2002). In Bordeaux, as these activities grew in importance, they by necessity began to be situated outside the walls. Texts and maps of the medieval city show the central core, surrounded by ramparts, and a whole series of specialized river ports outside the walls, such as the wedge around the new Faubourg des Chartrons that specialized in the export of products from the high country (wheat, wine, pastel). In Toulouse, medieval texts identify eight ports allowing boats to dock both inside and outside the walls of city.

As well during this period, a proto-industrial economy based upon hydraulic power (water mills, forges, etc.), emerged. Carriageway mills (built on stone wharves) were built in the twelfth century at Toulouse: Château Narbonnais, Daurade, and Bazacle mills (Mercié 2007) and at Agen: St-Caprais and St Georges mills (Lavaud 2017). They would be replaced by ship mills, called "nefs", which were floating mills anchored in the river.

Because the cities of the Garonne were commercially and economically turned toward the river, they were frequently at risk of flooding. In Agen, the construction of a bridge(s) across the river in the twelfth century was problematic because of flooding. Its construction would be put on hold several times over the next century (Ducarton and Lavaud 2017). In Toulouse, the covered and fortified bridges were unable to resist floods and floods. Among them, the Old Bridge (Pont viel) was rebuilt five times between 1258 and 1414 (Coppolani 1992).

In the Middle Ages, the waters of the Garonne should be considered as multi-functional. They were a source of food, of raw material for crafts, an energy source, a means of transportation, and an integral part of defenses (including moats around the ramparts of cities).

8.3.3 Developing City Riverfronts and Constraining the Garonne City (Modern Period Until the End of the Nineteenth Century)

The seventeenth and eighteenth centuries correspond to a pivotal period for the city-river relationship in the Garonne Valley. Since 1682 the river was linked to the Mediterranean Sea by the Canal du Midi, leading to increased river navigation activity everywhere. This in turn spurred the river cities to master the river for economic ends. To facilitate commerce, the old medieval cities took away defensive ramparts and gradually integrated the suburbs into the urban core. Faced with repeated floods on one side of the city, and growing trade on the other, the cities of Agen, Toulouse, and Bordeaux were doubly transformed between the seventeenth century and the first half of the nineteenth century.

In Toulouse, the Cours des Ormes, completed in 1601, served as both a dike protecting the Faubourg St-Cyprien from flooding and also as a vast promenade esplanade planted with a double row of trees. In Bordeaux, according to Leulier (2008), riverfronts and pavilions were substituted for leper stalls along the city walls in order to embellish the exceptional port site hugging the curve of the river. The inauguration of Place Louis XV (now Place de la Bourse) in 1743 opened the city up to the river. It began the metamorphosis of Bordeaux, wherein the city acquired between 1760 and 1850 a long riverfront of wharves, three monumental gates, and a public garden. “The riverfront of wharves is a sort of gigantic canvas”, writes Damas (1930: p. 25). The “canvas” corresponds to the pleasing view created by houses with uniform facades to which was added the construction of a fairly wide earth-filled quay encroaching on the Garonne. This practice of riverfront embellishment can be considered as “facade urbanism” (Harouel 1993).

In Toulouse during the same period, the Garonne was gradually corseted by a double line of quays, high on the right bank, and quay walls on the left bank. Within the quays, 3 ports were built in different forms: rectangular for the port of La Daurade, square for the port of Bidou, and circular for the port of La Viguerie (St-Cyprien). This architectural ensemble had a triple function, which is also found in Bordeaux. Firstly, it was a question of defending against floods; secondly, avoiding river erosion of the shoreline; and thirdly, of favoring trade while enhancing riverfront aesthetics. The dwellings built on the quays of Toulouse in red brick followed the idea of a uniform façade of houses like at Bordeaux (Valette and Carozza 2013). A little later, in the first half of the nineteenth century, the same scheme (quay/dike in red bricks) was taken up to embellish the island of Tounis further south of the city.

In Agen, the problem was somewhat different since it was first a question of modifying the gravel islands fronting the shoreline in order to improve navigation (Valette et al. 2011). The Gravier district, located near the river, was subject to sanitation improvement and drainage in the seventeenth and eighteenth centuries (gaining land on the Garonne to ward off the danger of flooding). Once modified, the gravel was planted with rows of abalones to embellish the riverfront (Valette 2017). Later, in 1840, the banks of the Garonne were definitively fixed in place by the establishment of a stone quay favorable to the development of port activities.

Finally, the modern period until the middle of the nineteenth century was a pivotal time for the urban river landscapes of the Garonne. Everywhere the banks of the river were encroached upon and rendered artificial through the construction of concrete and stone shorelines and quays, as well as beautifying movements focused on embellishing river facades. Many of today’s riverfront landscapes in Toulouse, Agen, and Bordeaux are directly inherited from this period. Moreover, those landscapes helped to form the identity of those places as riverfront cities.

8.3.4 Marginalization and Abandonment of the River Within the Urban Space (1850s-1990s)

At the turn of the twentieth century, river navigation all but disappeared from the landscapes of the Garonne Valley. The main consequence of this change in transportation was the abandonment of the port facilities of the river cities, essentially rendering them purposeless. This freed space was reinvented as car parks in the second half of the twentieth century. Nevertheless, navigation lasted longer in Bordeaux, which benefited from its role as an Atlantic port. Due to a lack of space, port activity migrated downstream from the city following the opening of wet docks in 1879 (Schoonbaert 2009). In 1927, the port of Bordeaux physically cut itself off from the urban core through the construction of water gates to create an artificial marine bassin apart from the river. In the second half of the twentieth century, as elsewhere in the Garonne Valley, the urban port facilities were gradually abandoned.

The urban development that began in the nineteenth century continued unabated throughout the twentieth century. It spilled over into the flood zones and gradually erased the early asymmetry of Garonne Valley cities. Toulouse became swollen on all sides, its area increasing by a factor of five. This growth included the left bank of the river, despite its having been submerged by the flood of 1875 (Valette 2002). At the beginning of the twentieth century, certain heavy and risk-prone industries were established on the banks of the Garonne, outside and to the south of the city including explosives, cartridges, the construction of combat aircraft, and the manufacture of fertilizers (by the Office National et Industriel de l'Azote, or ONIA). But, in the second half of the twentieth century, these areas would be incorporated into the growing urban tissue of the city (Valette 2016). In Bordeaux, a large population settlement, the Faubourg de la Bastide, began on the right bank and is linked to the construction in 1822 of a stone bridge connecting with the urban core. This district saw the development of a multitude of industrial activities around metallurgy, chemistry, as well as stations and military barracks (ZAC Bastide Niel 2014). The city of Agen has experienced similar trends, but somewhat later in the second half of the twentieth century with the urban agglomeration extending into the flood zone on the right bank (Agen South district) and on the left bank at Passage d'Agen (Valette et al. 2011).

Rather than being at the heart of the commercial and economic development of the city, the Garonne has become an element to be controlled, especially in relation to flooding. In the case of Toulouse, relations with the Garonne have been dominated by numerous initiatives to limit the disasters caused by floods and floods (dike construction). At the end of the 1960s, "the bed of the Garonne was enclosed in a system of concrete dykes from the Empalot bridges to the Blagnac bridge" (CUEIT 1993: p. 37). In the 1970s, the urban banks of the Garonne at Agen were encased in concrete on both the right and left banks. Between Bègles and Bordeaux, transportation routes (motorway A 631, 2–3 lanes) since 1974 closely follow the river's banks—creating a physical barrier between populations and the river.

It can generalized that during this period the towns and cities of the Garonne turned their back on the river, making the shoreline a "no man's land" of marginal spaces used as vast car parks or otherwise devoid of purpose. Since the beginning of the post-industrial period of the 1990s, however, this trend has begun to reverse itself. The Garonne has become an object of reconquest by municipal authorities in the name of improving quality of life through the creation of "natural" spaces.

8.4 The Revival of the Garonne in the City: Aesthetic, Cultural, and Festive Revitalization of the Riverbanks

Since the 1990s, the Garonne has been subject to initiatives aimed at improving the landscape aesthetics of its banks. Large cities such as Toulouse and Bordeaux have experienced broad movements to reapproriate urban river spaces, a trend which has spread to the medium-sized towns of the valley (Fig. 8.4).

8.4.1 The First Stages of the Revitalization of the Urban River Landscapes of the Garonne

Beginning in the 1990s, Toulouse and Bordeaux began several river improvement initiatives that resulted in a "river landscape metamorphosis". Toulouse was a leader in this trend (Valette 2002). One of the first signs of metamorphosis of the banks of the Garonne came with the interdiction of automobiles from the banks and the old ports in the early 90 s. As part of this new policy, 16 hectares of parks and promenades on the banks of the Garonne were consecrated in the decade leading up to the 2000s (Valette 2016). Coupled with the development of these green spaces was the rehabilitation of certain buildings and abandoned properties to create an aesthetic dynamism along the river (i.e., water tower gallery, Garonne theater, former tobacco factory, Bazacle space, art museum contemporary with the Slaughterhouses).

Bordeaux followed the lead of Toulouse a few years later. For decades the port of Bordeaux had been inaccessible to walkers, the waters of the river virtually invisible to pedestrians, hidden behind high gates and commercial sheds. "Until the mid-1990s", writes Martin-Herrou (2011:8) of the situation in Bordeaux, "any space that became available was targeted and turned into a car park". The first signs of river renewal in Bordeaux began with the removal of the tall gates surrounding the port in the summer of 1996, followed by the demolition of the disused sheds in the fall of 1999. Simultaneously, efforts were made to reduce the presence of cars and parking spaces, resulting in new public spaces dedicated to leisure. In the first decade of 2000, the Urban Community of Bordeaux launched a vast project to reclaim the quays through the widening of the sidewalks and new roads and tramways dedicated to leisure. This project was an integral part of the city's goal of having Old Bordeaux, in particular the Port of the Moon, classified as a UNESCO World Heritage Site in June of 2007. In the years since, the quays have become a festive gathering place (leisure, walking, sport, strolling, holidaying). Covering about 25 ha (4.5 km long, 80 m wide), the quays were redesigned around five thematic sequences, the best known of which is that of the "mirror of water" (Arc en rêve. Centre d'architecture 2009).

Elsewhere in the Garonne Valley, the renewal of the river banks is not so clearly traced. In the 1990s, Agen would invest in managing the riverfront, but the objective

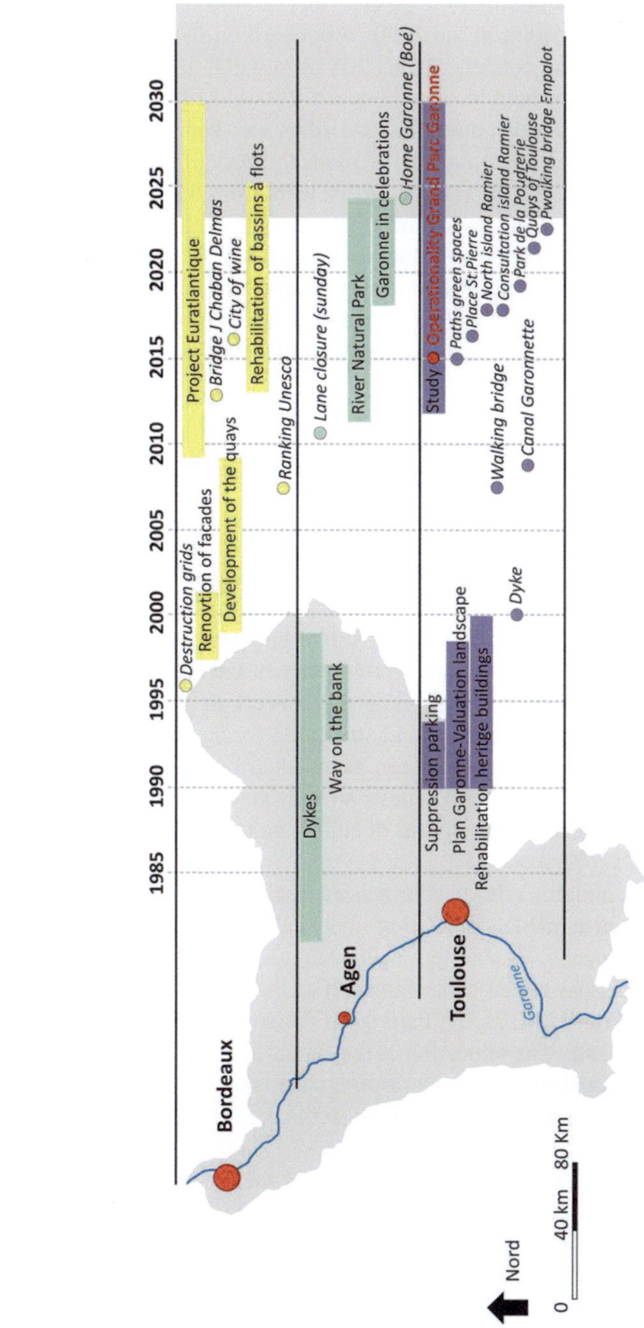

Fig. 8.4 Chronogram of riverbank revitalization operation in Toulouse, Agen, and Bordeaux (*Source* Valette 2023)

was not necessarily renewal but the construction of various dike and wall infrastructure to protect the city against flooding (Valette et al. 2011). Following trends dating back to the 1960s, flood protection in the 1990s was based on the artificialization of the landscapes by rebuilding concrete banks 205 m in width, dredging, and the construction of dikes. The dikes built in Agen between 1993 and 1999 vary in height and in shapes, some serving as walls, others as road dikes, etc. The Gravier dike, for example, allows pedestrians to contemplate the Garonne through different panoramic semi-circles. However, the spectacle presented to viewers from the top of the dike is far from the aesthetically pleasing panorama of renewal achieved in Toulouse and Bordeaux. At the foot of the dyke, an overflow car lane rudely interjects itself into the relationship between the citizens of Agen and the waters of the Garonne. Until the early 2010s, renewal initiatives remained confined to the old urban core, with the goal of showcasing the city's heritage facade (Rode and Valette 2019).

8.4.2 A Change of Scale in Renewal Initiatives

Since the beginning of the 2010s, there has been a marked change of scale in the renewal of the banks of the Garonne. The development of the Garonne at the inter-municipal level has become a major territorial issue. In Toulouse, the Grand Parc Garonne project was presented in 2012 with the aim of enhancing the banks of the river along approximately 32 km between the confluence of the Ariège and St-Jory (Lechner 2006). The basis of such initiatives is the concept of the Garonne as a natural and leisure corridor to be managed and structured for nearby urban centers (CUGT 2012). Four objective goals have been established by 2030: (1) preserve and enhance natural heritage, (2) develop pedestrian and cycle paths, (3) strengthen water-related uses, and (4) develop new spaces of culture and conviviality (Rode and Valette 2019).

Since 2015, various initiatives have commenced such as the development of soft paths to facilitate sustainable connections between different municipalities (Toulouse, Blagnac, Beauzelle, Fenouillet, Seilh, Gagnac, St-Jory). Along these connections multiple sites have been enhanced such as Place St-Pierre (2016), the north of the island of Ramier (2017), the Parc de la Poudrerie (2019), the historic quays of Toulouse (2021), etc. The relocation of the Île-du-Ramier exhibition center in 20xx DATE offers the possibility of transforming "the island into an urban park dedicated to nature, culture and leisure". This project is considered emblematic of future renewal initiatives for Toulouse.

In Bordeaux, following the success of the renewal of the quays district, similar initiatives have sprung up elsewhere. In 2009, the Bordeaux Euratlantique National Interest Operation (OIN) was defined, forming an arc between the St-Jean rail station and Bordeaux Lake (communes of Bordeaux, Bègles and Floirac). The goal of this vast ongoing project is to reduce the divides between the right left banks. The 738 ha targeted by this urban development project are distributed on both sides of the Garonne and joined together by two bridges. The first bridge was inaugurated in

2013 between the Bassins à Flots and the Bastide district (Jacques Chaban Delmas Bridge), while the Simone Veil Bridge will connect Bègles to Foirac (planned for 2024). As in Toulouse, several of these initiatives have been tied to larger municipal goals set for the year 2030: to modernize the St-Jean station, to develop a "Euratal-tique" business center, to develop housing, to develop green spaces on the river (50 ha planned), and to promote public transport. Everywhere, new buildings with contemporary architecture on both sides of the river have accompanied these projects. The "city of wine" in Bacalan, inaugurated in 2016, exemplifies this movement.

8.4.3 The Rebirth of a River and Its Banks

In recent years, the urban renewal of the banks of the Garonne has become a "territorial resource" (Carré and Deutsch 2015). The Garonne has clearly become central to urban management strategies at the intermunicipal level. This renewal is manifested in numerous festivities centered around the river.

Among these festivals, it is possible to note Toulouse Plages in Toulouse, Garonne en fête, and the river festival in Bordeaux (see Fig. 8.5). River festivals are also reflected in the erection of single-day, multi-day, and seasonal festivity tents (guinguettes) on the banks of the Garonne. Despite the length of time, each of these festivities marks a change in mentality and outlook toward the river in the city and a desire to reconnect with the Garonne.

This desire to reconnect with the Garonne is exemplified by recent trends in Agen. Since 2008 the problem of how to make the banks of the Garonne more accessible to Agenais has been under study. It is a question of rebuilding physical links between the city and the river, links that have been strained, in particular, by decades of flood protection works. Few development possibilities are available in

Toulouse (Source: Ph Valette) Agen (Source: Ph Valette) Bordeaux (Source: Ph Valette)

Fig. 8.5 Celebrating the Garonne—a symbol of the renewal of the urban river landscape (*Source* Toulouse Metropole; Agen Agglomeration; City Hall of Bordeaux)

the agglomeration of Agen, as the river banks are blocked off by autoroutes on both sides. Nevertheless, in 2010 an idea was proffered to close one of the riverbank lanes every Sunday to encourage walks along the river. Today, due to low attendance, the route is only occasionally closed during special events. The effort then focused in 2011 on the Agen-Garonne River Nature Park. The park is part of the repurposing of a former gravel pit site to the south of the town of Agen into a place of leisure (Lac de Passeligne). Soft connections for pedestrians and cyclists were created along the banks of the Garonne to connect the urban core with the new Park. Such developments represent the post-industrial reclamation of former industrial sites along the riverside for leisure purposes (Valette 2017). This urban reappropriation of the Garonne has also been manifested by the organization of festive events such as "Garonne en fêtes", in 2017. In addition to these examples of social reappropriation must be added examples of cultural reappropriation. The city of Boé will inaugurate in 2023 the Garonne House, where river heritage will be highlighted.

Such movements of renewal and reappropriation of the banks of the Garonne are spreading up and down the valley to the towns of Cazères, Muret, Portet-sur-Garonne, Blagnac, Gagnac-sur-Garonne, Marmande, Langon, and Langoiran. Since the signing in 2009 of the Garonne Plan, whose 4th goal is to "enhance the river and its cultural and landscape identity", there has been a broad transition toward viewing the Garonne as an environmental and landscape amenity. With this broader acceptance, the movement to reappropriate the Garonne has gone beyond its urban origins to encompass most of the valley.

8.5 Conclusion

In the Garonne Valley, the process of reclaiming the urban banks of the river has slowly spread, following the initiatives begun in Toulouse and Bordeaux in the 1990s. Since the 2010s, the scale of intervention has become intermunicipal. The Garonne has become a natural space within the city to be enhanced and rehabilitated. At the scale of the valley, linking such projects between different municipalities through soft pathways (bike trails and footpaths) has become an objective of intermunicipal cooperation. Overall, urban river bank renewal has been most successful in the old urban heritage core, while intermunicipal cooperation is still in its early days.

The Garonne is also becoming a marketing product, where many real estate projects in Toulouse and Bordeaux boast of their views of the river from their new complexes. All of these projects promote the river in the city, using the Garonne as a backdrop or visible horizon. The vast majority of such initiatives have focused on the banks of the river with the aim of making them more democratically accessible. But these initiatives also show a form of standardization, using similar devices from one place to another. Open river banks arranged in the form of stairs or steps are found at numerous sites. Even the recent success of the Bordeaux "Water Mirror", which heralded many of the most innovative attempts to enhance the urban appreciation of the Garonne, is not quite as new as many imagine it to be. The geohistory of urban river landscapes in the Garonne shows the rapid generalization of past riverbank management initiatives. The famous river facades of eighteenth-century Bordeaux and Toulouse being prime examples.

The urban river renewal that began in the urban cores of the major cities has gradually spread to the medium and small-sized towns. The banks of the Garonne, the quays, and the slipways of Langon were rehabilitated in 2018, in Langoiran in 2019, in Boé in 2021, and are currently in progress in Muret. Elsewhere, Garonne interpretation houses (spaces with cultural vocations) are opening their doors at Cazères (2019) and Boé (2023). Elsewhere, where the Garonne is not located immediately nearby, like in Portets or Cadillac for example, the desire is to make pontoons available for spending time on the river. In all these places, the objective is the social renewal with the river through various leisure activities. This form of renewal appears today as the dominant model, focused almost strictly on the river banks. On the other hand, ecological renewal in particular the question of the water quality in the Garonne—especially in connection with the possibility of bathing—is little mentioned. However, in the context of climate warming and the problem of urban heat islands, rivers in the city are a form of cool green zone that needs to be better integrated into urban climate change strategies. Combining the functions of green zone with the leisure activity of swimming would represent the ultimate renewal of urban populations with their river.

References

Arc en rêves. Centre d'architecture (2009) Les quais. Bordeaux 1999–2009. Editions confluences, 216

Arramond JC, Requi C, Vidal M (2004) La Toulouse des Volques Tectosages (Vieille Toulouse, Estarac, St-Roch). Guide de l'exposition Gaulois des pays de Garonne (IIe-Ier siècle Av J.C.), 92 p, pp 42–50

Barraud D, Dominique S, Maurin L, Sireix C (2009) Les origines de Bordeaux: de la protohistoire jusqu'à la fin du VIe siècle. Lavaud, S. (dir.). (2008), Bordeaux. Notice générale II. La formation de l'espace urbain des origines à nos jours, Atlas historique des villes de France, pp 17–39

Bergevin J (1992) Déterminisme et géographie. Ste Foy, Presses Universitaires de Laval, Hérodote, Strabon, Albert le Grand et Sebastian Münster, p 204

Carré C, Deutsch JC (2015) L'eau dans la ville. Une amie qui nous fait la guerre. L'aube, 319

Centre Urbain d'Initiation à L'Environnement de Toulouse (CUIET) (1993) La vie au bord des fleuves, 72

Communauté Urbaine du Grand Toulouse (CUGT). (2012). Grand Parc Garonne. Plan guide à l'horizon 2030, 108

Coppolani J (1992) Les ponts de Toulouse. Privat, 152

Damas P (1930) La façade de Tourny. Bordeaux, 123

Dolfus O (1985) Brèves remarques sur le déterminisme et la géographie. L'espace Géographique 14–2:116–120

Ducarton A, Lavaud S (2017) Agen communal (vers 1217-vers 1340). Lavaud S (dir.). Atlas historique d'Agen, Notice générale, tome 1. Ausonius Editions, pp 145–167

Friedberg C (1992) La question du déterminisme dans les rapports homme-nature. Jollivet M (dir.), Sciences de la nature, Sciences de la société. Les passeurs de frontières. CNRS éditions, pp 42–53

Harouel JL (1993) L'embellissement des villes. L'urbanisme français au XVIIIe siècle, Paris, p 335
Jacques P (2017) Agen Antique (Aginnum), du règne d'Auguste au début du Haut Moyen Age (30 A.C. à 500 P.C.). Lavaud, S. (dir.), Atlas historique d'Agen, Notice générale, tome 1. Ausonius Editions, pp 95–118
Lambert R (1982) Les crues de la Garonne. Ferro M (dir.), Une histoire de la Garonne. Paris, ramsay, pp 41–82
Lavaud S (dir.) (2017) Agen, Sites et monuments, Atlas historique des villes de France, 386
Lechner G (2006) Le fleuve dans la ville. La valorisation des berges en milieu urbain. Centre de documentation de l'urbanisme, Paris
Leulier R (2008) La ville embellie. Une transformation radicale de la ville : les embellissements de Bordeaux au XVIIIe siècle. Lavaud S (dir.) (2008). Bordeaux. Notice générale II. La formation de l'espace urbain des origines à nos jours, Atlas historique des villes de France, pp 171–225
Martin-Herrou A (2011) La reconquête du fleuve par la ville de Bordeaux. Portus Plus, 14
Mercié P (2007) Légendaires moulins à eau aujourd'hui disparus. Toulouse, Montauban, Albi. Editeur Pierre Mercié, 367
Pailler JM (dir.) (2002) Tolosa. Nouvelles recherches sur Toulouse et son territoire dans l'Antiquité. Ecole française de Rome, 601
Pelletier J (1990) Sur les relations de la ville et des cours d'eau. Revue De Géographie De Lyon 65:230–255
Poussou JP (2011) Bordeaux ou la ville sur les marais. Revue du Nord, Zones humides et villes d'hier et d'aujourd'hui : des premières cités aux fronts d'eau contemporains, N°26, Hors série, pp 205–217
Rode S, Valette P (2019) Par-delà les limites communales, le fleuve au cœur du projet de territoire métropolitain? Comparaison entre Perpignan et Toulouse. Sud-Ouest Européen, L'eau au service des territoires? Entre Valorisation Et Instrumentalisation 47:25–39. https://doi.org/10.4000/soe.5172
Rossiaud J (dir.) (2016) Villes et fleuve en Europe. Silvana Editoriale, 87
Schoonbaert S (2009) De la ville à l'urbanisme. Renouvellement et continuité de l'espace urbain de 1790 à 1914. Lavaud S (dir.), Bordeaux. Notice générale II. La formation de l'espace urbain des origines à nos jours, Atlas historique des villes de France, pp 227–290
Valette P (2002) Les paysages de la Garonne: les métamorphoses d'un fleuve (entre Toulouse et Castets-en-Dorthe). Université de Toulouse Le Mirail, Thèse de Géographie, p 554
Valette P (2016) L'espace fluvial de la Garonne à Toulouse. Un territoire en reconquête. Revue Espaces, Tourisme et loisirs, Dossier: usages récréatifs des espaces fluviaux: des enjeux à l'échelle des métropoles, no 333. Nov-Dec 2016:56–63
Valette P, Carozza JM, Boudou M (2011) Agen et la Garonne: géohistoire des politiques de protection contre les crues et les inondations. Revue Du Nord, Hors Du Lit: Aléas, Risques Et Mémoires 16:77–88
Valette P, Carozza JM (2013) Toulouse face à la Garonne: emprise de l'urbanisation dans la plaine inondable et géohistoire des aménagements fluviaux. GEOGRAPHICALIA, « Regiones fronterizas transpirenaicas », Universitat de Saragoza, 16. https://doi.org/10.26754/ojs_geoph/geoph.201363-64859
Valette P (2017) La place des cours d'eau à Agen. Regards géohistoriques sur l'imposante Garonne et ses petits affluents oubliés. Lavaud S (coord.). Atlas historique d'Agen, Notice générale, tome 1, Ausonius Editions, pp 61–84
ZAC Bastide Niel (2014) La Bastide, mutations historiques, livret expo, 28

Chapter 9
Management of the Turia River in Valencia (Spain): The Recent History of an Unfinished Metamorphosis

Iván Portugués Mollà

Abstract The management of the Turia River in Valencia (Eastern Spain) throughout the twentieth century and the early twenty-first century exemplifies the natural, social and even economic dynamics that usually affect Mediterranean urban riverbeds. However, this case has some particularities, such as the complete diversion of the river through a discharge channel and the conversion of the old riverbed into a linear urban park. This is the ultimate exponent of a profound environmental and urban metamorphosis that is still ongoing. This chapter identifies the different stakeholders involved in the governance of the Turia over the last 125 years, distinguishes the various stages of management, analyses the most evident territorial consequences of the hydrological policies and points out some aspects that are still pending in the fluvial scene.

Keyword River diversion · Landscaping · Renaturalization · Turia River · Valencia

9.1 Valencia and the Turia River, a Strained Relationship

Located in the center-east of the Iberian Peninsula, the city of Valencia (800.000 inhabitants in the urban core) has an eminently fluvial origin. Its foundation more than two thousand years ago took place on the banks of the Turia River (280 km long and a basin extension of 6.394 km^2)[1] and not on those of the Mediterranean Sea, at

[1] The Turia is a medium river by Spanish standards. Its main hydrological characteristics are its steep gradient (it originates at an altitude of 1,800 m above sea level) and its torrential nature. Although its average module barely reaches 11.4 m^3/s at the gates of the city, historically flood flows have exceeded 2000 m^3/s, especially during autumn storms (Carmona 1990).

I. Portugués Mollà (✉)
Department of Geography. Section of Physical Geography, University of Valencia (UV), Valencia, Spain
e-mail: ivan.portugues@uv.es

© The Author(s), under exclusive license to Springer Nature Switzerland AG 2024

J. Farguell Pérez and A. Santasusagna Riu (eds.), *Urban and Metropolitan Rivers*, The Urban Book Series, https://doi.org/10.1007/978-3-031-62641-8_9

that time a few kilometers away from downtown. In fact, the coastal occupation and the construction of the seaport did not occur until well into the Middle Ages.

The co-evolutionary relationship between Valencia and the Turia has always been very close. In the first place, the winding course of the river has defined the morphology of the old town. Almost until the diversion of the watercourse, and with the exception of sporadic constructions, the river was understood as the northern limit of the city. The riverbanks articulated its most noble promenades (Albereda, Petxina, Montolivet), then located on the outskirts. In that sense, the fluvial view of the northern façade has been a historical reference for illustrators, travelers and merchants (Fig. 9.1): the river and its parapets, the Gothic bridges, the medieval walls and bell towers offered a rich composition (Rosselló and Esteban 1999). Secondly, the Turia has fed an extensive agricultural ring (about 28 Km^2) through an intricate network of irrigation ditches and weirs of Arab origin, known as the Horta, which has sustained much of the metropolitan development. In addition, the stream provided water for human and livestock consumption, and the river bottom supplied aggregates for construction. Finally, the riverbed served as an esplanade that hosted public events, but also informal settlements and uses.

This overlap between a natural and a social system, both very dynamic, has not been free of conflicts, which became more decisive as the traditional uses of the river lost interest. On the one hand, the Turia was conceived as a threat due to its recurrent floods. That is why the authorities began its robust channeling five centuries ago. In contrast, Valencia evidenced a constant desire to conquer the river plain and increased its exposure to risk. On the other hand, the river began to be perceived as a nuisance

Fig. 9.1 Engraving by Alfred Guesdon (1856 ca) showing the praised fluvial façade of Valencia. On the right, the Albereda Promenade (*Source L'Espagne à vol d'oiseau* lithographic collection)

when the late industrial development environmentally and aesthetically degraded the river scene. It was also an inconvenient neighbor for the port, immediately north of the mouth. The tensions during the last century accelerated the denaturalization of the Turia that had been initiated sometime before. This is precisely the starting point of the present chapter.

The second part of the twentieth century meant the definitive rupture in the river-city relationship. Valencia woulds end up removing the Turia from its urban grid and annulling it as a hydrosystem in response to a major flood in 1957. From the 1980 onward, the abandoned riverbed was landscaped in line with emerging environmentalist demands. As a result, today the lower Turia branches into two channels without permanent flows: one has become a linear park and the other is a mere discharge channel. Although in the last fifty years the two watercourses have received unequal attention, both are in the focus of contemporary urban planning. These particularities give the Valencian case a uniqueness that deserves to be explored in depth.

9.2 Stakeholders in the Fluvial Scenario

The case of the Turia in Valencia exemplifies the interests involved in the management of Mediterranean urban rivers. In a territory not only subject to urban processes, but also to agricultural and port dynamics, river governance has involved multiple actors, both public and private. These stakeholders pursued objectives that did not always coincide. Management patterns have been linked to the socio-economic and political evolution of Spain, in general, and of the city, in particular; industrialization and hygienism, post-war misery and the long Franco dictatorship (1939–75), developmental technocracy or the democratization of participatory processes have been stages represented in the metamorphosis of the river. In turn, urban needs and deficiencies have always been reflected in the riverbed.

Many of the initial hydraulic proposals concerning the Turia had little impact or did not materialize due to economic insolvency, the development of the Spanish Civil War (1936–1939) or the succession of major floods. In other cases, discrepancies between administrations delayed execution. For their part, urban plans began to pay increasing attention to the potential of the riverbed as a structuring axis for future growth. The particularities of the fluvial regime, which resulted in a riverbed with a low flow and apparently overdimensioned, gave it an unquestionable strategic potential. The flood of 1957 forced a rethinking of all the projects, while the arrival of democracy in the mid-seventies brought about a change in the political paradigm.

9.2.1 The Júcar Hydrographic Confederation

Until the reclassification of the urban riverbed, the jurisdiction of the Turia River as it passed through Valencia fell exclusively to the Júcar Hydraulic Division (Júcar

Hydrographic Confederation since 1934), one of the twelve water bodies administered by the Spanish State.[2] With regard to the river guardianship issues, the Hydrographic Confederation (hereinafter JHC) maintained a permissive position that resulted in a profound deterioration of the river ecosystem and the urban landscape. In any case, it went unnoticed by the majority of the population, as the river acquired a peripheral role in the city.

As for the works policy, its management was characterized by partial hydraulic projects of little entity and municipal boundaries. In many cases they did not even go beyond the study phase. There were four basic areas of attention: (a) *sanitation*. In accordance with hygienist principles, the construction of a minor bed was proposed to redirect the low waters and to avoid diversions that would cause unhealthy puddles and bad odors; (b) *ornamentation*. The authorities recognized the degradation of the riverbed due to abusive uses and proposed its reintegration into the urban image, as in other Spanish cities[3]; (c) *flood management*. This is undoubtedly the issue that most concerned the authorities. To mitigate the risk, the widening and repair of the historic canalization and the dredging of a bed filled by sediments were constantly considered; d) *urbanization*. In the opinion of the technicians of the water agency, the actions contemplated in the previous points would allow the urban development of the riverbanks in accordance with the post-war urban development plans.

These objectives were included in the *Preliminary Project for the Channeling of the Turia River as it flows through Valencia* (García Labrandero 1949). Although little known, it was the most complete document of all those referring to the Turia in the first half of the twentieth century. However, only the extension of the parapets to the northwestern neighborhood of Campanar, historically surrounded by orchards but with expectations of approaching the riverbed, and some occasional works to stabilize the current, could be carried out. The 1957 flood deepened the idea of implementing a more decisive solution to the problem of urban flooding. As will be detailed later, the Turia would be diverted by means of a channel that came into operation in 1972 and would constitute the new Public Hydraulic Domain from then on. The Confederation took over the management of the new channel and at the same time delegated the administration of the old riverbed.

9.2.2 The Valencia City Council

The municipality of Valencia has played an outstanding role in the evolution of the urban bed of the Turia, although less than desired according to institutional documentation. The city insistently claimed ownership of the riverbed under a medieval

[2] The domain of the Júcar Hydrographic Confederation (42,735 km^2) covers a large part of the Eastern peninsula, between the basins of the Ebro, to the north, and the Segura, to the south. The competences of the basin agency were defined by the Spanish Water Law of 1879.

[3] As recognized by the technicians, the *Project for the Channelling of the Manzanares River* (Mendoza et al. 1942), implemented in Madrid, served as inspiration for the possible beautification of the Turia River.

right that recognized communal use for centuries. But these claims were initially unsuccessful, as they contradicted the provisions of the State Water Law of 1879. Thus, the municipal authority was reduced to the urbanization of the riverbanks, the improvement of the river walks and the construction of bridges to facilitate the late urban development of the north bank.

In any case, the local administration can be understood as the most influential pressure group in fluvial management, concerned by the distancing that was occurring between the river and the city. On the one hand, it constantly reported the poor condition of the riverbed and the permissiveness of the hydraulic authority in authorizing or ignoring abusive uses (Fig. 10.2). On the other hand, it demanded the reintegration of the river front in the urban landscape as well as the embellishment of the monumental hydraulic complex. Finally, it showed firm opposition to any attempt to remove the river in its entirety and detach it from its fluvial condition. Many of its claims, including the last one, were disregarded by the Confederation.

Once the river detour was completed, the municipality showed its interest in ensuring the ownership of the riverbed and defining suitable uses. This would prevent the implementation of ambitious developmental projects outlined from Madrid and would make possible the settlement of a green area increasingly demanded by the Valencians. Not without controversy, the transfer by the State took place by means of a royal decree on December 1, 1976. The resulting park has remained under municipal ownership to this day. Only specific sectors of general interest are managed by the regional government (Generalitat Valenciana).

9.2.3 The Port of Valencia

Although its participation in the affairs of the Turia has gone more unnoticed, the Port is one of the actors that deserves more attention. Its motivations were certainly forceful: the first, to move the river mouth away from the historic dock to avoid the costly annual dredging derived from sediment accumulation; the second, to eliminate its main physical obstacle in order to be able to expand to the south. Coinciding with the increase in export activity in the first third of the twentieth century, the various urban and hydraulic planning instruments took up the Port's pretensions. Almost all of them sketched a partial detour that only affected the mouth of the river.

During the sixties, the expansionist expectations of the Port were largely exceeded with the approval of the integral deviation of the Turia. The declaration of the Port Authority as an autonomous body in 1978 facilitated the extension of the installations to the new mouth and reaffirmed its strategic potential. In addition, it maintained the competences over the old mouth until 2017. In fact, in 2006 the Authority moved forward and buried the fluvial section that meets the sea, erasing forever the vestiges of the old estuary. The Port of Valencia is today a major infrastructure in the context of the western Mediterranean, although the continued enlargement of its facilities remains a source of controversy.

9.2.4 Private Initiatives

Already in the late nineteenth century, economists and developers showed interest in narrowing the river area and urbanizing its banks.[4] The profitability of a strip of land 120 m wide and about 10 km long, very close to the heart of the city, was very tempting. It is worth considering that urban plots were scarce in the crowded urban fabric. In addition, the high income of the Horta ring restrained the possible expansion of the city and its satellite towns. The urbanistic value of the riverbed has been evidenced in democracy by the housing development unleashed after its landscaping and improvement.

The industrial cluster established between the Pont de Ferro Bridge and the port at the beginning of the twentieth century also had a significant impact on the policy of river guardianship. Using the irrigation ditches as dumping channels, chemical, paper and alcohol factories polluted the final stretch, encouraged by legislative permissiveness. The CAMPSA oil complex in the final stretch had a double impact on the system: on the one hand, it experienced leaks that worsened water quality; on the other, it excessively restricted the mouth section and forced the JHC to redirect the estuary. During the 1980s, the relocation of these activities to newly created industrial parks was addressed.

9.2.5 Counter-Planning Movements

Both the weakness of the dictatorial regime at the gates of democracy and the sociological evolution of the Spanish population explain the appearance of the first counter-planning movements. Little by little, participatory and ecological processes were counteracting the environmental insensitivity and the privatizing desires of the Franco dictatorship.[5] During the early seventies, public movements in favor of the park led symbolic occupations, popular plantings or festive days on the riverbed. The professional associations drafted manifestos in that sense and some architect teams offered their own proposals. In 1973, when the media repercussion of the pro-park proclamations was already very evident, the Ministry of Public Works and Urban Planning showed an unprecedented interest in revising the highway project and taking on board the social demands.

It was not until 1976 that the gardening initiative would experience a great impulse. That year the *Pro-Cauce Commission of Citizen Entities* was formed, made up of professional associations, representatives of neighborhood movements and recreational entities. Under the slogan *The riverbed is ours and we want it green*, the

[4] This was the main purpose of the *Project for the diverting of the Turia River and the drainage of the Albufera lake*, (Llorens 1888), probably the first documented precedent.

[5] The Turia issue was not an isolated event in a context of developmentalism vortex. The most obvious example was the projection during the 1960s of a residential complex with hotels, golf courses and a marina in the nearby Albufera natural site, declared a protected area in 1986.

citizens' platform insistently claimed the municipal cession of the riverbed for its conversion into a park. It also warned of the more than probable destruction of the hydraulic heritage that the road construction would entail. The transfer of the riverbed at the end of the year certified the success of the popular mobilizations.

9.3 Evolution of the River Management

In terms of uses and hydraulic policies, the significant differences between the most central section and the peripheral ones highlighted a certain sectorization of the riverbed. In practice, there were up to three sections with a particular idiosyncrasy. The first, from the Rovella Weir to the San Josep Bridge, had a more agricultural character. The second, from that bridge to the Pont de Ferro Bridge, was eminently urban. Downstream, the river had an industrial profile (Fig. 9.2). Undoubtedly the central section always received better treatment. This zoning was not rigid, but underwent major changes starting in the sixties. Having lost its fear of the river, the city consolidated its expansions and jumped to the north bank. In the course of twenty years, the Turia became an axis of urban symmetry for the first time.

When dealing with the temporal evolution of urban-river governance, it is essential to highlight two major historical milestones for Valencia. The first was the flood of October 1957; the magnitude of the tragedy itself led to the detour of the river and,

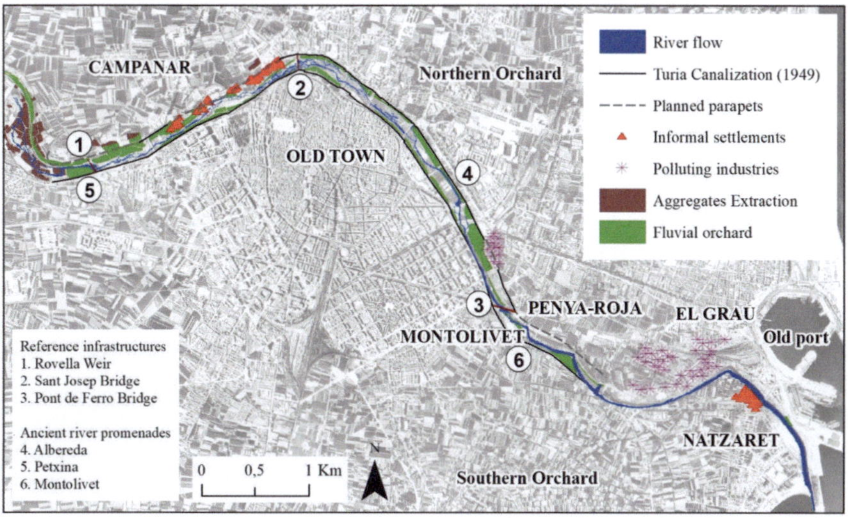

Fig. 9.2 Irregular uses in the Turia riverbed during the 1940s and early 1950s. City of Valencia. The orthophoto in the background belongs to the so-called American Flight of 1956 (*Source* Author)

subsequently, a profound metropolitan transformation. The second was the municipalization of the liberated riverbed, which facilitated its conversion into a large urban park. These two events define the three river scenarios described below.

9.3.1 The Secular Degradation of the Turia

Until the second half of the twentieth century, the Turia remained in a semi-natural state. Its flow was drained by the intensive agricultural exploitation of the metropolitan belt, but some returns of irrigation ditches and the connection to the sewerage system and the rainwater network ensured a minimum flow in the urban course, although it could dry out during severe summer droughts. The low waters drew intertwined streams that caused puddling and accelerated the putrefaction of the water. In contrast, episodes of flash-floods were frequent and had major social, economic and urban repercussions. This is why flood defense was not the exclusive concern of the basin organization, but of society in general.

Especially during the long post-war period, the Turia riverbed hosted numerous irregular uses in the same way as occurred in other Spanish rivers (Ollero 2002); the extraction of aggregates, the cultivation of family orchards or the irregular settlements made it the largest slum of Valencia (Fig. 9.2). All in all, the mouth of the river was the most affected sector due to the rise of modern industry. In this context of fluvial degradation, unsuccessful urban and sectorial planning considered the possibility of eradicating the problem at its root and diverting the Turia, either partially—from the Pont de Ferro Bridge—or totally—from the western boundary.[6]

A major flood in September 1949, which particularly affected the shantytowns inside the riverbed (there was talk of 9000 thousand dwellers), led to a change in the JHC's decision-making. Its management became more restrictive, which resulted in a notable improvement of the river's image, especially in the historical channeling. However, it had little impact on risk reduction. This was evidenced by the extraordinary episode of October 14, 1957, a real turning point in the hydraulic policy of the lower Turia. Due to heavy rains in the middle-lower basin, the flow recorded a peak of 3700 m^3/s and flooded three quarters of the city, which remained under mud for weeks (Portugués et al. 2016). The official victims were 84, 4000 houses were destroyed and the economic losses were estimated at 4400 million pesetas, 2.5% of the national income.

---

[6] The General Plan of Valencia and its Metropolitan Belt of 1946 or the *Preliminary Project for the Channeling of the Turia River* by the JHC (1949) proposed the rerouting of the final section in order to urbanize the riverbed and free the port. For its part, the 1947 Railway Development Plan (Berriochoa 1947) suggested the total detour to reorganize the railway network. Not by chance, its alternative for the new channel was practically identical to the one undertaken by the Southern Solution.

9.3.2 From the River Diversion to the South Plan

The 1957 event made it clear that partial hydraulic solutions were not enough. Therefore, only three months after the tragedy, the JHC presented the *Preliminary Project for the Defense of Valencia against the floods of the Turia River* (García Labrandero 1958). The document proposed three solutions: the Northern, the Southern and the Center one. The first two drew integral fluvial deviations in those directions; the central alternative maintained the natural course, but recommended an exhaustive dredging of the urban riverbed and the construction of a lamination reservoir in the middle-lower basin. Under Franco's authoritarianism, the technocrats opted for the so-called Southern Solution. In the summer of 1958, the drafting of the preliminary project was approved. The inoperative development plans of the forties, as well as similar hydraulic experiences in other international contexts,[7] inspired the formulation of the Southern Solution and contributed to its rapid gestation.

The planned diversion would start in the metropolitan municipality of Quart de Poblet, a few kilometers upstream from Valencia. At the embouchure, an automated weir (Assut del Repartiment) would be built to collect the flows of the Turia and distribute them through the profoundly modified network of irrigation ditches. The new riverbed would bypass the city along its southern perimeter up to its confluence with the sea at the Pinedo neighborhood, southeast of the capital. The channel would be 12 km long and 200 m wide, which ensured a hydraulic capacity of 5000 m^3/s. In order to adapt to the morphology of the terrain, three sections with different slopes and lengths were contemplated, most of them lined with concrete and with a breakwater bottom (García Labrandero 1961).

Nevertheless, the Confederation's engineers not only worked on solving the hydrological-hydraulic problem. Through a multisectoral plan, they also tackled the main urban planning deficiencies (outdated road and railway networks, insufficient rainwater drainage and sewerage systems, lack of urban zoning, etc.) (Martínez García-Ordóñez 1959). All this at a time of demographic growth fueled by an unprecedented rural exodus. Thus, the draft of the Southern Solution resulted in the so-called South Plan of 1961. The South Plan, in turn, was integrated into the 1966 General Plan, a revision of the 1946 metropolitan planning. The new urban framework reflected a desire for progress in line with the developmentalism prevailing in Spain at the time.

The discharge canal was built between 1965 and 1972 using abundant modern machinery (Fig. 9.3). The total cost of the work was estimated at 7000 million pesetas, of which 1500 million pesetas were spent on expropriations. With these figures, the Southern Solution is still today one of the largest hydraulic engineering works in the history of Spain. However, its effectiveness has not been empirically proven. Only in 1977 and 2000 did the canal carry significant floods, although much lower than in 1957.

---

[7] The bypass works executed on the Los Angeles River (California) in the first half of the twentieth century (Gumprecht 2001) were undoubtedly a benchmark for the Southern Solution.

Fig. 9.3 Appearance of the new Turia channel shortly after its inauguration (1973) (*Source* Júcar Hydrographic Confederation)

9.3.3 Municipalization and Greening of the Freed-Up Riverbed

Interventions in the natural riverbed did not cease despite the start of work on the new canal, as the Turia maintained its urban flow for more than a decade. The JHC focused on the most urgent hydraulic issues. First, the dredging of the bed and the reconstruction and enhancement of the parapets. Second, the construction of dykes to protect the factories and the maritime district of Natzaret, to the north and south of the estuary, respectively. As for the works of fluvial guardianship, the concession of plots of land for associations and private groups was the predominant feature from the seventies onward. The occupation of the riverbed by temporary sports and recreational facilities, compatible with the river's behavior but not very sensitive to the historical group of bridges and parapets, once again revealed the city's important deficits in terms of amenities.

As more or less spontaneous uses proliferated in the old riverbed, an intense territorial debate arose regarding its definitive uses. The South Plan itself had envisaged a West–East highway. Bounded by the canalization, it would connect the airport and the port, while compensating for the radiocentric layout of the urban road network. The 1966 Arterial Network project of the Ministry of Public Works included this proposal and ten years later the Integral Transport Study of the Valencia Area (1975–1976) continued to point in this direction. From the very beginning, society was reticent about the highway plan and demanded a large park to alleviate the lack of green spaces.

Once it received the old bed in 1976, the first democratic municipality assumed the ecological consigns. In 1978, it proposed the need to undertake a master plan that would include the new planning of the riverbed. In 1979, a participatory process was opened for the elaboration of the *Special Plan for the Urban Park of the Turia River as it passes through the city of Valencia* in order to provide some guidelines for the landscaping. However, in 1981 the city council opted to hire the services of the architect Ricard Bofill for the writing of a draft plan. This was the basis for the Turia's Interior Renewal Plan (*PERI of the Turia* in its Spanish acronym), which was approved in 1984 after lengthy negotiations between stakeholders. The increase of green spaces, the achievement of a unitary garden model, the valorization of the fluvial façade and the structuring of modern city growth were some of the objectives promulgated by the Plan, all without losing sight of the articulating role of water. It also contemplated a basic zoning with multiple services that were compatible with the green area. Although Bofill's studio had sketched a neoclassical garden, in the end a park model was imposed. Thus, the awarding of sections to different Valencian architectural firms distorted the initial unitary image.

The landscaping of the so-called Turia Garden began at the end of 1985 with investments from successive municipal governments and, occasionally, from the regional government. They did so in phases and at different rates. The radical transformation of the riverbed coincided with intense urban development on the banks, encouraged to a large extent by the General Urban Development Plan (GUDP) of 1988. The municipal planning gave a central role to the Turia and strongly advocated the revaluation of the riverbanks, solving part of the deficits in basic facilities and attracting the incipient tourist flows. To this end, it considered the implementation of more or less iconic facilities for cultural (Palau de la Música, inaugurated in 1987), sports (municipal athletics stadium 1991), recreational (Gulliver's playground 1991) or divulgative purposes (City of Arts and Sciences 1998–2009). Because of its magnitude (1.5 linear km), cost (1300 million euros) and international relevance (4 million visitors in 2022), the avant-garde complex of the City of Arts and Sciences deserves a specific mention.[8] In return, these invasive artifacts forced the definitive disconnection of the Turia stream, which contravened another of the main premises of the master plan.[9]

Between 2002 and 2004, work was carried out on the Capçalera Park (334,000 m^2) at the western edge of the riverbed, an area that had traditionally suffered from aggregate extraction and neglect by the authorities. Since 2008, an adjacent land parcel houses the Bioparc (107,000 m^2), a new generation zoo that reaffirms the authorities' intention to consolidate the area of the old riverbed as a tourist attraction.

[8] The complex, designed by Santiago Calatrava and Félix Candela, houses an IMAX cinema (Hemisfèric, opened in 1998), a museum (Museu de les Ciències 2000), an aquarium (Oceanogràfic 2002), an opera house (Palau de les Arts 2006) and a multipurpose space (Àgora 2009, now hosting the CaixaFòrum exhibition center).

[9] The premise of the preliminary plan was to make the current of the Turia compatible with the park. In this way, the new channel would function especially during floods, as in the case of the Venetian city of Padua (Zanetti 2013). The intense urban use of the riverbed, the construction of a new urban collector and the lack of pro-river positions led to the disappearance of the stream in 1983.

9.4 Territorial Footprint of Urban-River Policies

The river policies discussed above have had territorial manifestations at different scales that have shaped the current morphology of the Valencian area. In particular, two major spatial realities can be identified in relation to the Turia: the old riverbed and the new canal. In reference to both fluvial scenarios, ecologists, technicians and politicians are beginning to show interest in reversing a long process of artificialization that has been unattended for too many decades.

9.4.1 The Old Riverbed as an Urban Axis

Analyzed with a time perspective of 35 years, the success of the Turia gardening is undeniable. To this day, the old riverbed is home to 110 ha of urban green areas across the city and has become one of its identity signs. It is common to find people there strolling, playing sports or celebrating in groups. The environmental improvement of the riverfront has also been evident. The decontamination of the extinct industrial areas has been crucial. But this greening has had an impact that goes beyond the municipal boundaries. The Turia Garden is in practice the final part of a large metropolitan park that includes two protected natural areas (the Turia Natural Park and the Albufera Natural Park) managed by the regional government, a river park managed by the JHC, and the Horta, protected by a specific instrument for its landscape and ecosystemic value.

From an urbanistic point of view, the monumentalization of the riverbed by means of architectural landmarks constitutes an attraction in itself, while it has also been a vehicle for modern urban growth. Upstream it consolidated the Campanar neighborhood. Downstream, the city overcame the threshold of the Pont de Ferro Bridge and, over the decades, has moved closer to the sea. In particular, the City of Arts and Sciences revalued the neighborhoods of Penya-roja, to the north of the riverbed, and Montolivet, to the south, and encouraged developments that were not free of speculative dynamics (Fig. 9.4). It has also boosted the consolidation of recent neighborhoods such as Les Moreres, an inland extension of Natzaret.

In any event, it should not be forgotten that some of the considerations of the *PERI of the Turia* have not been fulfilled. On the one side, the construction by segregated sectors, the disparate priorities of alternate rulers as well as the fluctuating economic trends prevented obtaining a homogeneous whole. On the other, the fluvial condition of the city was condemned when the current of the Turia disappeared from the old riverbed. For its part, the hydraulic ensemble of Gothic bridges and parapets, surrounded by a thick mass of trees, has been decontextualized. Finally, the highway planned within the riverbed was transferred, at least functionally, to its banks. Therefore, high-intensity roads have distorted the traditional riverside promenades. In short, the park has not achieved the expected permeability and its evasive function has been slightly diminished.

Fig. 9.4 Surrounding area of the City of Arts and Sciences, a popular area for walkers and cyclists (*Source* Author)

9.4.2 The New Riverbed as a Non-Place

South Plan brought about profound hydro-geomorphological alterations. The main one was the practical annulment of the natural system. The total use of water in the embouchure weir, the use of cement for the lining and the disproportionate width of the canalization define the new Turia riverbed as an anodyne discharge channel virtually disconnected from the natural flow. But the territorial impact on the southern flank goes much further. The works involved the fragmentation of the productive agrosystem of the Horta, the replacement of the network of irrigation ditches, the restructuring of traditional roads, the segmentation of several municipalities and the destruction of a large part of the dispersed settlement of farmhouses of Arab heritage.

Landscape deterioration has been another obvious consequence. The new riverbanks defined a growth horizon that, under the framework of developmentalism, was urbanized at great celerity. The South Plan itself placed easements (drinking water infrastructure, sewers), highways and nuisance uses (Font de Sant Lluís railway logistics station, Vara de Quart industrial area, Pinedo treatment plant). Despite current conservationist ideals, in recent decades the new riverbanks have seen similar rates of occupation. The expansion of the ring road and the inclusion of more undesired facilities (power plants, council garages) and port uses on the mainland (industries, container depots, logistics platforms) magnify the footprint of the South Plan. In brief, the new fluvial domain is functioning as a large logistics zone. The reader will undoubtedly find a parallel with those dynamics of suburbanization that had affected the final stretch of the old riverbed from the forties onward. Paradoxically, they have been relocating toward the new one since the seventies.

At the same time, recent residential developments have been shaping a new riverfront without considering an overall treatment of the area. Using Augé's terminology (1993), this large piece of land can be defined as a true non-place. Gumprecht (2001) refers to the Los Angeles River as an invisible place in the eyes of the administrations and residents, but the expression is equally applicable to the case of the new Turia riverbed.

9.5 Remaining Challenges in the Fluvial Context

Despite the intense transformations of the river environment in the last half century, the Valencia-Turia binomial remains active. On the one side, the old channel continues to act as a catalyst for the urban development that now affects the old mouth. On the other, the new channel is beginning to be the object of attention at a time of new historical, territorial and ecological sensitivities. In any case, it seems certain that the future of the two watercourses involves both a rapprochement with the city and their ecological connectivity with the natural sites of the lower basin. All integrated in a sort of green corridor with two extremities that should finally reach the sea (Fig. 9.5).

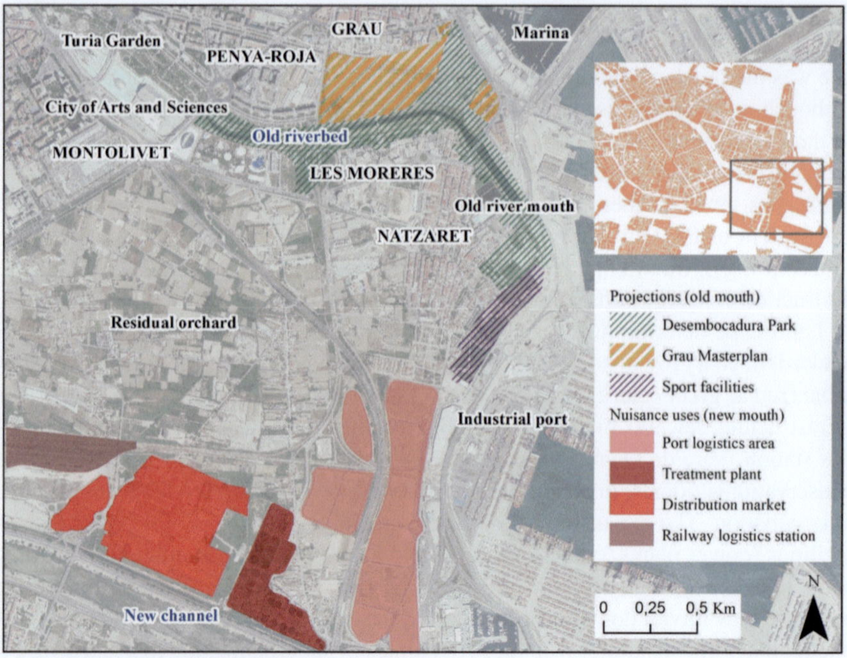

Fig. 9.5 Current and planned uses in both the old and the new mouth of the Turia River (Orthophoto of 2022) (*Source* Author)

9.5.1 Toward the Connection Between the Urban Park and the Seafront

As for the Turia Garden, its connection with the coastal front remains to be resolved. The cession in 2017 of the mouth—which was still Maritime-Terrestrial Public Domain—by the Port has been a necessary preliminary step. This favorable agreement for the city has recently made it possible to develop the ideas competition for the Desembocadura—river mouth in Spanish—Park (245,600 m^2), which in a first phase includes a green strip of 8.8 ha that follows the old river dell, reduced now to waste and rainwater collector. To this end, the port will remove the existing concrete embankments, while the municipality will have to guarantee a minimum water quality at the outlet of the sewer. As an appendix, in future phases the park will penetrate the old Natzaret seafront and traditionally battered by the port facilities. The final project will include sports facilities for the professional football team Levante UD and other services for the neighborhood.

However, the greatest interest of the planners is focused on the Grau district. More specifically on the 370,000 m^2 esplanade that decades ago housed industrial activity and, between 2008 and 2012, a Formula 1 circuit originally out of planning. In 2007 the municipal authorities put the Grau Masterplan out to public tender, which, following the premises of the 1988 GUDP, would weld the maritime districts to the urban fabric in the form of a large delta. The resulting designs showed a landscape with avant-garde vertical buildings, large green spaces and canals that, starting from the old river dell, diverged toward the marina. An avenue parallel to the bed connected the extension of the Albereda promenade with the waterfront transit. Since the change of local government in 2015, there have been several revisions subject to different perceptions about the suitability of the macroproject. The latest update, outlined as these lines are being written, shows a delta with 193,000 m^2 of green areas that subtracts weight to the roads and bets on pedestrian paths. On the other hand, the Turia canals have been discarded due to incompatibilities with the marina management. The main debate now revolves around the volume and priority use of the towers, which are expected to house 3000 dwellings.

Finally, the regional government has recently presented a metropolitan forest project that will connect the Turia Garden and the new channel, immediately upstream from Valencia. This will recover an abandoned area of about 50 ha that also used to be part of the natural riverbed and remained under the jurisdiction of the JHC.

9.5.2 The New Channel, a Metropolitan Ecological Corridor?

Undoubtedly, the policies of renaturalization of fluvial spaces promoted by the European Water Framework Directive (2000), a direct result of the new ecological demands, require the reversion of the enormous landscape and environmental affection of the South Plan. Thus, the channel remains today more or less isolated from the Turia Natural Park, only a few kilometers upstream, where the river is still in good health. In other words, the hydrological functioning is only activated during the episodes of autumn floods or reservoir releases.

In this context has emerged the non-binding study *New bed, New River* (Rivera 2018), drafted by a multidisciplinary team and promoted by the City Council. The preliminary project proposes to restore the fluvial ecosystem, guarantee the connection with the most representative sites of the Valencian plain (including the riverside forests of the Turia, the Horta and the marshes of the Albufera) and reestablish the transversal and longitudinal connectivity of the metropolitan municipalities. In short, it is a matter of mitigating the barrier effect of the new riverbed and recovering the river's leading role in the Valencian agglomeration. The case of the Los Angeles River itself, with similar pretensions, can serve as a good example of renaturalization. In this direction, the Hydrographic Confederation recently announced the restitution of an ecological flow in the new channel of 400 l/s in order to restore a certain ecosystemic dynamic. In any case, the matter is not without controversy. In the first place, the water authorities will have to decide where to draw this minimum flow in the context of increasingly prolonged droughts; treated water seems the most likely solution. Second, the engineers will have to tackle a landscaping that is compatible with the evacuation of floodwaters under the watchful eye of the citizens.

Acknowledgements This chapter has its origin in the Ph.D. thesis "The metamorphosis of the Turia River in Valencia (1897–2016): from urban torrential riverbed to metropolitan green corridor", defended at the University of Valencia in September 2017 and funded by the VALi+d 2013 program for researchers in training under the Generalitat Valenciana. Most of the technical documentation has been provided by the Júcar Hydrographic Confederation. I extend my gratitude to these institutions.

References

Berriochoa E (1947) 3ª Solución del Plan Berriochoa. Valencia: Speech given at an event of the Board of Railway Links

Carmona P (1990) La formació de la plana al.luvial de València. Geomorfologia, hidrologia i geoarqueologia de l'espai litoral del Túria. Valencia: Alfons el Magnànim Press

García Labrandero A (1949) Anteproyecto de encauzamiento del río Turia a su paso por Valencia. Valencia: Júcar Hydrographic Confederation

García Labrandero A (1961) Proyecto de defensa de Valencia contra las avenidas del río Turia. Valencia: Júcar Hydrographic Confederation

Gumprecht B (2001) The Los Angeles River. Its life, death, and possible rebirth. Baltimore: The Johns Hopkins University Press

Llorens J (1888) Proyecto de desviación del río Turia y desecación de la Albufera. M. Alufre Press, Valencia

Martínez García-Ordóñez F (1959) Una ciudad con futuro. In VV. AA. El futuro de Valencia-Conferencias. Valencia: Ateneo Mercantil Organization

Mendoza C, Luengo L, Herrera J (1942) Proyecto de canalización del río Manzanares. Manzanares River Commission, Madrid

Ollero A (2002) Ecogeografía del río Ebro. In De la Cal P, Pellicer F (eds) Ríos y ciudades. Aportaciones para la recuperación de los ríos y riberas de Zaragoza. Zaragoza: Fernando el Católico Institution-CSIC, 135–157
Portugués I, Bonache X, Mateu JF et al (2016) A GIS-based model for the analysis of urban flash floods and its hydro-geomorphic response. the Valencia event of 1957. J Hydrol 541:582–596
Rivera R (ed) (2018) Llit Nou-Riu Nou. Valencia: Valencia City Council
Rosselló VM, Esteban J (1999) La façana septentrional de la ciutat de València. Valencia: Bancaixa Foundation
Zanetti PG (2013) Acque di Padova. 150 anni del canale Scaricatore. Sommacampagna: Cierre Edizioni

Chapter 10
The Challenges of Urban Requalification of Rivers and Wetlands. The Case of Peri-Francilian Towns

Sylvain Dournel

Abstract Since the 1970–1980s, European urban stakeholders have been attaching importance to rivers and wetlands. The many projects related to their aesthetic and functional treatment illustrate the vivacity of this process, characterised by the notion of urban requalification. The aim is to remedy the recent downgrading of rivers and wetlands, to highlight their rich urban history and to enhance their cultural, economic and environmental potential. However, the content and the scope of these policies depend on the size and situation of the towns. In France, Lyon and Nantes have developed large-scale urban projects based on their metropolitan ambitions. The peri-Francilian towns have developed smaller but more diversified initiatives to attract residents and investors, emphasising their proximity to Paris and their local identity. This chapter focuses on the development potential of rivers and wetlands in this specific urban case, revealing the strategies employed and their contributions and impacts on the environment. This provides valuable lessons for an efficient urban requalification.

Keywords Medium-sized town · Public policy · Risk of standardisation · Rivers and wetlands · Urban requalification

10.1 Introduction

Since the 1970–1980s, European urban stakeholders have been attaching great importance to rivers and wetlands. There are numerous projects that focus on their aesthetic and functional treatment, such as the *Thames Gateway* in London, the *Isar Plan* in Munich and the *Maillage Bleu* in Brussels. In France, *Confluence* in Lyon, *Île de Nantes* and *Deux rives* in Strasbourg are among the most popular actions. Furthermore, several urban actions have been implemented on rivers and wetlands in medium-sized towns located near Paris, such as *Vallée idéale* in Amiens, *Presqu'île*

S. Dournel (✉)
CEDETE Laboratory, University of Orléans (FR), Orléans, France
e-mail: sylvain.dournel@univ-orleans.fr

© The Author(s), under exclusive license to Springer Nature Switzerland AG 2024
J. Farguell Pérez and A. Santasusagna Riu (eds.), *Urban and Metropolitan Rivers*, The Urban Book Series, https://doi.org/10.1007/978-3-031-62641-8_10

in Caen, *Loire trame verte* in Orléans et *Coulée verte* in Reims. The urban approach therefore concerns cities and towns with different situations. The generalization of urban actions for rivers and wetlands does not necessarily depend on the size and level of development of towns: the large and prosperous city is not always a centre of innovation, serving as a model for the small and modest town. On the contrary, the diversity of towns and cities that invested early in rivers and wetlands, illustrated in Fig. 10.1 for France, make a greater number of hypotheses: the interest of mayors for their rivers and wetlands, their capacity to develop innovative actions, the quality of the local environment, the hydrographic configuration, the availability of land, the local potentials, the level of involvement of other territorial authorities and associations. These few decades of urban actions on rivers and wetlands show a hybrid and heterogeneous process.

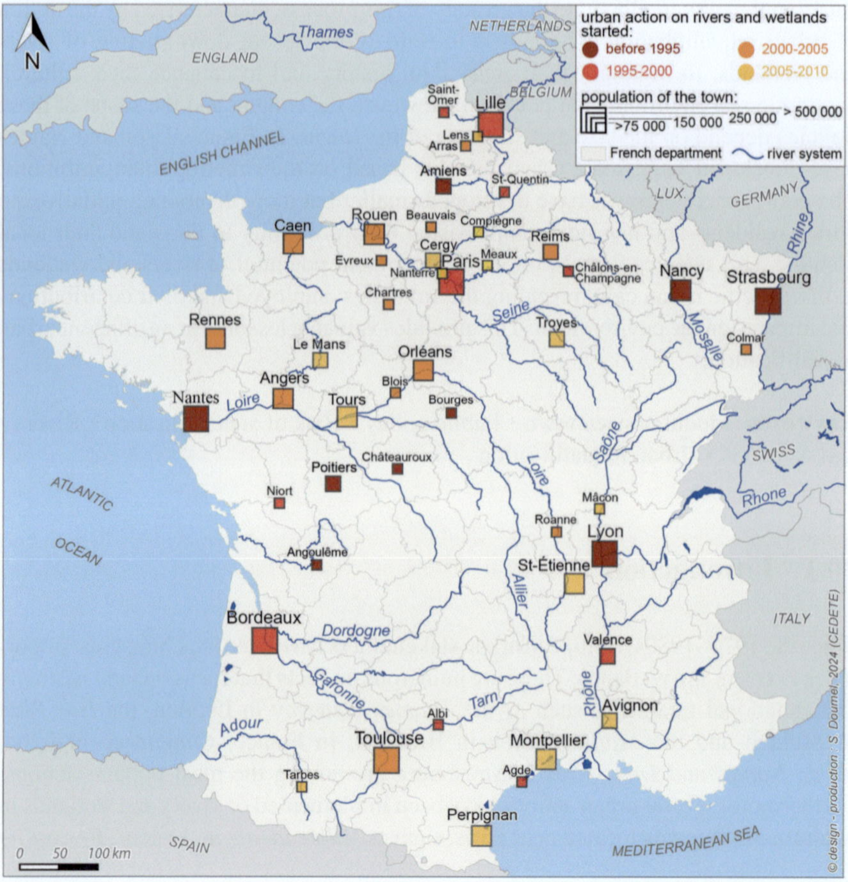

Fig. 10.1 The spread in France of urban actions to rivers and wetlands between the 1990s and the 2010s (*Source* Dournel 2023)

Beyond this diversity, three general lessons emerge from these actions:

- Remedy the frequent situations of marginalization and downgrading of rivers and wetlands: motorways or car parks along the banks, piped or recalibrated rivers, filled ponds or converted to landfill, marshes transformed into storage or commercial areas;
- Highlight their rich urban history: fisheries, market gardening, river transport, military hydraulics, craft activities;
- Enhance their important climatic, cultural, ecological, land, recreational and tourist potential.

The intersection of these three lessons reveals an approach that is simultaneously landscape, heritage and urbanistic in scope.

Beyond these hypotheses and general principles, what are the reasons of this unanimous enthusiasm of urban stakeholders for rivers and wetlands? In parallel, the aim is to characterise this urban approach to better understand its real scope. Considering the great diversity of cities and towns involved in this approach, is there one or several ways to act in favour of rivers and wetlands? This question can be asked in particular in the specific case of medium-sized towns located near Paris, as they are particularly active in the process. What are the contributions and limitations of projects developed on rivers and wetlands? What lessons can be learnt to make this approach more efficient?

This chapter is structured around this set of questions. The first part contextualises the enthusiasm of urban stakeholders for rivers and wetlands within current urban thinking. It then reveals several degrees of urban ambitions in the actions developed on rivers and wetlands. The second part studies strategies employed, contributions and limits of the urban approach through the specific case of peri-Francilian towns.[1] This provides valuable lessons for an efficient urban requalification.

10.2 Urban Stakeholder's Interest in Rivers and Wetlands: Favourable Urban Thinking, Varied Urban Ambitions

The first aim of this section is to provide a contextual and conceptual framework for the enthusiasm of urban stakeholders for rivers and wetlands. The aim is also to highlight a variety of urban and environmental ambitions in the same process.

[1] The term "peri-Francilian towns" refers to generally medium-sized towns, located close to Paris and the Île-de-France region. The peri-Francilian towns form a coherent urban network with a wealth of development potential on which the Paris metropolitan area can build. The good rail and motorway links, a growing number of business parks, wide availability of land, attractive property prices and a generally healthy environment are all factors that attract senior executives, companies and investors. As a result, there are many links between the Paris metropolitan area and the peri-Francilian towns.

10.2.1 Rivers and Wetlands at the Heart of Current Urban Thinking

The enthusiasm of urban stakeholders for rivers and wetlands reveals a paradigm shift in urban planning. The aim is to characterise this process in order to identify its meaning and scope.

10.2.1.1 Urban Rivers and Wetlands, Privileged Places for Implementing Urban Quality

The enthusiasm of urban stakeholders for rivers and wetlands is in opposition to modern urbanism, active from the 1930s to the 1970s in Western Europe. This trend in urban planning was in the extension of hygienist theories which condemned the harmful effects of urban wetlands on public health, leading to the removal of urban ponds, marshes and canals in the eighteenth and nineteenth centuries (Dournel 2010).

Marked by the innovations of the industrial revolution, the post-war economic dynamics and the needs for housing, modern urbanism promoted a rational, rigorous and standardized urban aesthetic, reflected in the monumental perspectives generated by the alignments of cubic buildings among important grassy surfaces (Auricoste 2003). This urban model, detached from any historical, cultural and environmental characteristics, caused the suppression of rivers and wetlands. Furthermore, the division of urban space into monofunctional areas (housing, transport, work, recreation) was opposed to the evolving and interdependent aspects of rivers and wetlands.

The economic crisis of the 1970s and the integration of social and environmental considerations introduced a paradigm shift. While modern urbanism was in crisis, culturalism and urban ecology became credible alternatives, in favour of rivers and wetlands. "The modern city is often built on filled or drained swamps. Visions of the postmodern city often involve the return of the repressed" (Giblett 1996).

Considering historical and cultural dimensions of towns, culturalism respects its proportions, integrating its architectural, morphological and environmental data (Choay 1996). Linked to urban history, rivers and wetlands present numerous legacies (Dournel and Sajaloli 2012): cultural and landscape policies have restored and protected moats, surrounding walls, cultivated marshes, religious buildings, watermills, canals, ports and bridges.

In the 1960s–1970s, urban ecology conceived the city as a living environment (Blanc 1998), emphasising its metabolism and biophysical components (Barles 2010), including rivers and wetlands. Within the framework of environmental charters, urban stakeholders have created green corridors to preserve landscape units and sensitive environments.

More recently, sustainable urban development complements both approaches (Haughton and Hunter 1994; Emelianoff 1999). This cross-cutting concept refers to a permanent project, attached to the historical, cultural and environmental dimensions of towns, in urban quality perspective (Laganier and Roussel 2000). In this context, the mayors wish to promote the potential of rivers and wetlands for urban nature, social diversity, local identity, urban renewal and territorial cohesion.

The urban project concept puts their wish into practice. Alternative to modern urban planning in reflection and action, this concept focused on the treatment of vacant spaces and discontinuities as well as on the valorisation of natural or cultural potentials. Thus, numerous urban projects are centred on the treatment of rivers and wetlands (Masboungi 2002). This qualitative concept focuses on landscapes, revealing a new conception of urban planning (Waldheim 2016).

10.2.1.2 *The Notion of Urban Requalification of Rivers and Wetlands*

Urban planning interest in rivers and wetlands describes a hybrid approach that needs to be defined in order to better understand its meaning and scope. Indeed, urban stakeholders want to manage areas with a high naturalness gradient which were not initially under their competence, and which are coveted by people in search of a better quality of life. Furthermore, these stakeholders use the tools and know-how of urban planning to implement an essentially landscape approach.

A review of projects focusing on rivers and wetlands brings out a panel of terms among which *reappropriation*, *reconquest*, *regeneration*, *rehabilitation*, *renaturation*, *restoration*, *blue grid*, *revalorisation* and *waterfront*. Most of them are built around the prefix "re-" and the suffix "-tion", indicating a reference to the past and a long-term dynamic action.

Restoration and *renaturation* stem from environmental management. The first aims to remedy damage to biodiversity and ecosystem dynamics. The second, more ambitious, aims to restore a natural reference state.

Regeneration, *waterfront* and *rehabilitation* are urban planning issues. *Regeneration* refers to the establishment of equipment to develop new functions in an area in economic decline. *Waterfront* expresses the specific regeneration of a harbour area. *Rehabilitation* implies a degradation of the building and a readaptation to current needs.

Blue grid is based on landscape ecology. The aim is to counteract the fragmentation of rivers and wetlands and integrates their ecological and landscape continuity into spatial planning. With *the green grid*, it allows animal and plant species to circulate and interact, contributing to the preservation of biodiversity.

Each of these terms refers to specific situations. None of them describes the framework of cross-cutting actions of the rivers and wetlands approach, translated into urban projects of various nature and size. Among the more cross-cutting terms used by urban stakeholders to justify their action, *reconquest* expresses, in a power relationship, the recovery of what has been lost. In a similar sense, *reappropriation* evokes a more political or land-related dimension. In economic terms, *revalorisation* refers to a new appreciation of a possession or an object. Thus, these terms are still too reductive in comparison with the multiple ambitions of the urban approach; they only describe very specific aspects.

On the other hand, the notion of *requalification* describes a new way of qualifying a territory, i.e. giving it a new quality (Dournel 2010). Firstly, the qualitative dimension reflects the landscape approach of the process. Quality research also brings together projects of various types and scales. In addition, a quality territory evokes careful urban planning, economic prosperity, high visitor numbers and a good state of the environment. Lastly, the notion, in its construction, evokes a reference to the past and the project principle. To better understand the nature and scope of the approach, it is appropriate to speak of "urban requalification of rivers and wetlands".

10.2.2 Rivers and Wetlands, Indicators of Varied Urban and Environmental Ambitions

After this contextual and conceptual reframing, the aim is to report, from two perspectives, on projects of varying types and scales, depending on the size, situation and local specificities of cities.

10.2.2.1 Structuring the Great Metropolitan Areas

In Europe, three metropolitan regions are among the most active in the requalification of rivers and wetlands: Greater London, the Randstad Holland and the Ruhr. All three have in common that they promote the structuring role of rivers and wetlands.

Influenced by the waterfronts of Boston, Baltimore and San Francisco, London began the regeneration of its docklands in the 1970s and 1980s (Bentley 1997). Depending on the opportunities, London stakeholders rehabilitated docks and warehouses, developed tertiary activities and created infrastructures, housing and urban parks (Michon 2008). The Thames has become the economic and environmental link of the metropolis. From the 1990s, local and national stakeholders extended this link to the river mouth, through the *Thames Gateway* project (Cohen and Rustin 2008; Beucher 2009). Applying the principles of urban renewal and regeneration, this 70 km long project develops accessibility and preservation of rivers and wetlands.

In the Netherlands, the popularity of water sports established marinas in the 1960s. With the *Randstadgroenstructur* in 1985, the state emphasised the structuring role of water bodies and waterways for agriculture, recreation and environment (Van Gessel 1985): surrounded by the major Holland cities, the *Groene Hart* (160 km^2) is an example of this (Kühn 2003). With the floods of 1995, the relationship with rivers changed. The national *Room for the River* programme applies integrated management of the flood areas (Wolsink 2006), which is expressed locally through various combinations between depoldering, widening riverbeds, creating retention areas, restoring wetlands, developing public spaces (Roth and Winnubst 2014). The town of Nijmegen, for example, has been reorganised around its river by opening a 2.5 km secondary branch of the Waal and developing an island urban park (Rădulescu et al. 2021).

In the Ruhr, regional and urban stakeholders structured the regeneration of the region by restoring rivers and creating wetlands. The *International Building Exhibition (IBA) Emscher Park* started the process in 1988 with the creation of the *Emscher Landscape Park* (300 km^2), the revitalisation of the Emscher River (83 km) and its tributaries (270 km) and the development of leisure activities around the canals (Weber and Konitzky 1993; Gruehn 2017). This planning policy networked several towns in the Ruhr area, including new urban areas such as the *Phoenixsee.* In the South of Dortmund, this 2.5 km^2 site is the result of 10 years of work transforming an industrial complex into a high value-added district, with soil decontamination, the creation of the lake (700,000 m^3 of water) and wetlands, landscaping, recreational facilities and residential development (Mihaila 2015).

The requalification policies conducted in London, Randstad Holland and the Ruhr are based on specific approaches, but have common structuring effects. The size of the areas concerned and the stakeholders involved, who come under the heading of urban planning but also regional planning, underline the national and even international importance of these policies. With a time lag, the actions conducted in the metropolitan regions of Brussels, Lille, Barcelona and Paris describe these same effect and challenge.

10.2.2.2 Revealing the Metropolitan Ambitions of Large Regional Cities

At a lesser level, the large regional cities active in river and wetland requalification aspire to become metropolises. In France, Lyon, Nantes and Strasbourg are examples of this ambition since the early 1990s.

In Lyon, the *Plan bleu*, an incentive and cross-cutting tool, initiated the requalification process. The urban stakeholders formalised it around projects for extending the urban centrality to the North, along the Rhône, with the *Cité internationale* (0.2 km^2), and, especially, to the South with the Saône-Rhône confluence (1.5 km^2). These new districts are home to a large number of offices buildings, housing units, services and structural facilities (Gérardot 2007). Both are structured and connected to other districts by a dense network of green spaces and urban parks along the rivers, including one designed around a basin. Located to the North-East of the city, the *Miribel-Jonage* park (22 km^2) completes this network by combining flood water expansion, outdoor recreation and biodiversity protection (Amzert and Cottet-Dumoulin 2000).

In Nantes, the regeneration of the industrial port island opposite to the city centre reflected the enthusiasm of urban stakeholders for the banks of the Loire. The creation of a dense network of roads and green spaces (0.7 km^2) facing the river, including a marina, preceded the construction of housing, offices and infrastructures (Masboungi 2003). The aim was to extend the city centre into this sparsely populated urban area of 3.4 km^2 by combining regeneration and urban renewal (Chasseriau 2004). The whole project is commensurate with the ambitions of a metropolis imagined along the Loire estuary, linking Nantes to Saint-Nazaire.

In Strasbourg, the requalification policy connected the city to its dense river system and to the Rhine, to the East. Urban stakeholders created the *Jardin des deux rives*, a 1.5 km^2 urban park organised around wetlands and a central footbridge linking France and Germany (Reitel and Moullé 2015). In addition, they developed the *Deux rives* urban project, focusing on the renewal of industrial and port wastelands along the Rhône-Rhine canal (Beyer et al. 2021). The aim is to develop 2.5 km^2 of housing, offices, facilities, green spaces and bridges. These actions reflect Strasbourg's ambition to become a cross-border urban area and to open up to the Rhine area.

The requalification policies of Lyon, Nantes and Strasbourg take advantage of rivers and wetlands to create central districts and central facilities. These actions again set these cities in their regional reality but also reveal their metropolitan ambition. With a time lag, the Bordeaux, Rennes and Toulouse projects have the same ambition. Comparing these actions with those identified in metropolitan areas, rivers and wetlands reveal urban and environmental ambitions of varying content and scope. What about medium-sized towns near Paris?

10.3 Rivers and Wetlands Requalification in Peri-Francilian Towns: Strategies Employed, Contributions and Limits

Several peri-Francilian towns are among the most active in France in the urban requalification process. By focusing on this urban sample, this section aims to identify the implemented approaches and to measure their urban and environmental scope.

10.3.1 The Peri-Francilian Towns: Urban and Environmental Similarities and Dissimilarities

The peri-Francilian towns are medium-sized and similarly situated, with many rivers and wetlands. However, their socio-economic dynamics and hydraulic characteristics differ. The result is a promising urban sample for assessing the challenges of urban requalification.

10.3.1.1 A Common Situation but an Urban Network with Contrasting Dynamics

The medium-sized towns surrounding the Île-de-France region are located in the area of influence of Paris and have metropolitan development potentials. Their rail and highway connections, business centres, land availability, real estate costs and quality of life attract executives, companies, structural facilities and cultural facilities. Describing these towns as "peri-Francilian" reflects the many links they have with Paris and its region. Nevertheless, these links vary in nature and intensity according to the demographic weight of these towns and to their distance to the capital.

On the first point, considering towns located less than two hours from Paris (average travel time by train and car), 19 towns have more than 75,000 inhabitants. However, this urban network is heterogeneous (Fig. 10.1):

- 4 are larger than 250,000 inhabitants: 3 have the status of administrative metropolis, two are regional capitals;
- 4 have between 150,000 and 250,000 inhabitants: one was regional capital until 2015;
- 11 are of more modest size, exercising the function of departmental prefecture.

The most populated cities have the most expertise and engineering: this is demonstrated by the many areas of intervention of their public establishment for intercommunal cooperation as well as the presence of urban planning agencies, mixed economy companies of planning, development councils, etc. The Paris region links with these towns, through inter-city cooperation policies.

On the second point, Mirloup (2002) mentions an active metropolisation within a polygon formed by Amiens, Reims, Orléans and Rouen and a passive metropolisation beyond (Fig. 10.1). Passive metropolisation refers to the vertical centre-periphery model, with low value-added activities deconcentrating from the Île-de-France to remote, sparsely populated areas. Active metropolisation refers to a horizontal model, to network organisation. The location of command functions and facilities (administrative headquarters, head offices, technology parks) in some peri-Francilian towns reflects their integration into the Paris metropolitan area.

10.3.1.2 *Importance and Diversity of Rivers and Wetlands in the Peri-Francilian Towns*

In addition to their common situation, all peri-Francilian towns are crossed by one or several rivers (Fig. 10.2); their original urban site depends on it: confluence site, crossing site…

Despite a common river identity and a certain geological homogeneity, due to their situation in the Parisian sedimentary basin, the peri-Francilian towns show hydromorphological contrasts. They are related to the length and regime of the rivers and the position of towns in their catchment area. Peri-Francilian towns have a great diversity of rivers (straight or meandering, narrow or wide), as well as old

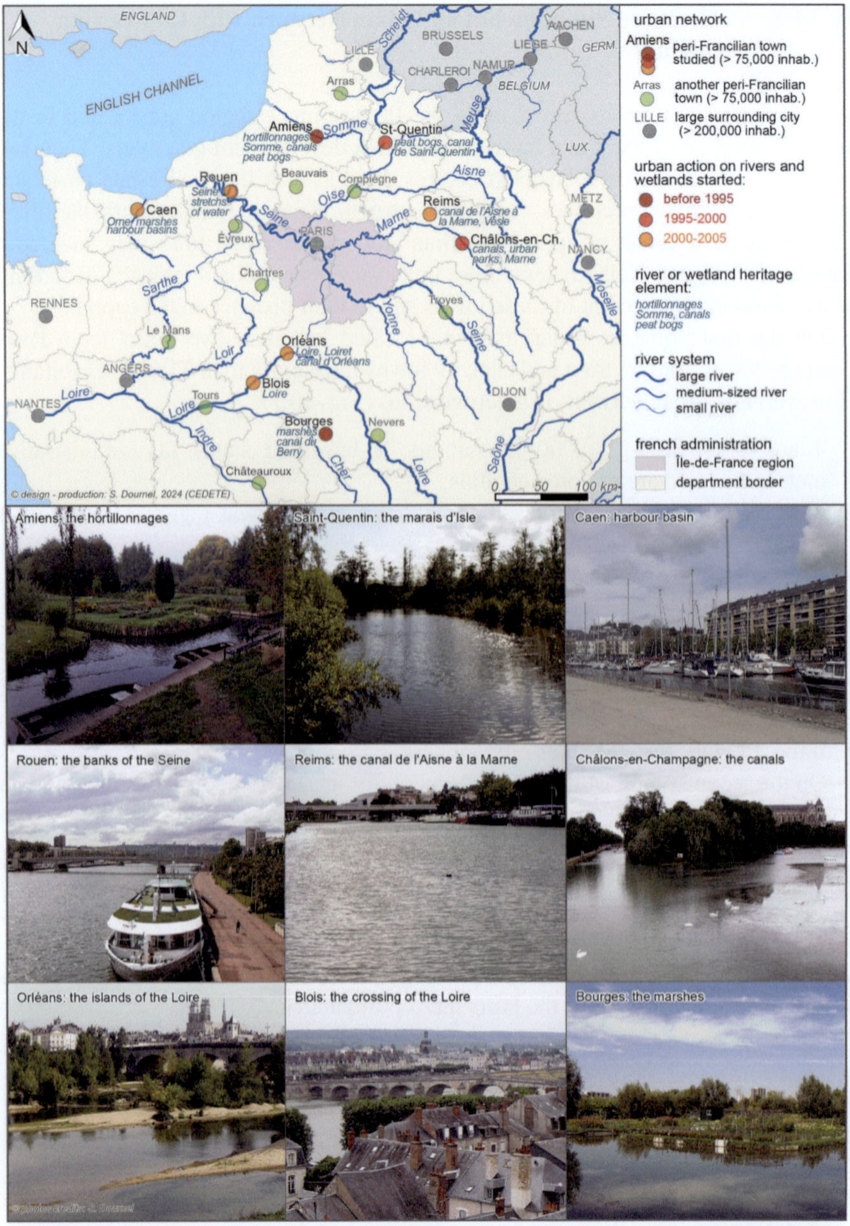

Fig. 10.2 Wealth and diversity of rivers and wetlands in the 9 peri-Francilian towns studied (*Source* Dournel 2023)

developments, responsible for the presence or absence of navigable waterways, quays and basins, drainage canals (Dournel 2010). These elements lead to linear, surface or mixed hydraulic configurations that have a greater or lesser impact on urban morphology or crossings, which can generate opposing banks (Fig. 10.2).

In addition, the peri-Francilian towns contain wetlands that vary in number, surface area and type (marshes, mudflats, peat bogs, ponds, riparian areas). Some of these wetlands are the result of major human activities, such as the marshes drained for market gardening (*hortillonnages* of Amiens, Bourges marshes) and water bodies created by gravel quarries (Loire Valley, Seine estuary).

The peri-Francilian towns therefore present a great diversity of river and wetland landscapes, resulting from physical and anthropic factors. Given the large number of towns involved, the study focuses on Amiens, Blois, Bourges, Caen, Châlons-en-Champagne, Orléans, Reims, Rouen and Saint-Quentin (Fig. 10.2). This selection is based on a combination of three criteria:

- the largest number of inhabitants;
- the towns with the most rivers and wetlands in terms of size and diversity;
- the most historically active in the urban requalification process.

10.3.2 Different Urban Approaches to Rivers and Wetlands

This sub-section examines the specificity of the peri-Francilian towns in the requalification of rivers and wetlands, based on the 9 towns selected. The urban approaches are varied, linked to different themes and heterogeneous urban planning tools.

10.3.2.1 A Diversity of Themes, Adapted to Local Challenges and Environments

The analysis of requalification policies is based on three complementary works:

- The study of communication tools: thematic brochures, municipal and inter-municipal newsletters, websites, social networks;
- Analysis of political project documents and technical studies, supplemented, if necessary, by semi-directive interviews with the administrative departments concerned;
- Conducting of field surveys.

This work highlights six main themes, each divided into 2 or 3 subcategories (Fig. 10.3).

In view of people's need for well-being, urban stakeholders place outdoor recreation at the heart of the requalification of rivers and wetlands. Each action is characterised by improved access, landscape treatments and (re)connections to local dynamics. The aim is to restore or create nautical bases and urban parks in wetlands, and to link them by means of pedestrian and cycle paths developed along the

Topic request		Amiens	Blois	Bourges	Caen	Châlons-en-Champagne	Orléans	Reims	Rouen	Saint-Quentin
Economy	Local agriculture	****	*	***			*	**		
	Tourism	****	***	***	**	****	***	**	**	***
	River navigation	**			***		*	***	***	***
Environmental management	Flood risks	*	****			*	*		*	
	Water and soil treatment	***			***	*	*		***	*
	Restoration of wetlands	***	**		***		**	***		***
Heritages	Natural heritage	****		**	**	**	*	**		****
	Cultural heritage	**	**	****	**	****	****	*	***	
Mobility	Place of the car				***		***	***		
	Soft modes of transport	**		**	***	**	***	***	*	*
	Crossings	***			*	***	**	***	***	
Outdoor recreation	Urban parks and nautical bases	****	***	***	****	**	***	****	***	****
	Walking and cycling paths	**	***	***	****	****	****	****	***	***
	Events	***	**	**	***	***	****	**	****	***
Urban construction	Urban densification	****		*	****		*	**	****	
	Urban facilities	****	*		****		****	*	****	***

****	***	**	*	
very high	high	medium	low	very low

Fig. 10.3 Diversity of themes mobilised behind a single urban requalification approach (*Source* Dournel (2023), according to Dournel (2010))

rivers. Recreational facilities and festive events (*Armada* in Rouen, *Loire Festival* in Orléans) enliven the network.

The development of pedestrian and cycle paths introduces the question of transport in towns exposed to saturation of the road network and, possibly, to opposing banks. The projects in Caen, Orléans, Reims and Rouen focus on the reduction of car use by removing expressways and car parks on the riverbanks, to jointly promote soft modes of transport such as walking, cycling and even river transport, and to build footbridges or bridges to remedy the isolation of neighbourhoods.

From an environmental point of view, flood risk management, water and soil treatment and the preservation of wetland ecosystems go hand in hand with some policies combining leisure and transport. Land-use regulations in flood-prone areas sometimes force mayors to develop functions other than housing. The *La Bouillie* project in Blois, for example, promotes wetland ecosystems, public spaces, agricultural activities and recreational facilities that are not vulnerable to flooding. Furthermore, towns with an industrial and port history, such as Amiens, Caen and Rouen, are introducing decontamination of degraded environments into their projects. As an extension of this, the restoration of wetlands is included in the requalification policies, particularly in Caen, Reims and Saint-Quentin.

Linked to this, the heritage values of rivers and wetlands constitute valuable levers for action. The image of nature in the city, accentuated by legal protection tools (Natura 2000 areas in Blois and Orléans, nature reserve in Saint-Quentin, Amiens and Reims, sensitive natural area in Bourges and Caen), supports any action that promotes river and wetland landscapes. In addition, the 9 studied towns are originally linked to water, which gives a historical dimension to their requalification project. The projects of Bourges, Châlons-en-Champagne and Orléans, which focus respectively on their market gardening, craft and boat heritage, have an important cultural and identity dimension.

Otherwise, the riverside areas host real estate operations and facilities that comply with the restrictions of flood risk prevention plans. Rivers stimulated architects to build collective housing, developed over large areas (*Caen Presqu'île*, *Rouen Flaubert*) or distributed over multiple small sectors (pericentral districts of Amiens). In both cases, these project managers work with water landscapes. At the same time, the reintegration of rivers and wetlands into the urban dynamics involves the creation of radiating facilities: recreational complexes in Saint-Quentin and Orléans, university centres in Amiens and Caen…

From an economic point of view, some initiatives support urban agriculture in wetlands, as in the *hortillonnages* of Amiens and Bourges marshes, or are trying to develop it again, as in the Vesle valley in Reims. Urban stakeholders also consider rivers and wetlands to be ideal places for sustainable tourism, as illustrated by the *Loire à vélo* in Orléans and Blois. The urban policies of Reims and Saint-Quentin are taking advantage of river tourism to boost traffic on their navigation channels; those of Caen and Rouen are taking advantage of this sector to diversify their port economy.

Behind a desire to structure their metropolitan region around rivers and wetlands, the policies pursued in Greater London, Randstad Holland and the Ruhr focus respectively on the economy and living environment, water sports and flood risks, outdoor recreation and quality of life. In the peri-Francilian towns, the wide range of themes covered makes for confusing reading. This is due to the variety of urban and environmental issues.

10.3.2.2 *Requalification Policies Using Heterogeneous Urban Planning Tools*

This variety of approaches is also due to the heterogeneity of the urban planning tools that are used. Here, there is no question of vast urban projects or ambitious development plans specifically created as in the metropolises and large regional cities.

Only the urban projects of Rouen and Caen are close to this scenario. *Seine-ouest*, including the *Rouen Flaubert* project, and *Caen Presqu'île*, cover 8 km^2 and 6 km^2 respectively, with several kilometres of riverbank. Begun in 2005 and 2010, they represent two to three decades of urban development. Managed by a local public company, they bring together the inter-municipal authorities, the riparian municipalities and the port authorities. These large-scale projects have involved the creation of concerted development zones.

To a lesser extent, the *Loire trame verte* project in Orléans and the *Coulée verte* project in Reims are among the key actions of the agglomeration projects, which were programmed in the 2000s for two decades of urban development. These two inter-municipal projects, which originated in the thinking of the member municipalities, bring together a multitude of actions in the public domain, focusing on wetlands.

In Amiens, the requalification project initiated at the end of the 1980s focuses solely on the municipal area. However, its approach is more cross-functional, directly linked to the urban renewal policy. Conceived on a neighbourhood scale, the urban actions are differentiated. The smallest projects are carried out by the local authority on public land while the larger ones are carried out in concerted development zones. Among these, the regeneration of the *La Vallée* neighbourhood, begun in the 2000s, covers 1.3 km^2. It involves a semi-public company and a local public company.

In the other peri-Francilian towns, urban development is punctual, using contractual agreements and partnerships with local public and associative stakeholders, classic urban planning tools and land pre-emption on a case-by-case basis. These are essentially opportunistic initiatives. Urban approaches generate few innovative urban planning approaches, such as the master plan of *Île de Nantes* (Masboungi 2003). Furthermore, urban development directly involves municipal or inter-municipal services of green spaces, urban planning, heritage, sports and leisure. Few of them benefit from territorial engineering support, such as the semi-public company *Lyon Confluence*, created for the urban project in 1999.

A comparison of the requalification projects in the 9 studied towns reveals a diversity of themes, reflecting local urban and environmental issues. The heterogeneity of the urban planning tools used also reveals requalification policies with varying ambitions, confirming contrasting urban dynamics.

10.4 Incomplete Urban Achievements and Similar Results?

After highlighting the diversity of approaches adopted by peri-Francilian towns, this sub-section looks at the impact of urban development on rivers and wetlands.

10.4.1 Often Incomplete Urban Achievements

The urban actions developed over the last 3 decades in peri-Francilian towns provide an assessment of the urban developments carried out on rivers and wetlands. Urban policies, with their varied approaches, have all created a dynamic around these environments. The popularity of the promenades, events and facilities developed along the quays confirms this observation. Riverfronts have even become showcases of towns.

However, requalification policies do not cover the entire river network, concentrating on the main river and neglecting secondary rivers (Rode 2017), or only concerning a specific section or bank as in Orléans. Furthermore, some intentions on rivers and wetlands are not systematically followed up by action. Urban players often prioritise building programmes and structural facilities to the detriment of environmental restoration projects, as illustrated by the regeneration of the *La Vallée* neighbourhood in Amiens. Some projects have also been scaled back, such as the *Parc de la Loire* in Orléans, imagined in 2002 on 300 hectares of riverbanks, gravel quarries and farmland. Although this park was to include a historical interpretation centre for Orléans, tourist accommodation and an environmental discovery centre, only this third aspect was finally retained in 2018, in a reduced version.

These examples show the vulnerability of requalification policies, which is linked to six constraints (Dournel 2019):

- administrative: too many stakeholders and regulations;
- climatic: damaging flood or low-water episodes;
- financial: high costs of environmental restoration;
- land: actions limited by the fragmented land parcels in wetlands and the state-owned or private status of rivers;
- political: discrepancy between the short-term elective period and the long-term project implementation period;
- socio-cultural: various representation systems and uses of rivers and wetlands.

To overcome many of these constraints, urban stakeholders try to extend their field of competence, to acquire land or to develop coordination actions. Nevertheless, urban achievements are often incomplete.

10.4.2 Reading and Understanding the Potential Uniformity of Rivers and Wetlands

Although the peri-Francilian towns studied have adopted a variety of approaches to the treatment of rivers and wetlands, these approaches are in fact very hierarchical (Fig. 10.3). The nine towns favoured four fields of application: pedestrian and cycle paths, urban parks and nautical bases, tourism, and events. These were followed by urban facilities and cultural heritage. Soft modes of transport, natural heritage, urban densification, crossings and navigation are all well used, but only by specific towns. The other fields of application relate to local environmental issues.

This hierarchy highlights a risk of uniformity of rivers and wetlands, which can be measured by similar landscaping and urban functions.

Strengthening access to rivers and wetlands through the systematic development of urban parks and pedestrian and cycle paths can transform wetlands into classic urban green spaces (Fig. 10.4). Should rivers and wetlands really become public open spaces, recreational areas and tourist destinations? The ecosystems, environmental dynamics and socio-economic activities found there prove the contrary. Rivers and wetlands are hybrid environments, anthropized and natural, open and closed, public and private, requiring integrated and differentiated public action.

The aesthetic treatment of old and new building frontages confirms the risk of uniformity. In the first case, urban stakeholders accentuate the mineral character of the quays, linked to the proximity of the old town centre, by developing esplanades or promenades that are often similar, using the same urban furniture, the same vegetal compositions and the same lighting systems. In the second case, building projects use the same architectural codes (Fig. 10.5), often with little connection to the shapes, materials and volumes of the surrounding buildings. When they are not transformed into urban green spaces, wetlands are not intended to host building alignments.

In theory, the risk of standardisation can be correlated to three factors:

- the concept of nature in the city that is contemplative, aesthetic and accessible;
- urban renewal, often interpreted as the densification of vacant spaces;
- the concept of the living environment, promoted by territorial marketing.

In practice, the risk of standardisation can be correlated to four factors:

Fig. 10.4 Grassed areas and artificial plantations to the detriment of the ecological and landscape diversity of wetlands (*Source* Dournel (2023): Blois, Orléans and Amiens)

Fig. 10.5 Wetland rivers prey to uniform architectural styles (*Source* Dournel (2023): Caen, Amiens and Châlons-en-Champagne)

- urban actions confined to the competences of municipalities and inter-municipalities;
- the culture of public space, the obsession with accessibility to environments and the possibility of communicating about them;
- tourism and heritage developments which, while highlighting local characteristics, use the same techniques of staging and animation;
- reference to media actions that lead to imitation, despite the local characteristics.

Although the peri-Francilian towns play an important role in the requalification of rivers and wetlands, their approaches are varied but refocused on specific fields of action. With urban issues taking precedence over environmental issues, there is a palpable risk of standardisation, which is further exacerbated by the situation on the outskirts of Paris.

Compared with other types of towns that are active in the approach, the urban players emphasize the promotion of their requalification policy, highlighting the availability of land, the attractive cost of property and the quality of life generated by rivers and wetlands. For major urban projects, they promote structuring facilities and major events with a view to integrating their town into the Parisian metropolitan system. At the same time, urban stakeholders combine requalification policies with major communication and promotion campaigns to change people's perception of the town. The aim is to bring it out of the shadow of Paris and give it some meaning, by emphasising local identity and heritage. In addition, the challenge is to remedy the diffuse and disorganised nature of urban areas by maximising the potential for territorial coherence and solidarity of rivers and wetlands.

Considering these three issues, there is indeed a way to act in favour of rivers and wetlands in the peri-Francilian towns. The proximity of Paris has an impact on the content of requalification projects. Other regional factors are also important, such as the proximity of the seafront in Caen and Rouen and, to a lesser extent, the Loire axis in Blois and Orléans, which brings local nuances to the aesthetic and functional treatment of rivers and wetlands.

10.5 Conclusion

Rivers and wetlands in towns and cities are therefore ideal places for expressing the urban quality sought in the implementation of sustainable development principles, which explains the unanimous enthusiasm of European urban stakeholders for these environments. However, the urban and environmental ambitions of urban policies change according to the size and situation of the towns. This has been demonstrated in three specific urban categories, including peri-Francilian towns. The study, which focused on nine of them, revealed the strategies employed, and the benefits and limitations of the requalification projects. Considering the risk of standardisation of these environments, generated by global factors as well as factors specific to the situation on peri-Francilian towns, valuable lessons were drawn for more efficient urban requalification of rivers and wetlands.

Firstly, urban requalification is a constantly evolving project, a permanent framework for action to improve the quality of rivers and wetlands based on environmental, landscape and urban planning considerations (Bush and Doyon 2019). This cross-cutting approach requires differentiated and integrated public actions to address the multiple challenges identified, from the local to the global. It also requires flexible and reversible urban development in order to adapt constantly to changing environmental dynamics and socio-economic needs (Desse et al. 2017; Kaplan and Kaplan 2005). At the same time, urban requalification requires a long-term territorial diagnosis to identify the hybrid nature of rivers and wetlands, the current and past uses and the resulting urban development and systems of representation (Carcaud et al. 2019; Carré 2011; Dournel 2013, 2021; Rivière-Honegger et al. 2014). Finally, it is a collective and shared project, bringing together public, voluntary and private stakeholders, as well as owner and user representatives. In this way, urban requalification expects the mayor to take a new form of public action (Theys 2017), emphasising his ability to create a general dynamic, devise an innovative mode of governance and coordinate action in accordance with the principles of complementarity and subsidiarity. These are the challenges of the urban requalification of rivers and wetlands.

References

Amzert A, Cottet-Dumoulin L (2000) Du “sauvage” à “l’inaltérable”: les conditions sociales de création d’un espace naturel en milieu urbain: le cas du parc de Miribel-Jonage. Géocarrefour 75(4):283–292

Auricoste I (2003) Urbanisme moderne et symbolique du gazon. Communications 74:19–32

Barles S (2010) Society, energy and materials: the contribution of urban metabolism studies to sustainable urban development issues. J Environ Planning Manage 53(4):439–455. https://doi.org/10.1080/09640561003703772

Bentley J (1997) East of the city. The London Docklands story. Pavilion, London

Beucher S (2009) Londres 2012, événement phare ou projet de ville durable? | London 2012, major event or sustainable city? Bulletin de l’Association des Géographes Français 3:312–323

Beyer A, Héraud JA, Rossano F, Steiner B (2021) De la ville-port à la métropole fluviale. Un portulan pour Strasbourg. Autrement, Paris
Blanc N (1998) 1925–1990: l'écologie urbaine et le rapport ville-nature. L'espace Géographique 4:289–299
Bush J, Doyon A (2019) Building urban resilience with nature-based solutions: how can urban planning contribute? Cities, 95. https://doi.org/10.1016/j.cities.2019.102483
Carcaud N, Arnaud-Fassetta G, Évain C (eds) (2019) Villes et rivières de France. CNRS éditions, Paris
Carré C (ed) (2011) Les petites rivières urbaines d'Île-de-France. Piren-Seine, Agence de l'eau Seine-Normandie, Paris
Chasseriau A (2004) Au cœur du renouvellement urbain nantais: la Loire en projet. Norois 192:71–84
Choay F (1996) L'allégorie du patrimoine. Le Seuil, Paris
Cohen P, Rustin MJ (eds) (2008) London's turning. Thames Gateway: prospects and legacy. Ashgate, Aldershot
Desse RP, François A, Holvoet M, Sawtschuk J (2017) Adapter les territoires aux changements climatiques: transition urbanistique et aménagement de l'espace. Norois 245:7–13
Dournel S (2010) L'eau, miroir de la ville: contribution à l'étude de la requalification urbaine des milieux fluviaux et humides (Bassin parisien, Amiens, Orléans). Doctoral thesis in Geography, Planning and Environment, Orléans: Université d'Orléans, https://theses.hal.science/tel-00925925
Dournel S (2013) Les zones humides à Amiens et Orléans. Reconstitution et transmission des paysages au défi d'une histoire tourmentée. In Galop, D. (edS) Paysages et environnement. De la reconstitution du passé aux modèles prospectifs. Besançon: Presses Universitaires de Franche-Comté, 347–356, https://books.openedition.org/pufc/43390
Dournel S (2019) L'eau, vecteur de projets communs, à contrecourant d'approches sectorielles, concurrentielles et autocentrées. Gestion et valorisation des milieux fluviaux dans le département du Loiret. Sud-Ouest Européen 47:69–91. https://doi.org/10.4000/soe.5308
Dournel S (2021) Les hortillonnages, des héritages emboîtés | The Hortillonnages, Interlocking Legacies. In Art and Jardins Hauts-de-France (EdS) Hortillonnages Amiens. Festival international de jardins. Amiens: Éditions Arts and Jardins Hauts-de-France, 10–19, https://hal.science/hal-03260355
Dournel S, Sajaloli B (2012) Les milieux fluviaux et humides en ville, du déni à la reconnaissance de paysages urbains historiques. Urban Hist Rev 41(1): 5–21. https://doi.org/10.7202/1013761ar
Emelianoff C (1999) La ville durable, un modèle émergent. Géoscopie du réseau européen des villes durables (Porto, Strasbourg, Gdansk). Doctoral thesis in Geography, Orléans: Université d'Orléans
Gérardot C (2007) Fleuves et action urbaine: de l'objet à l'argument géographique. Le Rhône et la Saône à Lyon, retour sur près de trente ans de « reconquête » des fronts d'eau urbains. Doctoral thesis in Geography and Planing, Lyon: Université Lumière Lyon 2
Gessel van PH (1985) The structure of green areas in the urban agglomeration of the Western Netherlands. The Dutch approach: planning a Randstadgroenstructuur. Landscape Urban Plan 18(3–4):257–263, https://doi.org/10.1016/0169-2046(90)90013-R
Giblet R (1996) Postmodern wetlands. Culture, History, Ecology. Edinburgh University Press, Edinburgh
Gruehn D (2017) Regional planning and projects in the Ruhr Region (Germany). In Yokohari M, Murakami A, Hara Y, Tsuchiya K (eds) Sustainable landscape planning in selected urban regions. Science for Sustainable Societies. Tokyo: Springer Japan, 215–225
Haughton G, Hunter C (1994) Sustainable cities. Regional studies association. Routledge, London
Kaplan R, Kaplan S (2005) Preference, restoration and meaningful. Action in the context of nearby nature. In Barlett PF (ed) Urban place: reconnecting with the natural world. Cambridge: MIT Press, 271–298

Kühn M (2003) Greenbelt and Green Heart: separating and integrating landscapes in European city regions. Landsc Urban Plan 64(1–2):19–27. https://doi.org/10.1016/S0169-2046(02)00198-6
Laganier R, Roussel I (2000) La gestion de l'écosystème urbain pour une ville durable. Bulletin de l'Association des Géographes Français 2:137–161
Masboungi A (ed) (2002) Projets urbains en France I French Urban Strategies. Le Moniteur, Paris
Masboungi A (ed) (2003) Nantes: la Loire dessine le projet. La Villette, Paris
Michon P (2008) L'opération de régénération des Docklands: entre patrimonialisation et invention d'un nouveau paysage urbain. Revue Géographique de l'Est 48(1–2). http://journals.openedition.org/rge/1104
Mihaila M (2015) Transforming the built landscapes—initiatives on city cultural sustainability: Phoenix project Dortmund. In ICAR2015 International Conference on Architectural Research, Bucharest: Proceedings Books
Mirloup J (ed) (2002) Régions périmétropolitaines et métropolisation. Presses Universitaires d'Orléans, Orléans
Rădulescu MA, Leendertse W, Arts J (2021) Metamorphosis of a waterway: the city of Nijmegen embraces the River Waal. Environ Soc Portal, 29. https://doi.org/10.5282/rcc/9357
Reitel B, Moullé F (2015) La resémantisation de la ligne frontière dans des régions métropolitaines transfrontalières: le Jardin des 2 Rives à Strasbourg et la place Jacques Delors à Lille. Belgeo 2. https://doi.org/10.4000/belgeo.16527
Rivière-Honegger A, Cottet M, Morandi B (eds) (2014) Contribution of stakeholder perceptions to managing aquatic environments. ONEMA, Paris
Rode S (2017) Les berges fluviales secondaires: des marges urbaines à résorber? Bulletin de l'Association des Géographes Français 94(3):472–488
Roth D, Winnubst M (2014) Moving out or living on a mound? Jointly planning a Dutch flood adaptation project. Land Use Policy 41:233–245. https://doi.org/10.1016/j.landusepol.2014.06.001
Theys J (2017) Prospective et recherche pour les politiques publiques en phase de transition. Nature Sciences Sociétés 25:84–92
Waldheim C (2016) Landscape as urbanism: a general theory. Princeton University Press, Princeton
Weber P, Konitzky A (1993) L'Exposition Architecturale Internationale I.B.A. Emscher Park (Ruhr). Hommes et Terres du Nord 2:85–90
Wolsink M (2006) River basin approach and integrated water management: governance pitfalls for the Dutch space-water-adjustment management principle. Geoforum 37(4):473–487. https://doi.org/10.1016/j.geoforum.2005.07.001

Chapter 11
Eliminate, Ignore or Integrate Urban Rivers? Challenges and Opportunities for the River Huatanay (Cusco, Peru)

Sisko Fernando Rendón Cusi, Juan José Zúñiga Negrón, and Albert Santasusagna Riu

Abstract The river Huatanay, located more than 3000 m above sea level, is one of Peru's most polluted rivers. Moreover, for most of its approximately 30-km length, the river is urbanized, as it runs through Cusco, a city of more than 400,000 inhabitants. In this chapter, we seek to highlight the transformations, both historical and recent, that have led to the river's current condition and stress the part played in this by the failure to implement territorial planning resolutions and to adopt instruments of water management that might promote the river's sustainability. Although a number of promising initiatives were taken, especially during the last third of the twentieth century, which saved the Huatanay channel from elimination, the current state of the river is critical: its water quality is deplorable, and there are no projects on the table to seek its urban integration. Certain stretches of the river have been channelized, but this work has been constrained by the city's express ways, and the Huatanay finds itself relegated to playing a secondary role in Cusco's urban fabric. In the second part of this chapter, we identify a number of important questions that might serve as the basis for a river recovery strategy to be implemented by Cusco's public authorities.

Keywords Polluted rivers · Strategic planning · Territorial development · Huatanay river basin · Peru

---

S. F. Rendón Cusi (✉) · J. J. Zúñiga Negrón
Faculty of Engineering and Architecture, Andean University of Cusco, Cusco, Peru
e-mail: sisko.rendon@vrin.uandina.edu.pe

J. J. Zúñiga Negrón
e-mail: jzunigan@uandina.edu.pe

A. Santasusagna Riu
Department of Geography. GRAM (Grup de Rercerca Ambiental Mediterrània) and Institute of Water Research (IdRA), University of Barcelona (UB), Barcelona, Spain
e-mail: asantasusagna@ub.edu

© The Author(s), under exclusive license to Springer Nature Switzerland AG 2024
J. Farguell Pérez and A. Santasusagna Riu (eds.), *Urban and Metropolitan Rivers*, The Urban Book Series, https://doi.org/10.1007/978-3-031-62641-8_11

11.1 Introduction

The Andean city of Cusco, administrative capital of the region, and a UNESCO World Heritage Site since 1983, faces a series of significant challenges attributable to its demographic growth and chaotic urban expansion (Branca and Haller 2021). The last population census, conducted by the Peruvian national institute of statistics (the INEI), reported a total of 437,538 inhabitants, the result of the sustained demographic growth of the last few decades. This rising population has historically contributed to increasing pressure being placed on the city's river, the Huatanay (Esquivel and Apaza 2017), which has become the city's main sewer for its wastewater. In response, the public institutions of Cusco (above all the regional government, the provincial municipality and the district councils) have, for several decades now, sought to solve this problem, albeit with limited success. Indeed, the management of the river Huatanay continues, to this day, to represent a severe, persistent environmental problem (Gil 2021).

The environmental deterioration of the river Huatanay is due, first and foremost, to pollution by the solid and liquid waste generated by the city of Cusco (waste that, unfortunately, does not get treated adequately). A second contributing factor is the alterations to the geodynamics of the river basin, which is subject to significant erosion in the rainy season. Finally, another of the main threats to the river's sustainability is the deficient nature of public management. Faced by the deplorable situation of the river Huatanay, the obvious question that needs to be addressed is what can be done to ensure its correct management. The question poses obvious challenges and various solutions have been brought to the table. Indeed, since the late nineties, a number of interventions have been made to treat the city's solid waste, stabilize the riverbed and its banks, improve governance and create greater environmental awareness among the city's inhabitants, the *cusqueños* (Guzmán 2015). However, to implement these actions requires considerable economic investment to construct essential infrastructure, including wastewater treatment plants and river flood defence walls, and to sponsor citizen awareness campaigns.

The main reason why the initiatives implemented to date have failed can be attributed to insufficient financing, which has jeopardized the sustainability of these river management projects. However, this is not the sole cause as an important contributory factor has been the failure to apply existing regulations with any rigour and the general disregard shown for the social component in the programs, projects and activities aimed at recovering the waters of the river Huatanay. Thus, the aim of the present study is to analyse the problem of Cusco's river and to propose solutions for consideration by the local public authorities that can improve its management. To do so, we have undertaken a bibliographic review of working documents and institutional reports and conducted field visits to the study area.

11.2 The Impact of Deficient Territorial Planning on Cusco's Water Management

Peru's national environmental policy, in its primary goal of promoting the conservation and sustainable use of the Republic's natural resources and biological diversity, seeks first and foremost the integrated management of its water resources and the effective organization of the use and occupation of national territory by means of what it defines as ecological-economic zoning (or EEZ). It also seeks to adapt the population to climate change and to establish measures of mitigation aimed at sustainable development. Integrated river basin management for the sustainable management of water resources, in accordance with the territorial planning policy, is also included among current policies and regulations; yet, in practice, it has yet to be defined and implemented.

According to the EEZ established in the Cusco region, water-related aspects referring to river basin headwaters and drainage basins are not spatially represented, despite the fact that the Amazon Cooperation Treaty (1978) considers the relationship between the water, land and natural units in these geographical zones as being of fundamental importance. Consequently, the water management problems presented by the river Huatanay are not solely the result of natural phenomena, but also of the transformation and urbanization of riverside areas. It is evident that the goals have not been achieved, since it has proved impossible to harness river basin management in any effective fashion to territorial planning, despite the efforts attempts made in this regard by the management of the drainage basins.

To implement the aforementioned policies, the Peruvian State created an autonomous institution under the auspices of the Ministry of Agriculture responsible for hydrological planning so as to ensure adequate water management, the protection of the country's water basins and the controlled use of the resource. Thus, the National Water Authority (hereinafter, ANA in its Spanish acronym) emerged, and which, according to the country's hydrological plan, seeks to guarantee that the "management of water resources be participatory, with achievable and verifiable objectives; and, harmonized and coherent with national, sectoral and regional policies" (ANA 2016:14). Against this backdrop, ANA developed a set of hydrological planning instruments that might guarantee the sustainability of the resource in the country, promote multisector projects aimed at improving the spatial and temporal distribution of resources, as well as the availability of natural sources to harmonize supply and demand, while also ensuring their quantity and quality so as to secure sustainable development at the national level. However, in the territorial planning at the regional level, sufficient importance has not been attached to the water resource, with the result that its absence from the plans impacts the management of this resource.

Attempts at guaranteeing the efficient management of Peru's water resources are not new. More than half a century ago, a new territorial organization of water was introduced based on the creation of micro-regions made up of small drainage basins in which efforts were made to undertake joint actions with different state sectors (Velasco 2013). However, the outcomes were disappointing and projects of this nature would not be financed again, highlighting once more the abandonment of regional planning centred on the country's drainage basins. One of the problems complicating the management of the drainage basins is the administrative political boundaries, since there is an overlap between the natural and political limits: the management of a drainage basin implies an administrative compromise between the natural components and the economic and social systems that control natural resources, which is not easily achieved in the case of Peru (Cordero 2011).

Romero (2008), moreover, highlights the importance of taking into account the relationship between human populations and the environment and, in particular, water. In the most highly urbanized areas of Peru, the pressure caused by demographic growth and rising population densities has increased the demand for water for domestic use, while in rural areas extractive activities have been largely responsible for the increased demand for the resource. This explains the need to take water resources into consideration in territorial planning, and in most cases this occurs at the scale of the drainage basins and requires detailed information and precision mapping. ANA seeks to find a balance between supply and demand, and between the quantity and quality of the water resource, promoting its efficient use to ensure its sustainability (ANA 2016). Consequently, the water authority actively participates in the territorial planning process via the drainage basin system and hydrological planning within the framework of broader national, sectoral and regional policies, with the active participation of the sector's stakeholders. However, the role that ANA plays is insufficient and disjointed from that of other sectors that have a part to play in this process. The case of the river Huatanay highlights the absence of this integrated perspective for the planning of water resources at different scales and for providing them with effective territorial instruments.

11.3 Cusco's River Huatanay: An Unresolved Problem

The river Huatanay, which crosses the city of Cusco from west to east (Fig. 11.1), ranges in width between 6 and 20 m and has a largely irregular discharge regime, being heavily dependent on the rainy season. The river, born from the confluence of two rivers, the Huancaro and Saphy, is 30-km long and occupies a catchment of approximately 500 km^2. The Huatanay forms part of the larger river Vilcanota basin, into which it flows near the village of Huambutio, in the Lucre district of the province of Quispicanchi (ANA 1999). In its turn, the river Huatanay is joined by various tributaries, the Marashuayco and Tankarpata on its right bank and the Huacoto on its left.

The severest problem faced by the river Huatanay is one of environmental pollution, the result of the direct discharge of wastewater without any type of treatment (Flores 2022). Both banks of the river have also become, along various stretches, sites for clandestine rubbish dumps and illegal livestock farming (ANA 1999; Quispe et al. 2022). According to Mendívil (2002) and Carreño (2005), the quality of the river

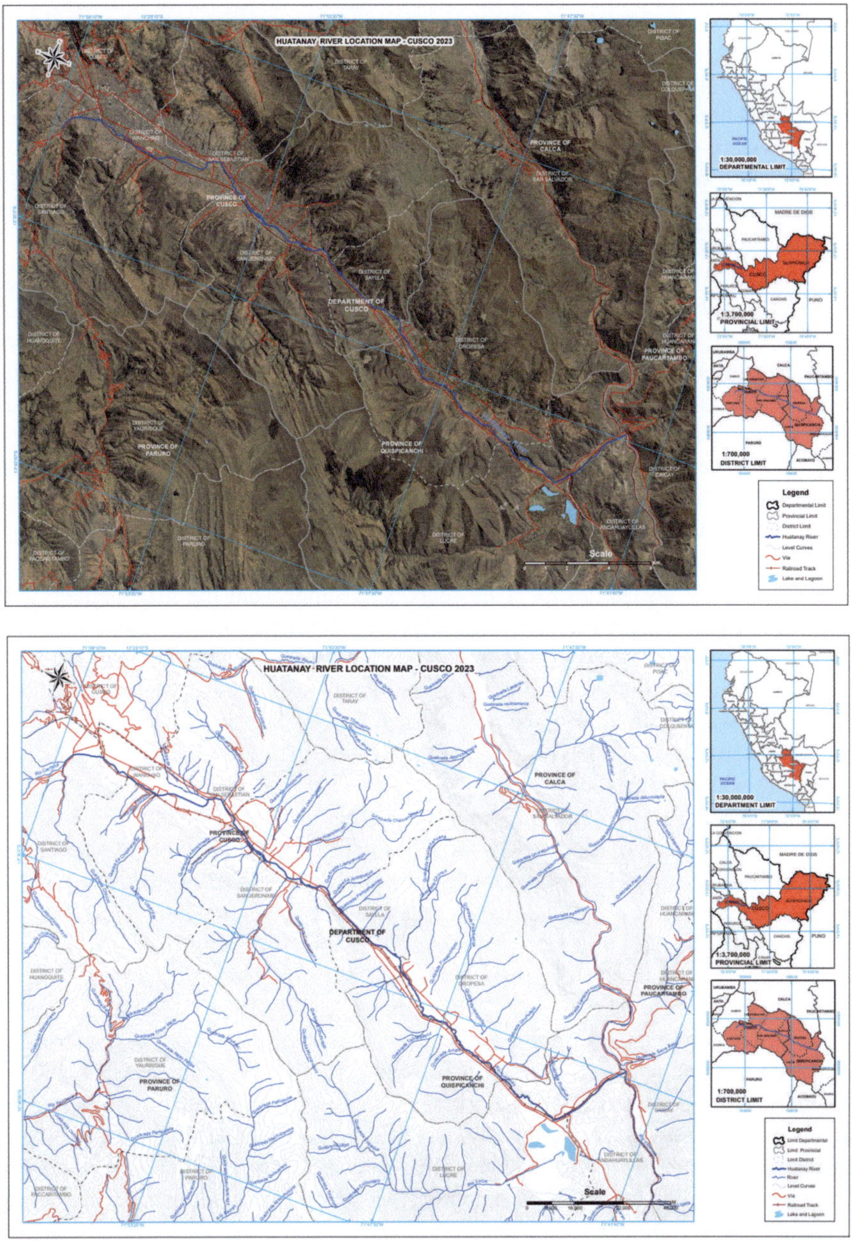

Fig. 11.1 Map showing the location of the river Huatanay in the province of Cusco (*Source* Juan José Zúñiga Negrón)

water is also affected by the transportation of solids from the erosion of agricultural systems along its banks and the constant occupation of marginal strips, which in some zones, due to the accumulation of waste produced by house building, even chokes the natural channel. Similarly, there are multiple points at which waste is discharged from privately owned factories, a practice to which the municipalities, who have responsibility for their management, turn a blind eye. This has resulted in highly deficient wastewater treatment and considerable negligence in the management of the drainage and sewage systems.

The deficient management of the river places those *cusqueños* who occupy its two banks in serious danger year after year. During the rainy season, normally concentrated between the months of November and March, the Huatanay frequently overflows, causing material damage and significant economic losses to the homes built on its floodplain (Camino 2012; Paredes 2019). Moreover, along the river course there are various stretches exposed to a high risk of slippage and land erosion that require the urgent stabilization of the river banks (Benavente et al. 2011; Vargas and Serna 2020). As confirmed by the Institute of Water and Environmental Management (hereinafter, IMA in its Spanish acronym), water management in the Huatanay basin is severely limited and deficient, not only in relation to the physicochemical quality of the water, but also in terms of its sources of supply and the management and treatment of rainwater, and domestic, industrial and hospital wastewater (Mendívil 2002). The rapid urban expansion undergone by Cusco over the last half century, which has far outstripped existing urban development plans, has generated a high demand for water and, in turn, the need to treat large volumes of wastewater.

Another matter of relevance is the provision, in the prevailing regulations governing water policy (2009 Water Resources Law—Nº 29,338), of a protected riverbank space—the so-called *faja marginal* or marginal strip (referred to in Spain as the "public domain of the rivers")—that has to be respected in all of the country's rivers. Article 74 of this Law states that "in the lands immediately adjacent to natural or artificial channels, a marginal strip of land shall be reserved for the purposes of protection, the primary use of the water, free transit, fishing, patrol roads and other services." But the lack of any regulations governing the management of water resources in the regional planning of Cusco has given rise to problems in the application of and compliance with this article. This is primarily the result of the absence of any corresponding mechanisms for its application in the municipalities and in the entities of the different State sectors. In many cases, the ambiguity inherent in the institutional roles and functions hinders further the effective application of these regulations, in addition to the lack of mechanisms that might facilitate the interrelation of functions and deployment of actions (Mendívil 2002). All in all, the absence of territorial planning in Cusco negatively impacts the regions' water management.

Finally, the complexity inherent in the diversity of institutional actors—both public and private—involved in Cusco's water management cannot be ignored. Water management is not only the responsibility of ANA, but also of other actors, including the water service provider (SEDACUSCO S.A.). However, this company limits itself to meeting the demand for water and takes no responsibility for the treatment of surface runoff and wastewater that have triggered alterations in the environmental quality and quality of life of the *cusqueños* (Mendívil 2002).

11.4 Ideas for the Future of the Urban Stretch of the River Huatanay in Cusco

In what follows we highlight a number of issues that are, we believe, of interest to the public authorities of Cusco for the implementation of a strategy of revitalization for the river Huatanay. The Huatanay remains, today, an overground river, although the space it is allowed to occupy is limited and its waters polluted. In full awareness of the critical nature of these circumstances, we propose a series of short- and medium-term measures that might be adopted to reverse (albeit only partially) the current situation and which specifically address three major problems: the river's especially severe environmental pollution, its physical instability and vulnerability and its deficient management model (which results, in turn, in little environmental awareness on the part of society as a whole). However, we first wish to acknowledge the various projects that have been carried out, especially since the nineties, to manage the river's urban stretch as it flows through Cusco, many of them financed by European and Spanish international aid funds. Thanks to the work of such organizations as IMA (mentioned above) and the Guamán Poma de Ayala Centre (a non-governmental development organization), several specific actions have been implemented in various urban stretches of the river. One example of this is the channelization of the Huatanay as it flows through the neighbourhood of La Quispiquilla (Fig. 11.2), an action involving the artificialization of the channel and interpreted and presented, by the city authorities, as an action for the "recovery" of the river Huatanay.

Within this proactive framework—developed from an eminently academic and technical perspective—a key question to bear in mind are the provisions made in the municipal budget for drawing up a strategic local development plan that seeks to revitalize the river space and which is fully adapted to the complexity of the city's environment and current needs. To be effective, such a plan would need the full backing of, and to be led by, Cusco's local authorities, with the participation and representation of all territorial stakeholders of the urban stretch of the river Huatanay, above all the residents of the homes and buildings immediately adjacent to the river. This plan should be technically ambitious and identify achievable short- and medium-term goals, both as regards the improvement of environmental quality and social involvement. The plans implemented to date have taken a largely sectoral perspective, but specific actions cannot be taken without understanding the needs of the whole city system. The city authorities need to promote an integrated discourse as regards the river that encourages citizens to be informed and involved in improving the quality of the river ecosystem. Here, the academic community could play a key role, especially those public and private universities with faculties of environmental engineering, architecture and related disciplines. Collaboration between the public authorities and the university community could have excellent outcomes in terms of raising citizen awareness of the value of the river Huatanay and, consequently, of promoting a co-responsible decision-making process. Open information sessions for the general public, workshops for the collaborative drawing up of proposals and environmental awareness sessions for the young are just some of the strategies that

Fig. 11.2 Channelization of the river Huatanay in Cusco (La Quispiquilla neighbourhood, San Sebastián district) (*Source* Jessica Harte 2020)

might be deployed within the framework of this plan. In addition, actions might also be taken to recover historical visions and memories of the river's heritage related to the Inca culture, as a means of strengthening ties with the ancestral identity of the *cusqueños*.

However, it should be borne in mind that the main problem of the river Huatanay is not that of an attitude of contempt or neglect on the part of the *cusqueños*, but rather the absence of economic investment in instruments that might improve the citizens' day-to-day relationship with the river space, as well as a lack of understanding of the many risks they run in their relationship with the river's waters, especially those related to pollution. Although there may be stretches along which waste concentrates, the existence of bulk rubbish dumped by the local population is not common along its entire length: it is by no stretch of the imagination an open-air landfill, as might be the case of other urban rivers around the world. The main environmental problem is low water quality and poor risk management, which contribute to generating a cycle of negative perception of the river environment that worsens the more degraded the river becomes (Fig. 11.3). The outlook is not at all promising. The main source of pollution is, as discussed above, the urban discharge that is poured into the river as it flows through the city, as well as that derived from specific sectors, including the construction and demolition industries. The absence of widespread wastewater treatment in Cusco makes major changes impossible; thus, improving the physicochemical quality of the water has to be the priority. The building of new sewage treatment plants, as well as the replacement and improvement of discharge systems, eliminating direct dumping into the river, are critical measures if the city hopes to imagine a decent future for the Huatanay. Likewise, the citizens need to be made aware that contaminated water cannot be used for farming, which currently proliferates in Cusco's urban periphery.

Finally, a window of opportunity is currently opening up for a possible physical reconciliation at least between the river and the city, due to the construction of the new international airport in the town of Chinchero, 20 km outside Cusco. This new infrastructure, which should be operational by 2025, will replace the Teniente Alejandro Velasco Astete airport, located in the city of Cusco itself. The building of a new international airport is as an obvious economic opportunity and should also serve to decongest the city of Cusco, as the current airport is embedded in its urban fabric. However, the local administration has yet to address the significant challenge of stripping an area of approximately 130 ha, which includes a 2-km stretch of the river Huatanay, of its urban function. This stretch of the river runs from a lateral avenue (La Costanera) and connects with a parallel express way of the same name as it crosses the airfield transversally. The left bank of this stretch has been left unbuilt, home until now of the airport's runways. We believe that it is an excellent opportunity to reclassify the land as a new green space.

This reclassification would be a great step forward with regards to two essential concerns for the urban configuration of Cusco. On the one hand, it would provide a

Fig. 11.3 On the left, partially channelized stretch of the river Huatanay, as it skirts the Huancaro market, to the southwest of the city. On the right, a residential stretch also near the market, with untreated waste being pumped directly into the river (*Source* Albert Santasusagna Riu, June 2023)

response to the urgent need for new green spaces, since it would allow the current ratio of square metres of green space per inhabitant (significantly lower than the 9 m^2/inhabitant recommended by the World Health Organization for all urban populations) to be increased and, in turn, would serve to structure the city beyond its historic centre, with a new linear space in the form of a large central park. On the other hand, this reclassification would give rise to the environmental, aesthetic and landscape re-evaluation of urban river spaces, linked for decades to processes of new recreational uses and citizen enjoyment. In practice, it could represent the creation of a new urban centrality and, therefore, the opportunity to use this stretch of the river Huatanay as a new sociable public space for citizens.

11.5 Conclusions

The river Huatanay is a paradigmatic example of an urban river affected by a series of physical and social variables that combine to leave it in its current critical condition. In the course of this chapter, we have shown that the absence of an effective territorial plan for the region's water resources has a direct impact on the management of such rivers. Although various government authorities, operating at different scales, have as their goal the correct planning and management of water and urban river spaces, in practice they present notable deficiencies in terms of their organization and the actual implementation of regulatory norms. The administrative structure fails to meet its objectives and, like a house of cards, collapses, giving rise to the serious prejudicial scenario that the citizens of Cusco face in their day-to-day relationship with the river Huatanay.

Due to its polluted waters and the implicit risks derived from its poor management and planning, the river Huatanay today constitutes a problem rather than an opportunity. It is for this reason that at various moments in its recent history, the decision has even been taken to eliminate the river completely. A poorly managed river is the focus of endless problems, and in the case of the Huatanay the biggest problem is the pollution derived from the uncontrolled dumping and waste that accumulates along its urban stretch. Not having an adequate water purification system, nor a sufficient urban sewage system, means the pollution combined with the impermeabilization of the surface creates a situation that is especially hazardous and destructive during episodes of heavy rain, resulting in floods that impact above all the poorest strata of society. Furthermore, the presence of buildings in flood zones and the absence of river flood protection systems considerably increase the vulnerability of much of the city, giving rise to the worst possible scenario imaginable: a river urbanized practically right up to its riverbed without an adequate risk management system.

In the case of the river Huatanay, it is vitally important that this vicious circle of adverse, negative conditions be broken. To do so, it is essential that the city—beginning with the local public authorities—starts to rethink the technical and scientific discourse used to date for managing the river Huatanay and finds a way of identifying global solutions based on a fully integrated approach that avoids the biases of a sectoral perspective. Taking advantage of the enormously significant land use change represented by the construction of a new airport outside the city, Cusco is presented with a great opportunity to re-plan a space that will be stripped of its previous function. This, together with the implementation of local development strategies, environmental education and citizen participation should serve as a catalyst of change for the river and encourage the city of Cusco to imagine the Huatanay it wants in the future, despite the many uncertainties that doubtless await at the global scale. We believe that Cusco should and can position itself as an example of best practices were it to champion this process of change within the framework of the Global South.

Acknowledgements This study has benefitted from the scientific and economic support of the CONCYTEC ProCiencia project: "Characterisation of territorial planning policy in the adoption of innovative methods in science and technology for territorial management in the Cusco region" (*Proyectos de Investigación Aplicada en Ciencias Sociales*, N° PE501078255-2022-PROCIENCIA). It has also received funding from grant 2021SGR00859 awarded by the Agency for the Management of University and Research Grants (AGAUR) of the Catalan Government of the *Generalitat* (SGR 2021–2024).

References

Autoridad Nacional del Agua [ANA] (1999) Delimitación de la faja marginal del río Huatanay. Resumen ejecutivo. Lima: Administración Técnica del Distrito de Riego

Autoridad Nacional del Agua [ANA] (2016) Planificación hídrica en el Perú. Lima: Ministerio de Agricultura y Riego

Benavente CL, Delgado GF, Fidel L (2011) Evaluación del río Huatanay en el tramo Puente Agua Buena y Urbanización Cachimayo. Distrito de San Sebastián, región Cusco. Informe Técnico N° A6443. Lima: Instituto Geológico, Minero y Metalúrgico

Branca D, Haller A (2021) Cusco: profile of an Andean city. Cities 113:103169

Camino GA (2012) Rainfall drainage: a case study of Cusco city. In Strecker EW, Huber WC (eds) Global Solutions for urban drainage: proceedings of the ninth international conference on urban drainage, Held in Portland, Oregon, September 8–13, 2002. Reston: American Society of Civil Engineers

Carreño R (2005) Geological and geomorphologic relationship of the sub-active landslides of Cusco Valley, Peru. In Sassa K, Fukuoka H, Wang F, Wang G (eds) Landslides. Berlin: Springer

Cordero E (2011) Ordenamiento territorial, justicia ambiental y zonas costeras. Revista De Derecho De La Pontificia Universidad Católica De Valparaíso 36:209–249

Esquivel J, Apaza R (2017) La modernización de la ciudad y su salubridad: la canalización del Cusco a principios del siglo XX. Summa Humanitatis 9(1):110–168

Flores AH (2022) Influencia en la gestión de aguas residuales de la ciudad del Cusco en la calidad del río Huatanay, año 2022. Universidad César Vallejo, Lima

Gil JE (2021) Deterioro y pérdida de los ríos urbanos. El Antoniano 134(1):1–29

Guzmán E (2015) Valoración económica de mejoras en los servicios ambientales en el contorno del Río Huatanay, Cusco-Perú. Informe Técnico N° A1-PBCus-T2–01–2014. Cusco: Consorcio de Investigación Económica y Social (CIES)

Harte J (2020) Gestión del agua. Recuperación del río Huatanay en Cusco. Programa Internacional de Cooperación Urbana Unión Europea-América Latina y el Caribe, Lima

Mendívil R (2002) Gestión del agua en la cuenca del río Huatanay y la concertación para el tratamiento de problemas ambientales. Lima: Instituto de Manejo de Agua y Medio Ambiente

Paredes JC (2019) Derecho a gozar de un ambiente equilibrado y adecuado para el desarrollo de la vida. Estudio del caso del río Huatanay (Cusco). Cusco: Universidad Nacional de San Antonio Abad del Cusco

Quispe V, Oros W, Felix Z (2022) Educación ambiental y manejo de residuos sólidos en Cusco. Ciencia Latina. Revista Multidisciplinar 6(3):2800–2807

Romero G (2008) Componente de la gestión de riesgos para el ordenamiento territorial de la ciudad de Calca. Distrito Calca, Región Cusco, Perú. Calca: Centro de Estudios y Prevención de Desastres (PREDES), Welthungerhilfe, Prevención de Desastres en la Comunidad Andina (PREDECAN)

Vargas CG, Serna MA (2020) Condiciones de habitabilidad de viviendas aledañas a la cuenca de ríos: caso Huancaro-Cusco. Yachay. Revista Científica Cultural 9(1):530–542

Velasco O (2013) Perú. La difícil construcción de una república para todos. Universidad de San Martín de Porres, Lima

Part III
Perception

Chapter 12
The Role of Bluespaces for Well-Being and Mental Health. Rivers as Catalysts for the Quality of Urban Life

Ana Bonifácio

Abstract The recognition of urban rivers' relevance on urban environment and urban planning and political decision-making was achieved in the late twentieth century. However, "concrete curtains", roads, or heavy traffic still block many of these natural elements. With the COVID-19 pandemic, mental illness problems gained preponderance in health but also in new urban intervention projects. However, regarding this theme, some taboos persist, and these new interventions do not systematically apply approaches linked with mental well-being. Urban regeneration efforts to revitalise waterfronts have been essential for citizens' quality of life. It is known that regardless of the river's route, shape, or size, it can act as a catalyst for population well-being. Therefore, with higher public consciousness, bluespaces are a collective urban asset that provides comfort and promotes greater spatial justice. Starting from the foundations of public space creation, through the lenses of neurourbanism and urban planning and designing, this chapter intends to open the discussion to a new approach of planning that recognises bluespaces as a critical factor in the population's quality of life and, particularly, their mental health.

Keywords Urban planning · Urban health · Mental health · Public space · Bluespaces

12.1 Introduction

The opportunity to integrate the present analysis in a publication dedicated to urban and metropolitan rivers from the perspective of geomorphology, urban planning, and perception—especially because these sub-domains intersect—seems particularly relevant when different global challenges converge on the health of populations.

A. Bonifácio (✉)
Associate Laboratory TERRA, Centre of Geographical Studies, Institute of Geography and Spatial Planning, University of Lisbon, Lisbon, Portugal
e-mail: ana.bonifacio@edu.ulisboa.pt

Institute of Physiology, Lisbon School of Medicine, University of Lisbon, Lisbon, Portugal

© The Author(s), under exclusive license to Springer Nature Switzerland AG 2024
J. Farguell Pérez and A. Santasusagna Riu (eds.), *Urban and Metropolitan Rivers*, The Urban Book Series, https://doi.org/10.1007/978-3-031-62641-8_12

The United Nations (UN) states that «six out of every ten people in the world expected to reside in urban areas by 2030, rising to 83% by 2050» (United Nations Human Settlements Programme [UN-Habitat] 2022, p. 9). Thus, urban health seems particularly relevant when transdisciplinary concepts, such as "neurourbanism" (Adli et al. 2016, 2017) from the domains of neuroscience, psychiatry, or psychopathology or "neuropolis" from the perspective of urban sociology (Fitzgerald et al. 2016), are being sought to be put into practice.

The emancipation of these concepts meets the evidence brought by scientific research studies that claim the need to promote a close collaboration between health and territory professionals (Adli et al. 2017) since several emotional and mental disorders such as anxiety, stress, or depression have been identified in urban settings for several decades now (Burton 1990; Peen et al. 2010; Montgomery 2013). The cost of seeking the density of large cities for their variety of leisure activities, rich cultural life, better access to employment, and anonymity is reflected in health (Adli 2011).

The study of urban health has been growing in academia, and researchers with different backgrounds and specialisations draw attention to its implications, the impact factors of the built environment and ways to create evidence-based knowledge on how it affects people's psychophysiological health. More recently, we are witnessing a discussion and several exploratory research works regarding using different types of technologies, e.g. mobile EEG, eye tracking devices, and other wearable devices, that can collect and measure emotional responses to urban environmental stimuli.

On the one hand, neurourbanism seeks to investigate the biological bases of mental states and disorders to improve the quality of life in urban environments (Pykett et al. 2020; Buttazoni et al. 2022; Ancora et al. 2022) and on the other hand, there is a growing awareness of the importance of subjective concepts such as quality of life, well-being, or happiness in urban space.

Several efforts have been made to find definitions and measurement models for these concepts, and this search will undoubtedly continue, as it is still in its embryonic phase. Due to its multidimensional characteristics, quality of life covers the following domains: physical, material, social, and emotional well-being, development and activity (Felce and Perry 1995). Therefore, the concept of well-being is intrinsically associated with quality of life, and it is known that its evolution has accompanied the changes in lifestyles (visible in the last hundred years) mainly associated with the increased power of control over one's destiny (Santasusagna Riu 2021). Happiness, a term often used as a synonym for life satisfaction, is based on the influence of the cultural dimension of each person's context (Delle Fave et al. 2016).

In the context of subjectivity, urban space should become an objective inducer of quality of life and, therefore, it is necessary to foster a paradigm shift in the policies of reconversion, requalification, or renovation of waterfronts (installed since the end of the last century) by adding the mental health layer.

How can bluespaces contribute to blurring the negative impacts of urban life on mental health?

Before moving on to the hypotheses for answering this question, it is essential to illustrate the process of political and citizen awareness of the importance of riverside fronts as natural elements of urban territories that, in the past, took place due to the scarcity of financial resources and the lack of mobilisation public policy for urban development. In response to a changing landscape, urban planning integrated large-scale territorial revitalisation projects initially implemented in the United States, then in northern Europe and, at the end of the twentieth century, in the northern Mediterranean (Carrière and Demazière 2002).

12.2 The Case of Portugal and Medium-Sized Cities' Waterfronts. A Particular Public Policy of Cities at the Turn of the Millennium

In the late twentieth century, Portugal witnessed transformations in its economic structure and impacts on territorial development. This situation led to a reconfiguration of urban centres. The concept of quality of urban life transcended the mere provision of services and basic infrastructure. It has come to depend on the central role of public spaces and their ability to enhance environmental elements (Queirós and Vale 2005) whilst at the same time seeking to respond to the greater purpose of promoting urban transformations capable of increasing territorial attractiveness and competitiveness (e.g. Smyth 1994).

By the 1990s, Portugal experienced substantial economic growth, which improved the living conditions of the Portuguese population. At the same time, the revitalisation of the eastern area of Lisbon for EXPO'98, the Lisbon World Expo, became the catalyst for the emergence of a critical awareness of the waterfronts' importance.

These issues of the urban environment and city sustainability won significant attention, particularly after the 1992 Rio Conference, and were reinforced at the European level through the Green Paper on the Urban Environment (Commission of the European Communities [CEC] 1990). As a result, the need to establish new framework policies emerged, globally and locally, with a specific emphasis on integrating environmental protection and urban planning. These policies aimed to foster sustainable development by promoting the relationship between environmental and ecological preservation and urban enhancement.

Since 1994, the European Union has prioritised specific urban policy programmes for medium-sized cities. These programmes focused on addressing the importance of urban centres in the national urban system and promoting urban environmental requalification projects, mainly on "inherited" city areas (Portas et al. 2003).

The introduction of the National Economic and Social Development Plan (Ministério do Equipamento, do Planeamento e da Administração do Território [MEPAT] 1998), aligned with this dominant orientation, highlighted the importance of prioritising actions related to the urban environment. These actions were intended to improve urban spaces' environmental and social quality. It became clear that city intervention should go beyond day-to-day urbanistic management and incorporate the environment, sustainability, and urban planning into a strategic dimension (Correia et al. 2000).

Within this context, the Portuguese government recognised the need to implement a policy to strengthen the urban network system of medium-sized cities in Portugal through the POLIS—Programme for Urban Requalification and Environmental Valorisation of Cities (Correia et al. 2000).

This Programme meant to respond to the strategy enshrined in the Regional Development Plan (2000–2006), financially delimited by the III Community Support Framework, which outlined a programme of urban requalification operations inspired by the major operation that preceded EXPO'98, an international major scale event. That process brought a profound transformation to the eastern area of the city of Lisbon, as other successful international best practices.

Like the initiatives that had significant media repercussions, such as the 1992 Barcelona Olympic Games, EXPO'98 focused on the theme of the oceans and catalysed a city policy centred on the reinvention of the territory and a paradigm shift towards urbanism and practices distinct architectural features (Carrière and Demazière 2002).

That was the challenge of POLIS, designed to play a demonstrative role in Portuguese cities' public policies, like what happened with the Tagus riverfront. In alignment with EU and national strategic guidelines for territorial requalification, the primary objective of POLIS was to enhance the quality of life in cities through requalification operations based on environmental and heritage aspects. These operations were developed through public commitment between local authorities and the central administration.

The Council of Ministers Resolution nº26/2000 in which was published the Programme, states «(…) give great relevance to the strategic role that cities assume in the reorganisation of the territory and the importance that the quality of the environment and the correction of urbanistic errors have for that underlined strategic role». The following aims were defined as the leading specific objectives of POLIS, under the goal of improving the quality of life of citizens:

- develop large integrated urban requalification operations with a robust environmental valorisation component;
- develop actions that contribute to the requalification and revitalisation of urban centres, that promote the multi-functionality of these centres and reinforce their role in the region they are located at;
- to support other requalification actions that would improve the quality of the urban environment and enhance the presence of structuring environmental elements such as a river or coastal fronts;
- to support initiatives to increase green areas, promote pedestrian areas, and restrict car traffic in urban centres (Correia et al. 2000).

In this context, 18 Portuguese cities were selected to carry out integrated strategic planning processes and develop specific projects to be implemented on the ground in the shortest possible time.

The POLIS programme supported several city initiatives that responded to the pre-defined aims, including having a waterfront to enhance as a structuring element of the territory. These cities should also serve as a demonstrative example for the subsequent phases of the Programme in other cities, even with different management models.

The municipalities had the opportunity to produce a reliable integrated operation anchored in areas of intervention previously conceived and strategically diagnosed by the respective municipalities in the pre-defined time frame of 2001–2006. The competent entities chose territories where operations could be carried out in striking and structuring elements of cities, such as riverside fronts (rivers or coastlines) and their surrounding public spaces, with the capacity to act as social and cultural triggers and of "positive contamination" for the remaining areas of the urban territory.

One characteristic differentiating the POLIS programme was its focus on public participation, surpassing the legal public debates mandated by territorial management instruments. It entailed implementing a comprehensive communication strategy by establishing local monitoring committees, conducting awareness-raising initiatives, disseminating information through local media outlets, and organising engagement sessions with associations and citizen groups, particularly mobilising the population affected by the programme.

Having sea fronts or urban riverfronts as their motto, most of the interventions achieved their objectives through the implementation of actions that distinguish these natural elements in the daily lives of citizens, contradicting the natural phenomenon in many cities in the anterior decades that grew with their backs turned to the rivers. «For years, cities have overlooked their riverfronts; for hubs of economic activity, many city riverfronts have become neglected in the post-industrial era—forgotten, blighted eyesores» (Roe and McCay 2021: 50).

The river is often perceived as an obstacle to the organic expansion of the city, changing potential developments in the territory mitigated with river transport and the construction of crossings such as bridges or tunnels. Conversely, the city can act as a barrier to the river's natural flow, and urban planning has not consistently addressed this dynamic. It has often struggled to effectively manage the inherent risks and uncertainties associated with the river whilst failing to harness the diverse environments and unique interactions fully it offers (Silva and Pinto 2010).

The interventions conducted during that specific period, sometimes disrupted by various obstacles, yielded several positive outcomes, which included: the affirmation of cities on a regional scale, the establishment of a network of notable public spaces centred around water as a significant element, the enhancement of the value of historical urban areas, the strengthening of the urban ecological framework, the improvement of internal urban connectivity systems, and the establishment of functional anchors.

Leiria was one of the Portuguese cities whose portrait was, at the time, quite illustrative of the set of urban problems that accumulated at the end of the twentieth century in cities of the same dimension in Portugal and Europe. Extensive urbanisation, emptying of the historic centre, environmental disqualification of its river (in the case of Leiria, the river Lis), degradation of the natural and built heritage, lack of public transport, and excessive use of individual transport, amongst others, were problems that the interventions of POLIS, whilst not intending to solve them entirely, came to mitigate their impacts.

The expansion of the city of Leiria took place along the natural course of the river. However, it was found that «(…) the Lis River was losing its status as a structuring waterline for the morphological and economic development of the city through a general abandonment and even a disrespect for its characteristics as a natural element. (…) At the end of the last century, the city grew with its "back turned" to its river» (LeiriaPolis, Sociedade para o Desenvolvimento do Programa Polis em Leiria [LeiriaPolis] 2007:23).

Several studies that evaluated the results and impacts of the programme's implementation led to the conclusion that the paradigm of fruition and the experiences of the population have changed, mainly due to the creation of new functional valences (sports, leisure, and stay) and public spaces qualified that added the possibility of walking and crossing smoothly (on foot or by bicycle) the entire city's urban centre, along the river Lis. Improving the quality of life, the conquest of a sense of belonging, and mainly, the promotion of socialisation and the increase in physical activity (often anchored in sports open-air activities) was a noticeable added value of the direct results of the program's implementation (Figs. 12.1 and 12.2).

There are new health challenges in Leiria and other urban territories with waterfronts, mainly those related to urban and mental health. Bluespaces and rivers, in particular, must be treated as "crown jewels" at a time when water acquires vital importance for Humanity.

Fig. 12.1 Rossio de Leiria and the River Lis in its most urban section before the intervention (2002) and after (2006) (*Source* Leiria City Council, LeiriaPolis archive)

Fig. 12.2 An urban river area in a naturalised section before the intervention (2002) and after (2004) (*Source* Leiria City Council, LeiriaPolis archive)

12.3 Urban Health-Induced Public Policies: The Point of no Return

According to Peter Bosselmann (2009), the relationship between a city and water is inherently unstable, driven by human well-being and survival considerations. Acknowledging that water systems shape cities brings a sense of optimism because understanding those systems and their changes enables informed planning of new cultural landscapes, considered "commons" representing shared heritage even in large metropolitan areas.

The authenticity of a city is closely tied to its location within a particular territorial morphology. The city's form evolves in response to natural systems, influencing and altering them over time, rarely eradicating them. Rediscovering previously modified biological systems has sparked increasing interest amongst urban planning and design professionals. This interest should drive them to explore the unique characteristics of each city and seek creative solutions that draw inspiration from natural systems when designing new cultural landscapes (Bosselmann 2009).

Where do we stand today, more than two decades after the beginning of the third millennium, thinking, planning, and valuing waterfronts?

In Portugal, in Europe, or on an increasingly voluble planet, despite the distinct territorial contexts and global socio-economic asymmetries, we are today under a reality that is witnessing an exponential increase in natural disasters—in the course of the twentieth and early twenty-first centuries—and that places climate change on the side of causes and problems (Zêzere 2019). The climate crisis is one of the aspects that cannot fail to be relevant to health, immediately perceptible in the wake of disasters, in the loss of human lives, or increasingly in the generalised loss of quality of life.

Another crisis, the 2020–2023 COVID-19 pandemic, put the attention on mental health problems, which contributed, on the one hand, to encourage the fading of a persistent social stigma, even though, in everyday life, the return to the "old normality" was more immediate than an announced extension of what was then called the "new normal"; on the other hand, the measures to contain the disease that were adopted exposed the urgency to study in depth the articulation of this phenomenon with different geographical contexts and models of life. The pandemic increased outdoor exercises like walking or biking in nearby forest areas or coastal regions when these activities were allowed. In fact, urban green parks and riverside areas experienced a significant rise in visitors (Franco and Marques da Costa 2021; Louro et al. 2021a) and became a vibrant spots of urban life.

The importance of contact with green and bluespaces was thus underlined, emphasising their role in nurturing communities' resilience to the stress induced by the virus threat and the resulting physical constraints. In addition, these spaces were considered alternative places for physical activity and social interaction (World Health Organization [WHO] 2021). As a non-elastic territory, the public space needed to guarantee the imposed physical distances, reinforcing its importance due to the need to reach it.

In addition to these crises, we cannot forget other relevant "pandemics" which accelerate the need to produce more updated and better-equipped urban public spaces: the digital transition and artificial intelligence (translated into the market to reformulate ways of working, for example); the macro-economic crises; the geopolitical crises and the wars; or the forced migration movements.

To exemplify the importance of urban public space from a physical and functional point of view, Remesar evokes, for example, the case of two squares in Paris and Nancy, referring to the fact that they are open and multifunctional spaces as mechanisms for articulating two different parts of the city, separated from each other. In the centre of each square, there is a monument and other additional elements, such as fountains and architectural details; around it, the monumental vertical plane of the building façades, the green of the gardens, or the open blue of the river (Remesar 2020).

Borja and Muxi stated in the 2000s that public space represents the history of a city «because it is the place where all social and political activities are imprinted, and the place of all discussions and decisions that affect the collective» (2003, p.15). From the global challenge of the pandemic, what can we add to the place that, being everyone's, also affects the character, behaviour, and health of each individual?

The public space of urban areas must respond to more than just the needs imposed by increasing the longevity of human life or promoting better conditions for physical activity.

From a governance perspective, public space must be understood as a privileged stage for social, economic, environmental, cultural, and open-access responses. Using public space (mobile or immobile, perennial or performative contents), a substitute for the antidepressants or anxiolytics prescribed at the health centre may still be far from reach. However, it is also evident in the scientific literature and in the interdisciplinary experiences that support it—and if possible, in a transdisciplinary way by strengthening the teaching of neurourbanism or disciplines such as environmental psychology—that it is conceivable to develop solutions that promote health and well-being, as reflected, in WHO Sustainable Development Goal No. 3 "Ensure healthy lives and promote well-being for all at all ages".

Despite the arguments that these crises have brought to the fore in recent decades, human health—mainly urban health and, from there, mental health—are transported to a "point of no return" about public policies and political decisions. Gone are the improvement interventions that had as their main objective the enhancement and integration of waterfronts—often accompanied by interventions for the environmental recovery of water quality and biodiversity—to increase physical activity by introducing pedestrian routes, bike lanes, playgrounds, or maintenance fields.

12.4 Health, Urban Health, and Mental Health

As stated in the Constitution of the WHO, «[h]ealth is a state of complete physical, mental and social well-being and not merely the absence of disease or infirmity» (WHO 1946). This definition implies that health goes beyond simply not having specific undesirable conditions. It encompasses a broader and more positive state (Evans et al. 1990).

To frame the evidence demonstrated by the studies in the literature review which has already addressed this issue, most of them admit a set of five groups of factors that impact human health (also called "health determinants" not only "social determinants of health") which are assigned different names (depending on the source) but which we can designate as follows: (1) Environmental factors; (2) Individual behaviour; (3) Medical care; (4) Genetics and biology; and (5) Social and economic factors (Fig. 12.3).

Regarding urban health, the concept also encompasses elements such as "urban governance", "population characteristics", "natural and built environment", "social and economic environment", "food quality", and "emergency health services and management". These factors can either positively or negatively influence urban health, including the health of individuals within the urban context (WHO 2010).

Mental health is a multifaceted issue with no singular cause influenced by various factors that can affect an individual's overall mental well-being. According to the WHO, «Mental health is a state of well-being in which an individual realises their abilities, can cope with the normal stresses of life, can work productively and can contribute to his or her community» (WHO 2018:1).

Regarding the environmental factors to which the authors mentioned above attribute a maximum influence of 10% on health, we highlight the aspects that relate directly to urban health (with multiple impacts on physical and mental health), collected following a narrative review of the literature, developed as part of the eMOTIONAL Cities project (Bonifácio et al. 2023).

We focused on the urban context's built and natural environment that can impact physical and mental health positively and negatively, underscoring the importance of planning and policy measures that consider health perspectives (Fig. 12.4).

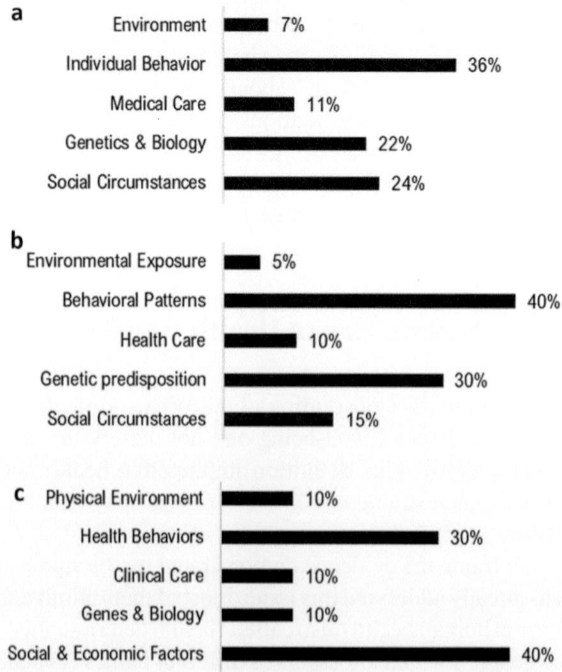

Fig. 12.3 Collection of charts of impact factors groups in human health according to **a** GoInvo (2020); **b** Awosogba et al. (2013); and **c** Pizzorno cited by Weeks (2017). The graphics represent the impact of each set of factors on one's health, with the designations attributed by each author (*Source* Adaptation by the author 2023)

According to the research carried out, we intend to detail the potential effects of bluespaces (and their different typologies) on the mental health and well-being of citizens because, as Roe and McCay state, «(…) exposure to urban water settings (bluespace) can help improve mental wellbeing; the benefits include increased subjective wellbeing, a reduced risk of depression and stress alleviation» (2021:41), in an attempt to theoretically sustain the materialisation of the concepts of Healthy Cities and Livable Communities in the context of political decision (Louro et al. 2021b). This enables practical application across various territorial scales and formulation of distinct yet complementary policies.

12.5 Bluespaces: The Knowledge Gap Exists

Urban bluespace is often a combination of water and green landscapes and has often been grouped with greenspaces. However, despite the possible "overlap", contact with bluespaces adds a layer of greater sensoriality in experiences (e.g. through the sound produced by moving water), which may result in different benefits compared to greenspaces (Haeffner 2017).

The evidence regarding the health benefits of natural environments is highly significant for health professionals, planners, and policymakers. They are crucial in translating the available evidence into interventions and policies that promote and support

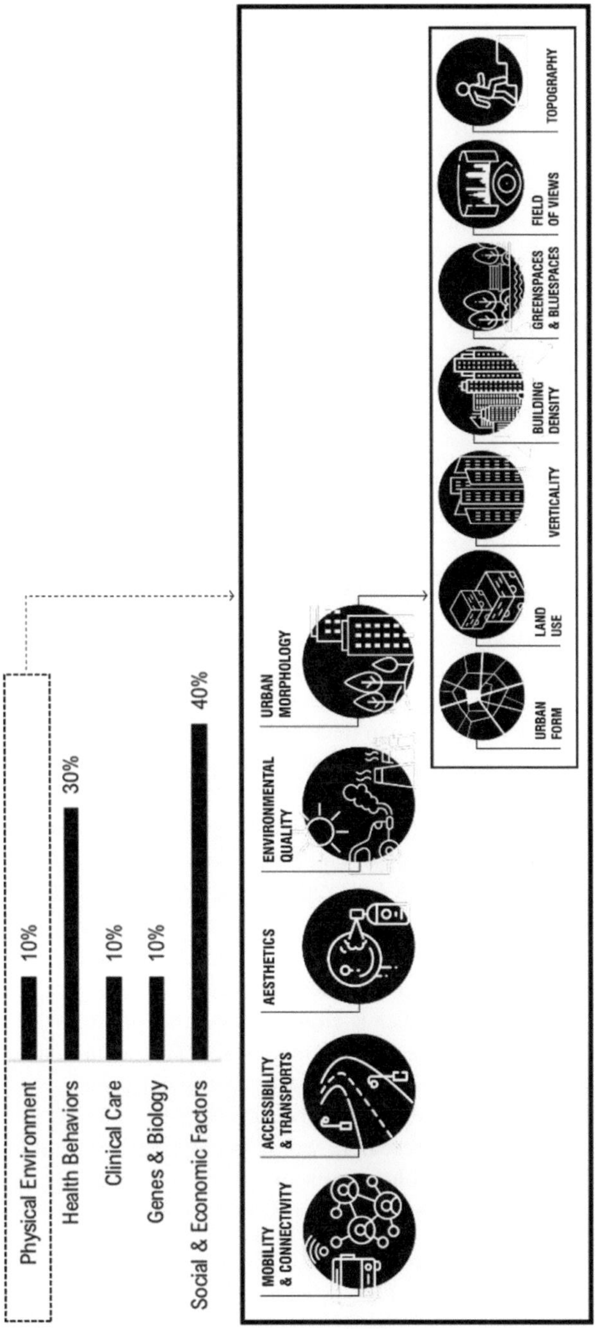

Fig. 12.4 The environmental factors that impact health are included in the "Physical Environment" results, related to the urban context (*Source* Author 2023)

health and well-being. Most studies in this field have focused on the health benefits of green spaces, as seen in numerous systematic and non-systematic reviews (Gascon et al. 2017). In smaller numbers, there are also those studies that have been exploring the health impacts of bluespaces, which stand out for their prominent water features and accessibility to citizens either through physical proximity, distance, or virtual experiences (Grellier et al. 2017).

Whilst it is crucial to enhance our understanding of the connection between health and bluespaces through scientific methods, the evidence suggests a significant potential to serve as a resource for mental health. They can support emotional well-being and contribute to resilience against heat stress and climate change. The systematic review conducted by Gascon et al. (2017) examined the evidence of the health benefits of bluespaces. At the time (2017), only twelve quantitative studies were identified, illuminating the potential mental health benefits of residing near or intentionally visiting bluespaces. Today, mainly as a result of post-pandemic research, there are new studies researching mental health and the use of bluespaces, yet they are still too limited.

The difficulty in doing a more objective-driven study and measuring people's mental health based on exposure to water bodies is multiple. One of the issues concerns the typological diversity of these spaces, whether natural or artificial. Consequently, the public spaces that surround them (banks, beaches, wetlands, linear parks, etc.) also have distinct characteristics that provide diversified types of use/fruiting, some with the possibility of access to water and others with only visual access.

In the literature review of Gascon et al. (2017), it was highlighted that the limitation based on estimating bluespace exposure is related to the variety of tools used, as it can either be based on land cover map measurements to determine the percentage at the level of the census area unit or distances between water bodies and specific points such as schools or residential areas.

Another limitation is based on the measured results. Some refer to mental health, well-being, life satisfaction, happiness, and psychological distress (Alcock et al. 2015; Brerenton et al. 2008; Huynh et al. 2013; MacKerron et al. 2013; Nutsford et al. 2016), and others are more oriented towards behavioural issues or depression or perceived anxiety and their relationship with medication taken by users (Amoly et al. 2014; Rogerson et al. 2016; and Triguero-Mas et al. 2015).

Moreover, the design of the studies is either longitudinal or cross-sectional, and only one is supported by a pre-/post-observational study (Rogerson et al. 2016). The results take time to obtain despite having the same instruments, such as the General Health Questionnaire, for comparison and conclusion.

Therefore, accordingly to our research, we have pinpointed several knowledge gaps regarding the influence of bluespaces on people's mental health, and the reasons can be summarised by the lack of consistent scientific evidence about this relationship.

12.6 Conclusions to Open Research Gates

Whilst some studies have indicated a connection between living close by and using outdoor bluespaces and mental health and general well-being, the available number of studies remains insufficient, and there is variability in their designs and results. Indeed, a methodological protocol for scientific evidence still needs to be improved.

However, recognising the qualitative methodological approach within this research domain has proven critical, as it already highlights several advantages of experimenting with bluespaces (Foley and Kistemann 2015). Qualitative research investigates aspects that quantitative analysis often ignores, such as assessing perceptions, sensory exposure, and individual experiences. These qualitative dimensions help explain the associations identified by studies based only on quantitative data.

This particular form of research remains valuable in generating insights for policy formulation or intervention initiatives related to bluespaces in urban environments. However, as Roe and MacCay (2021) noted, health outcomes are often overlooked and not given due importance in driving urban transformations regarding blue city projects.

How people interact with bluespaces can change in different cultures and climates worldwide, resulting in distinct meanings, functions, and effects on health and well-being. As a result, it is crucial to consider the inherent "cultural bias" of existing and future studies when examining this relationship (Gascon et al. 2017).

Scientific research has highlighted the importance of establishing metrics that can be used in an academic context and in a governance logic that policymakers can understand. Specifically, there has been increasing emphasis on measuring, evaluating, and categorising factors related to the impact of urban environments on citizens' mental well-being. This issue has emerged as a critical theme in urban health debates as health, or better, illness is a significant burden on the public budget, and health issues are increasing in the context of climate changes in urban areas.

Therefore, we must continue developing and implementing strategic and specific actions to tackle health issues caused by urban lifestyles. There is urgent to systematise the advantages of being exposed to natural elements such as rivers in cities. We will have to place at the centre of a discussion the role of riverside territories and how they can mitigate the impacts of other collateral damage (such as the increase in heat waves in urban environments), stress, depression, and anxiety, or how they can improve happiness, well-being, and quality of life.

Acknowledgements The author gratefully acknowledges the possibility of developing this investigation within the scope of the eMOTIONAL CITIES Project, to Paulo Morgado—Centre of Geographical Studies (GEG), Institute of Geography and Spatial Planning (IGOT), University of Lisbon—for his support, sharing of ideas and scientific knowledge. The eMOTIONAL CITIES Project received funding from the European Union's Horizon 2020 research and innovation program under subsidy contract no. 945307. More information at https://emotionalcities-h2020.eu/.

References

Adli M. (2011) Urban stress and mental health. Cities, health and well-being. Urban age—a worldwide investigation into the future of cities. LSE Cities:1–3.

Adli M, Berger M, Brakemeier EL, Engel L, Fingerhut J, Gomez-Carrillo A, Stollmann J et al (2017) Neurourbanism: towards a new discipline. The Lancet Psychiatry 4(3):183–185. https://doi.org/10.1016/S2215-0366(16)30371-6

Adli M, Berger M, Brakemeier EL, Engel L, Fingerhut J, Hehl R, Stollmann J et al. (2016) Neurourbanistik–ein methodischer Schulterschluss zwischen Stadtplanung und Neurowissenschaften. [Neurourbanism - a methodical alliance between urban planning and neurosciences] Die Psychiatrie 13(02):70–78

Alcock I, White MP, Lovell R, Higgins SL, Osborne NJ, Husk K, Wheeler BW (2015) What accounts for Englands green and pleasant land'? A panel data analysis of mental health and land cover types in rural England. Landsc Urban Plan 142:38–46. https://doi.org/10.1016/j.landurbplan.2015.05.008

Amoly E, Dadvand P, Forns J, López-Vicente M, Basagaña X, Julvez J, Alvarez- Pedrerol M, Nieuwenhuijsen MJ, Sunyer J (2014) Green and blue spaces and behavioral development in Barcelona schoolchildren: the BREATHE project. Environ Health Perspect 122(December (2)):1351–1358. https://doi.org/10.1289/ehp.1408215

Ancora LA, Blanco-Mora DA, Alves I, Bonifácio A, Morgado P, Miranda B (2022) Cities and neuroscience research: a systematic literature review. Frontiers in Psychiatry, 13. https://doi.org/10.3389/fpsyt.2022.983352

Awosogba T, Betancourt JR, Conyers FG, Estapé ES, Francois F, Gard SJ, Yeung H et al (2013) Prioritizing health disparities in medical education to improve care. Ann N Y Acad Sci 1287(1):17–30. https://doi.org/10.1111/nyas.12117

Bonifácio A, Morgado P, Peponi A, Ancora L, Blanco-Mora DA, Conceição M, Miranda B (2023) Musings on neurourbanism, public space and urban health. Finisterra 58(122). https://doi.org/10.18055/Finis29886

Borja J, Muxi Z (2003) El Espacio Público: Ciudad y Ciudadanía [Public Space: City and Citizenship]. Ed. Electa, Barcelona. ISBN: 84-8156-343-9 84-7794-904-2

Bosselmann P (2009) Cities shaped by water, common ground for a metropolitan landscape. In Saraiva MG (Coord) (2010). Cidades e Rios: Perspectivas para uma relação sustentável [Cities and Rivers: Perspectives for a sustainable relationship]. Parque Expo, Lisboa. ISBN: 978-972-8106-51-5

Brereton F, Clinch JP, Ferreira S (2008) Happiness, geography and the environment. Ecol Econ 65:386–396. https://doi.org/10.1016/j.ecolecon.2007.07.008

Burton I (1990) Factors in urban stress. J Sociol Soc Welfare 17(1), Article 5. https://doi.org/10.15453/0191-5096.1928

Buttazzoni A, Doherty S, Minaker L (2022) How do urban environments affect young people's mental health? A novel conceptual framework to bridge public health, planning, and neurourbanism. Public Health Rep 137(1):48–61. https://doi.org/10.1177/0033354920982088

Carrière JP, Demazière C (2002) Urban planning and flagship development projects: Lessons from EXPO 98 Lisbon. Plan Pract Res 17(1):69–79. https://doi.org/10.1080/02697450220125096

Commission of the European Communities (CEC) (1990) Green Paper on the urban environment: communication from the Commission to the Council and Parliament. COM(90) 218 final. Brussels

Correia FN, Cruz JB, Martins RB, Liberato P, Morbey L (2000) POLIS, Programa de Requalificação Urbana e Valorização Ambiental de Cidades [POLIS—Programme for Urban Requalification and Environmental Valorisation of Cities]. Ministério do Ambiente e do Ordenamento do Território (MAOT). Lisboa

Evans RG, Stoddart GL (1990) Producing health, consuming health care. Soc Sci Med 31(12):1347–1363. https://doi.org/10.1016/0277-9536(90)90074-3

Fave AD, Brdar I, Wissing MP, Araujo U, Castro Solano A, Freire T, Soosai-Nathan L et al (2016) Lay definitions of happiness across nations: the primacy of inner harmony and relational connectedness. Front Psychol 7:30. https://doi.org/10.3389/fpsyg.2016.00030

Felce D, Perry J (1995) Quality of life: its definition and measurement. Res Dev Disabil 16(1):51–74. https://doi.org/10.1016/0891-4222(94)00028-8

Fitzgerald D, Rose N, Singh I (2016) Living well in the neuropolis. Sociol Rev 64(1_suppl):221–237. https://doi.org/10.1111/2059-7932.12022

Foley R, Kistemann T (2015) Blue space geographies: enabling health in place. Health Place 35:157–165. https://doi.org/10.1016/j.healthplace.2015.07.003

Franco P, da Costa EM (2021) Atividade física no quotidiano familiar das periferias.: Uma visão a partir de Rio de Mouro–Sintra [Physical activity in the family dailylife of the peripheries. A view from Rio de Mouro—Sintra]. Finisterra, 56(116):183–203. https://doi.org/10.18055/Finis20067

Gascon M, Zijlema W, Vert C, White MP, Nieuwenhuijsen MJ (2017) Outdoor blue spaces, human health and well-being: a systematic review of quantitative studies. Int J Hyg Environ Health 220(8):1207–1221. https://doi.org/10.1016/j.ijheh.2017.08.004

GoInvo (2020) Determinants of Health. USA. https://www.goinvo.com/vision/determinants-of-health/#references

Grellier J, White MP, Albin M, Bell S, Elliott LR, Gascón M, Gualdi S, Mancini L, Nieuwenhuijsen MJ, Sarigiannis DA, van den Bosch M, Wolf T, Wuijts S, Fleming LE (2017) BlueHealth: a study programme protocol for mapping and quantifying the potential benefits to public health and well-being from Europe's blue spaces. BMJ Open 7:e016188. https://doi.org/10.1136/bmjopen-2017-016188

Haeffner M, Jackson-Smith D, Buchert M, Risley J (2017) Accessing blue spaces: social and geographic factors structuring familiarity with, use of, and appreciation of urban waterways. Landsc Urban Plan 167:136–146. https://doi.org/10.1016/j.landurbplan.2017.06.008

Huynh Q, Craig W, Janssen I, Pickett W (2013) Exposure to public natural space as a protective factor for emotional well-being among young people in Canada. BMC Public Health 13:407. https://doi.org/10.1186/1471-2458-13-407

LeiriaPolis, Sociedade para o Desenvolvimento do Programa Polis em Leiria (2007) Intervenção do Programa Polis em Leiria [Intervention of the Polis Programme in Leiria]. Leiria

Louro A, Marques da Costa N, Marques da Costa E (2021b) From livable communities to livable metropolis: challenges for urban mobility in Lisbon metropolitan area (Portugal). Int J Environ Res Public Health 18(7):3525. https://doi.org/10.3390/ijerph18073525

Louro A, Franco P, Marques da Costa E (2021a) Determinants of physical activity practices in metropolitan context: the case of Lisbon Metropolitan Area, Portugal. Sustainability, 13:10104. https://doi.org/10.3390/su131810104

MacKerron G, Mourato S (2013) Happiness is greater in natural environments. Glob Environ Chang 23:992–1000. https://doi.org/10.1016/j.gloenvcha.2013.03.010

Ministério do Equipamento, do Planeamento e da Administração do Território (MEPAT) (1998) Plano Nacional de Desenvolvimento Económico e Social 2000–2006 [National Plan for Economic and Social Development 2000–2006]. Secretaria de Estado do Desenvolvimento Regional. Lisboa

Montgomery C (2013) Happy city: transforming our lives through urban design. Penguin UK. ISBN: 9780141047546

Nutsford D, Pearson AL, Kingham S, Reitsma F (2016) Residential exposure to visible blue space (but not green space) associated with lower psychological distress in a capital city. Health Place 39:70–78. https://doi.org/10.1016/j.healthplace.2016.03.002

Peen, J., Schoevers, R. A., Beekman, A. T., & Dekker, J. (2010). The current status of urban-rural differences in psychiatric disorders. Acta Psychiatrica Scandinavica, 121(2) (pp. 84–93). https://doi.org/10.1111/j.1600-0447.2009.01438.x

Portas N, Domingues A, Cabral J (2003) Políticas Urbanas. Tendências, estratégias e oportunidades. Fundação Calouste Gulbenkian. Lisboa. ISBN: 972-31-1061-X

Pykett J, Osborne T, Resch B (2020) From urban stress to neurourbanism: how should we research city well-being? Ann Am Assoc Geogr 110(6):1936–1951. https://doi.org/10.1080/24694452.2020.1736982

Queirós M, Vale M (2005) Ambiente urbano e intervenção pública: O Programa POLIS [Urban environment and public intervention: The POLIS Programme]. Actas do X Colóquio Ibérico de Geografia. Lisboa

Remesar A (2020) From drinking fountains to promenades. Water as artistic medium? On the waterfront. The International on-line Magazine on Waterfronts, Public Art, Urban Design and Civil Particiapation 62(1):3–84. ISSN: 1139–7365

Roe J, McCay L (2021) Restorative Cities: urban design for mental health and wellbeing. Bloomsbury Publishing. ISBN: HB 978-1-3501-1287-2

Rogerson M, Brown DK, Sandercock G, Wooller J-J, Barton J (2016) A comparison of four typical green exercise environments and prediction of psychological health outcomes. Perspect Public Health 136:171–180. https://doi.org/10.1177/1757913915589845

Santasusagna Riu A (2021) El bienestar en el ultimo siglo: un gran salto adelante. Atlas de los Países en Busca de la Felicidad. [Well-being in the last century: a great leap forward. Atlas of the Countries in Search of Happiness]. Larousse, 32–33. ISBN: 978-84-18473-78-4

Silva JB, Pinto PJ (2010) As cidades fluviais de Portugal Continental, Métricas, Tipologias e alguns dilemas segundo a leitura RiProCity [The river cities of mainland Portugal, Metrics, Typologies and some dilemmas according to RiProCity reading]. In Saraiva MG (Coord). (2010). Cidades e Rios: Perspectivas para uma relação sustentável [Cities and Rivers: Perspectives for a sustainable relationship]. Parque Expo, Lisboa. ISBN 978-972-8106-51-5

Smyth H (1994) Marketing the city: The role of flagship developments in urban regeneration. Taylor and Francis. ISBN 0-203-97595-2

Triguero-Mas M, Dadvand P, Cirach M, Martínez D, Medina A, Mompart A, Basagaña X, Gražulevičienė R, Nieuwenhuijsen MJ (2015) Natural outdoor environments and mental and physical health: relationships and mechanisms. Environ Int 77:35–41. https://doi.org/10.1016/j.envint.2015.01.012

United Nations Human Settlements Programme (UN-Habitat) (2022) Tomorrow Today Together, Delivering the New Urban Agenda. Nairobi. https://unhabitat.org/sites/default/files/2022/04/nua_tomorrow_today_together_digital_a.pdf

Weeks J (2017) Integrative Scientist Pizzorno: "Toxicity is the Primary Driver of Disease" in HuffPost. https://www.huffpost.com/entry/integrative-scientist-piz_b_11353632

World Health Organization (WHO) (2018) Fact sheets on sustainable development goals: health targets. Mental Health. World Health Organization Regional Office for Europe, Copenhagen

World Health Organization (WHO) (2021) Green and blue spaces and mental health: new evidence and perspectives for action. WHO Regional Office for Europe. Copenhagen. ISBN: 978-92-890-5566-6

World Health Organization (WHO) (1946) International Health Conference, New York, N. Y., June 19-July 22. Report of the United States Delegation, including the Final Act and Related Documents. Washington

World Health Organization (WHO) (2010) Why urban health matters (No. WHO/WKC/WHD/2010.1). World Health Organization

Zêzere JL (2019) Riscos e Desastres em tempo de Alterações Climáticas. Transversalidades: Fotografia Sem Fronteiras 2019 [Risks and Disasters in times of Climate Change. Transversalities: Photography Without Borders 2019], pp 106–110. Centro de Estudos Ibéricos. ISBN 978-989-8676-21-4

Chapter 13
Three Rivers and Different Approaches of Urban Riverscapes in Zaragoza City: Hydromorphology, Memory, Perception, and Planning

Alfredo Ollero, Laura Albero, Pedro Boné, Jaime Díaz-Morlán, Valeria N. Pirchi, and Eberval Marchioro

Abstract A comparative analysis is developed on the characteristics, perception, and management of the Ebro, Gállego, and Huerva Rivers in the urban and peri-urban space of Zaragoza. These three rivers present different hydrogeomorphological styles, are appreciated in very different ways by the population, and have been the subject of different planning strategies over time, until the current situation in which they compete for European funds for their restoration. Zaragoza is a fluvial city, and its local analysis can be extrapolated to other cases of urban-fluvial interaction. Hydrogeomorphological parameters are worked on, the history of floods and the evolution of management measures are updated, and a public perception survey has been carried out, as well as interviews with experts and municipal agents. All this leads to establishing a forecast of urban-fluvial behaviour with proposals in the context of environmental change in the short term.

A. Ollero (✉) · L. Albero · V. N. Pirchi
Department of Geography and Regional Planning (Climate, Water, Global Change, and Natural Systems Research Group), University of Zaragoza, Zaragoza, Spain
e-mail: aollero@unizar.es

V. N. Pirchi
e-mail: vpirchi@unizar.es

P. Boné
TYPSA Aragón, Zaragoza, Spain
e-mail: pbone@typsa.es

J. Díaz-Morlán
Department of Urban and Territorial Planning, Engineering and Architecture School, University of Zaragoza. Atalaya Territorio, S.L, Zaragoza, Spain
e-mail: atalaya@atalayaterritorio.com

E. Marchioro
Department of Geography, Federal University of Espírito Santo (UFES), Vitória, Brasil

© The Author(s), under exclusive license to Springer Nature Switzerland AG 2024
J. Farguell Pérez and A. Santasusagna Riu (eds.), *Urban and Metropolitan Rivers*, The Urban Book Series, https://doi.org/10.1007/978-3-031-62641-8_13

Keywords Hydrogeomorphology · Environmental change · Urban planning · Risk management · River restoration

13.1 Introduction and Study Area

In the NE of the Iberian Peninsula, the city of Zaragoza is the capital of Aragón and it has a population of about 700,000 inhabitants. In an area of confluence between basque, iberian, and celtiberian groups, and built from an iberian city called *Salduie*, it was founded as a roman colony 2037 years ago by Emperor Augustus, thanks to its excellent location at the confluence of the Ebro, Gállego, and Huerva Rivers, and promoted as an administrative and logistical space, qualities that it has maintained through history (Pueyo et al. 2017).

The city was born and grew alongside the three rivers, of different dimensions and characteristics, which have configured three very different fluvial landscapes in the urban space: the Ebro River, big and integrated, the Gállego River, freer and peri-urban, and the Huerva River, the smallest and most punished by human interventions. They are also three fluvial landscapes differently perceived and appreciated by the population of Zaragoza and subjected to different models of public management. The fluvial urban panorama is completed with an artificial course, the Aragon Imperial Canal, which goes through the city but is not considered in the present study. There are also some ephemeral streams blurred in the city and its surroundings, which disappear as channels.

Zaragoza is located in the centre of the Ebro basin (Fig. 13.1), in a depression enclosed by mountain ranges, with a continental Mediterranean climate, an average annual rainfall of 320 mm and an average temperature of 14.6°C. This produces a strong evapotranspiration and water deficit, especially in summer. The rivers of the city provide flows from the upper Atlantic basin and the western Pyrenees (Ebro River), from the Pyrenean peaks (Gállego River) and the Iberian range (Huerva River), creating a privileged humid space (fluvial oasis) in contrast to the steppe and semi-desert landscapes of its surroundings.

The old town of Zaragoza was built on one of the Pleistocene terraces of the right bank of the Ebro River, formed by thick and strongly cemented gravel (Longares and Peña-Monné 2013), but its current urban area extends along both banks of the Ebro River, in confluence with the Huerva and Gállego Rivers. Floods of the Ebro River have constantly affected the city since antiquity. River channels are geographical spaces associated to necessities, trajectories, tendencies, evolutions, technologies, and cultural dynamics of the social groups. The interactions between physical, anthropic, and immaterial elements produce transformations that condition the future of rivers (Pueyo et al. 2017). The cultural and ideological changes and the new sensibilities add a complexity and mutability that should help refine the tools and indicators to improve river channels, being useful for new positive scenarios.

In this context, the present study analyses the role, vision, present, and future of the rivers of Zaragoza in the urban framework as a socio-environmental and management

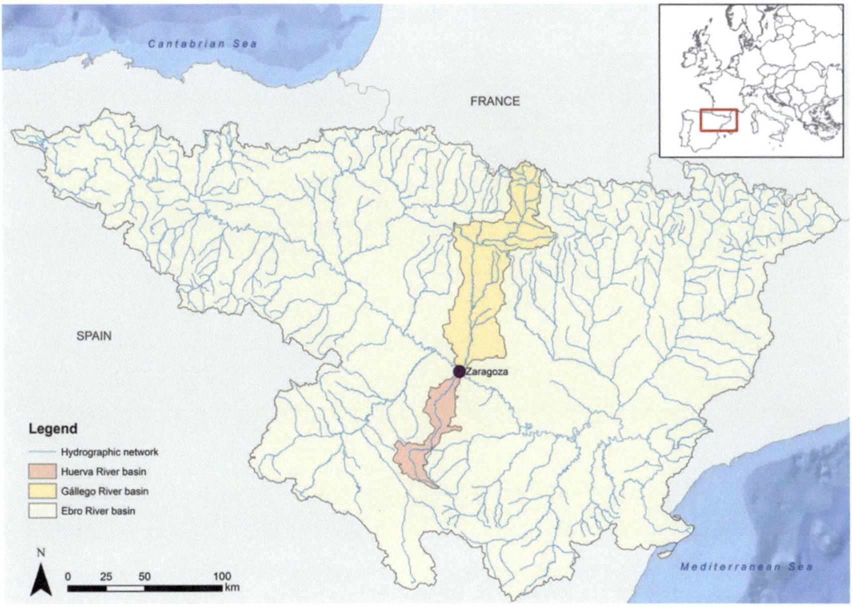

Fig. 13.1 Ebro, Huerva, and Gállego River basins in NE Iberian Peninsula. Prepared by the authors, based on information from the Ebro Basin Authority

diagnostic. The urban area has been defined as the included inside the ring road Z-40 (Fig. 13.2), so the studied segments of the three rivers begin at their respective bridges of the above-mentioned infrastructure, concluding in the fluvial confluences.

13.1.1 The Middle Ebro River

Zaragoza is located in the middle stretch of the Ebro and in its free meandering reach, which extends from Logroño to La Zaida over 345 km (Ollero 1992). Its extensive floodplain has a maximum width of 6 km downstream of Zaragoza. Floods are very frequent and have generated continuous changes in the channel through history, starring a unique fluvial dynamic, of great value on a peninsular scale. However, the complex fluvial dynamics of the Middle Ebro River were considerably conditioned in recent decades by the basin global change, especially by human actions such as regulation of flows or defence against floods (Ollero et al. 2021a). Different researchers (Ollero 1992; Magdaleno 2011; Ollero et al. 2015; Díaz Redondo et al. 2018) concluded that throughout the second half of the twentieth century, there was a significant transformation of the fluvial functioning due to anthropic factors. It can be stated that the "Anthropocene" manifests itself in the Middle Ebro River through a great acceleration of human intervention between 1950 and 1990. Hydrological

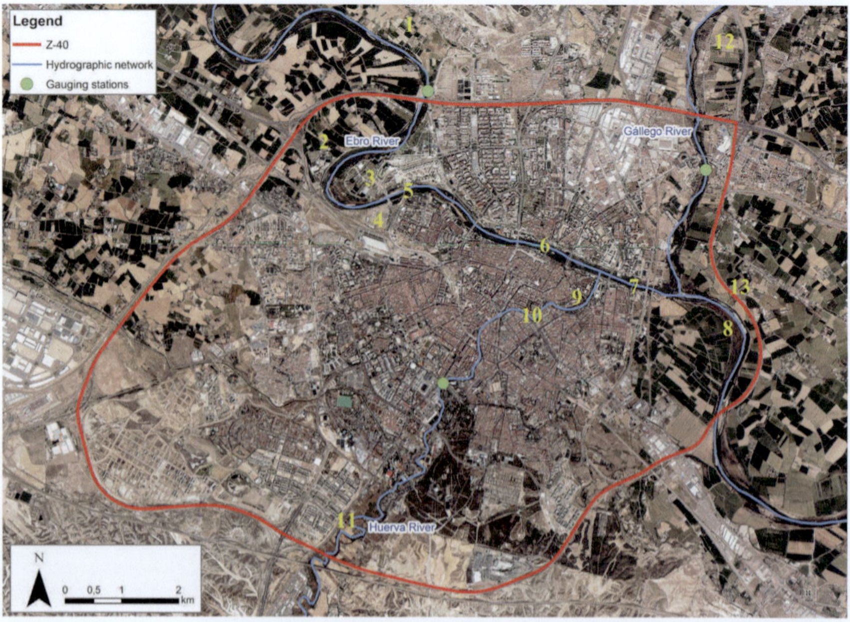

Fig. 13.2 Zaragoza city and the urban reaches of Ebro, Huerva, and Gállego Rivers. Bases: current PNOA orthophoto and GeoPortal SITEbro. Prepared by the authors. Location of some sites indicated in the text: 1. Juslibol, 2. Almozara, 3. Expo 2008, 4. Parking South, 5. Pavilion-Bridge, 6. Piedra Bridge, 7. Vadorrey Weir, 8. Cantalobos Forest, 9. Bruil Park, 10. Constitución Blvd., 11. Junquera Fountain, 12. Montañana, 13. Movera

data, continuous since 1950, and aerial images allow us to clearly verify this process (Ollero 2010). The basin area of the Ebro River at Zaragoza is 40,434 km^2 and the average discharge is 230 m^3s^{-1} (Ollero et al. 2021a), which implies a specific discharge of 5.7 $l\ s^{-1}km^{-2}$.

The Ebro River in Zaragoza locally reduces the width of its floodplain to 2.9 km, and increases the slope of the channel, from a value of 0.00092 mm^{-1} in the old town, compared to 0.00045 mm^{-1} in the river reach upstream. This fact implies a faster and more effective flood evacuation in the urban area (Ollero 1992).

13.1.2 *The Lower Huerva River*

The Huerva River has a drainage basin of 1020 km^2, narrow and elongated, with its headwaters in the Iberian range. It has a very low discharge that is compensated in the city by the drainage of 10 hm^3yr^{-1} from the Imperial Canal, so that the urban Huerva River mainly carries water from the Ebro River, reaching an average discharge of 3.03 m^3s^{-1}, which implies a specific discharge of 2.97 $l\ s^{-1}\ km^{-2}$. Its floods are infrequent, in part due to their regulation (Las Torcas and Mezalocha reservoirs), and generally occur in spring (Mateo and Sánchez-Navarro 2011).

The channel is meandering, with some pronounced curves in the upstream segment of the city (sinuosity rate of 1.68, with a slope of 0.00426 mm^{-1}). In urban area, between the Z-40 bridge and the beginning of the underground central section, slope slightly increases (0.00493 mm^{-1}) and sinuosity remains high (1.49). Underground section is 1.2 km long; from its beginning to its mouth in the Ebro, sinuosity is reduced to 1.23 due to urban channelling, and slope rises to 0.006 mm^{-1}.

13.1.3 The Lower Gállego River

Gállego River drains a basin of 4020 km^2 and reaches 200 km in length. It offers a great variety of fluvial landscapes and constitutes a corridor of great environmental value that traverses various morphostructural units from the Pyrenean axial zone to the centre of the Ebro Depression. Lower Gállego River valley opens in gypsum and salt Miocene formations and develops terraces embedded in deformations due to dissolution of underlying substrate (Peña et al. 2011, 2020), which originates alluvial thickening and superimposition of terrace levels of different chronology (Benito et al. 1996, 2010). The evaporitic substrate barely emerges, despite the current incision processes (Marcos 1991; Marqués 2018).

In this lower valley, with an average slope of 0.0034 mm^{-1} and a sinuosity rate of 1.23, the channel was wide braided until 40 years ago, but it is currently single and sinuous. The river easily creates and cuts meanders, with a very remarkable dynamic in each flood despite the regulation. The incision is currently very significant between the Urdán weir and Montañana, reaching values of up to 7 m since 1960 (Ferrer-Boix 2010; Martín Vide et al. 2010), caused by old gravel extractions and the weir effect. Average discharge is only 12.5 m^3s^{-1}, which implies a specific discharge of 3.1 $l\,s^{-1}\,km^{-2}$, after diverting a large part of its water to Monegros Canal and other irrigation systems (Ollero et al. 2004).

13.2 Material and Methods

General measurements of river channels have been made over the most recent orthoimage (2021 PNOA) and have been processed and mapped with ArcMap (ArcGIS 10.2). Field measurements were made in previous projects (Ollero, dir. 2005; 2018; Martín-Vide, dir. 2018a, 2018b). The sediment mobilization analysis using Helley-Smith method was carried out in the Gállego and Ebro Rivers, but not yet in the Huerva, so in this case, it has been estimated based on the particle size, discharge, and slope values. For hydromorphological assessment, the IHG index has been applied (Ollero et al. 2021b).

Discharge data have been obtained from the gauging stations managed by the Ebro Basin Authority situated in Zaragoza: 9311/A011 station in Ebro River, 9089/A089 in Gállego River and 9216/A216 in Huerva River. For Gállego and Huerva Rivers, series are short, so it is necessary to resort to other stations upstream: Mezalocha (9105/A105) in Huerva R. and Ardisa (9012/A012) and Zuera (9209/A209) in Gállego R.

The study of perception, diagnosis, planning, and proposals has been based on the experience of the authors, as well as on online surveys on the population and on interviews with 15 experts in environmental and fluvial urban management.

The open surveys on the population have been carried out since January 2023 and only the first 100 responses received have been used for work. These responses show an appropriate age, gender, and education distribution. Amongst other aspects, it was requested to attribute two adjectives to each river and to rank the three rivers by their beauty and scenic value in the city, their integration in urban framework, their naturalness or degradation, their accessibility and possibilities of citizen use, and by their security against floods. It was also requested to point out the problems of each river, to mention floods and events remembered, to give their opinion on recent actions, and to propose improvement measures for each river.

In the interviews with the 15 experts, they were asked about the main problems of Ebro, Gállego, and Huerva Rivers in the urban area. They were asked to give their opinion on actions and plans developed in each river in the last three decades, and they were encouraged to suggest measures or proposals for each river.

13.3 Results

13.3.1 Hydromorphology and Flood Memory

In Table 13.1, clear differences are seen in the hydromorphological dimensions between the three rivers. The similarity amongst rivers of the specific power is very relevant, since the slope compensates discharge. This similar power explains how the average sediment size is also similar in the three rivers. However, this gravel size material is moved by very different flows, also in response to the different slopes (Table 13.2).

Floods of all three rivers never overlap since they have different hydromorphological sources and their origin is quite distant. Therefore, there are no synergetic processes. Ebro River in flood penetrates the final segments of both tributaries a few hundred metres.

In the city, there are no observable limn marks that help remember flood episodes. From 100 people surveyed, 85 remembered some flood of Ebro River, although they generally did not guess the moment the events took place. However, only two people remembered some flood of Gállego River, and another two people remembered some flood of Huerva River. So, the memory of events is, generally, poor and almost exclusively focused on Ebro River.

Table 13.1 Dimensions and main hydromorphological characters in the three rivers

River	Channel length in the city (km)	Low-water width/ bankfull width (m)	Slope (m m^{-1})	Bankfull discharge (m^3 s^{-1})	Specific stream power bankfull (W m^{-2})	Sediment size average (mm)	Discharge moving gravel (m^3 s^{-1})
Ebro	10.03	~80/150	0.0007	~1600	73.17	37.8	~450
Huerva	10.12	~6/16	0.0056	~20	68.60	43.2	~15
Gállego	3.29	~20/90	0.0031	~200	67.51	39.5	~60

Table 13.2 Most relevant floods in the three rivers

River	Main floods date	Discharge m^3 s^{-1} (gauging station)	River	Main floods date	Discharge m^3 s^{-1} (gauging station)
Ebro	05/02/1952	3260 (Zaragoza)	Huerva	15/05/1971	56 (Mezalocha)
	29/05/1956	2744 (Zaragoza)		11/06/1975	70 (Mezalocha)
	16/12/1959	2790 (Zaragoza)		02/06/1977	80 (Mezalocha)
	02/01/1961	4130 (Zaragoza)		08/05/2003	140 (Zaragoza)
	16/11/1961	2570 (Zaragoza)		31/01/2020	34 (Zaragoza)
	12/11/1966	3154 (Zaragoza)	Gállego	24/08/1942	1300 (Ardisa)
	06/01/1968	2494 (Zaragoza)		01/06/1979	1314 (Ardisa)
	19/01/1981	2939 (Zaragoza)		18/12/1997	735 (Ardisa)
	05/02/1978	3154 (Zaragoza)		24/11/2003	626 (Zaragoza)
	16/01/1979	2581 (Zaragoza)		03/04/2007	384 (Zuera)
	19/01/1981	2939 (Zaragoza)		21/10/2012	639 (Zuera)
	08/02/2003	2832 (Zaragoza)		24/11/2016	395 (Zuera)
	02/03/2015	2448 (Zaragoza)		11/04/2018	400 (Zuera)

With the application of the IHG index (Fig. 13.3), a negative evolution of hydromorphological indicators is observed in the three rivers in the second half of the twentieth century, whilst in the twenty-first century, only Gállego River has achieved some improvement.

	1956	2000	2023
Ebro River	moderate	poor *regulation, channelization*	poor *weir, dredging*
Huerva River	poor *channelization*	bad *contraction*	bad
Gállego River	good	poor *regulation, extractions*	moderate

Fig. 13.3 Simple diagram of the evolution of hydromorphological state (IHG index) in the three rivers and critical impacts. Prepared by the authors

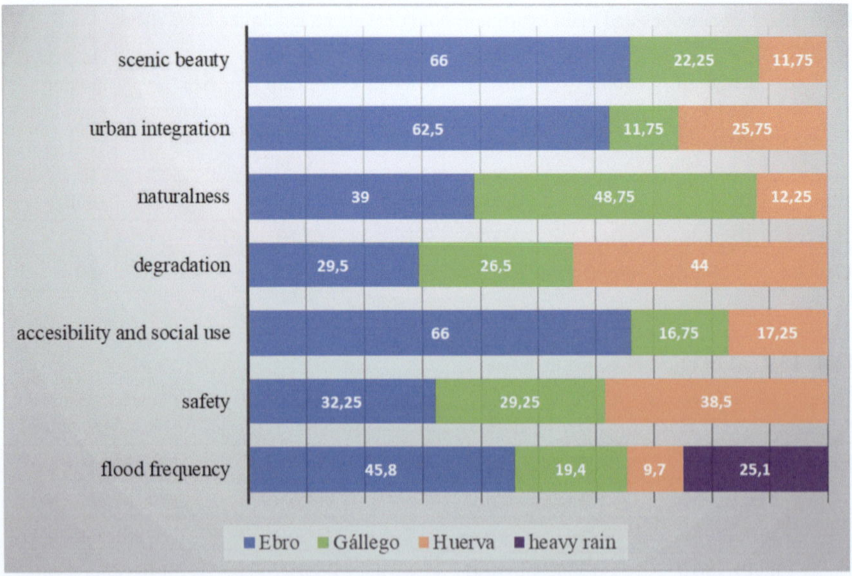

Fig. 13.4 Social perception (% responses) from the conducted surveys. Prepared by the authors

13.3.2 Perception, Diagnosis, and Planning

According to the social perception of the citizens of Zaragoza (Fig. 13.4), Ebro River is the best integrated in the city, Gállego River is the most natural, and Huerva River is the most damaged. This perception is generally consistent with the diagnosis of the experts and with the urban planning in relation to its rivers as it will be detailed in the following subsections.

Floods are mostly associated with Ebro River, mainly because of its great mediatic repercussions as it affects all the Middle Ebro River. Urban episodes caused by intense rain in situ are considered in second place (Fig. 13.4).

(a) *Ebro River, the Most Integrated*

Citizen perception highlights the greatness of Ebro River, its scenic beauty, its integration in the city, and its accessibility and social use. Between the characterizing adjectives of Ebro River, many people agree, so few words appear in the diagram (Fig. 13.5). The most frequent adjectives are mighty, big, long, broad, and important, highlighting the dimensions of the river and its main role in the city and region. The riverside plan of the International Exposition 2008 implied a very notable approach of citizens to the river, which has been maintained over time (Pellicer 2015; Pellicer and Sopena 2019). That action had a hybrid model: it was neither total urbanization nor restoration, and its main management is as a park. But there has been a notable increase in biodiversity in some banks of Ebro River, in line with what happened in

Fig. 13.5 Word cloud about the Ebro River from the first 100 surveys. Prepared by the authors. Software: https://www.jasondavies.com/wordcloud/

other Iberian cities (Molina et al. 2020), highlighting the abundance of waterfowls, wader birds, and insectivores.

The current problems of the Ebro, according to the surveyed people opinion and the interviewed experts, are the following: (i) the persistence of an interventionist mentality and management in the urban reach, which considers the river an annoying entity, to which is added as determining factor the architectural-urban character of great part of its banks; (ii) a citizen perception that does not value the natural state of the river, and considers the sediments and vegetation in bars and islands as dirty and dangerous; (iii) scarcity of space to release energy in the frequent floods, due to river and phreatic constraining (floodable left bank urbanized); (iv) reminiscences of negative aspects of Expo 2008, such as the elimination of the orchards, the dredging for navigation, or the fixing of banks; (v) citizen complaints about the general maintenance of promenades and urban furniture along the entire riverbank; (vi) deficient operation of the water treatment plants, with a system of collectors that causes direct discharges into the channel at high flows, proliferating solid elements not retained in the grids; (vii) overflow of combined sewer overflows (CSO) in rainy episodes, with an undersized sewerage network and alleviation points affected by floods; (viii) proliferation of macrophytes, algae, the invasive *Azolla filiculoides* (Fig. 13.8a), the big fish *Silurus glanis*, and *Simulium* sp., *Aedes albopictus,* and other insects with effects on health, associated with the high load of nutrients and nitrates from agricultural areas; (ix) poor condition of the surface water mass and groundwater, which accumulates heavy metals; (x) several inadequate or poorly located infrastructures in the channel: Vadorrey Weir, Pavilion-Bridge, and unused piers, as well as clear DPH (see Table 13.4) occupations with flagrant cases of exposure such as a restaurant-leisure centre (Fig. 13.8d). The problem with the Pavilion-Bridge

(Fig. 13.8c), designed by Zaha Hadid, is very remarkable because of its location and characteristics, with a big base in channel which has altered water and sediment flows, causing the closure of the left arm and a notable incision and rising erosion on the right riverbank (Martín-Vide et al. 2022). The Vadorrey Weir (Fig. 13.8b), aiming to achieve as stable watersheet in the city (Gutiérrez-Serret 2002), has been the work with the greatest environmental impact in urban Ebro River, it has been much discussed since its beginning and with different management avatars, it meant an evitable expense for an inappropriate navigation and a permanent fluvial damage (Ollero and Briz 2018).

(b) *Huerva River, the Most Punished*

Citizens and experts coincide in a mainly negative perception. The Huerva is a small river, confined, channelized, contaminated, and denatured, highlighting its underground reach in the city centre. Its confinement derives from urban development itself, and it extends upstream Zaragoza (in Cuarte and Cadrete municipalities), where the orchards have disappeared due to recent industrial and urban developments. This has led to a DPH kidnapping assimilated by the PGOU (see Table 13.4), which makes any future intervention difficult and limited. The drastic space scarcity and the absence of hydromorphological dynamics have made the incorporation of river in the city as a backbone natural element impossible, resulting in a little accessible river for the people. There are associated green areas, from Junquera Fountain to Bruil Park, but they are elevated over the river and unconnected. The channel is accompanied by strips of vegetation with autochthonous species and high presence of exotic ones, highlighting *Ailanthus altissima* in the urban reach and *Arundo donax* upstream. Social perception also highlights the presence of garbage and debris, with a general image of dirty river, neglected and forgotten, which include some reach with visible wastewater collectors. The experts add problems as the lack of sediment supply and transport, the phreatic nonexistence in the urban reach, the absence of connectivity, the great amount of CSO due to the sewer under sizing, and the historic pollution by heavy metals from valley industries. Flood risk is not perceived due to the confinement, and none of the 100 people surveyed remembered the important May 2003 flood.

Planning in Huerva River consisted in the implementation of hard defences for decades until the PDH of 2010–2015, which proposed a more modern urban vision, prolonged in the most natural reach upstream. The PDIVZ (2017) offered a more interesting and integrated environmentalist vision that has not been carried out (Fig. 13.6).

In 2023, Huerva River has obtained finance from the PRTR (NextGeneration EU). Actions include "fluvial and landscape restoration" favouring natural processes of the river and the conservation and development of biodiversity in the urban area, alongside citizen accessibility and security (Gallardo 2023). As it happened in the Expo 2008 with Ebro River, the work focuses on really integrating Huerva River in the city and bringing citizens closer to the fluvial space. To renature, ecosystem services and the concept of restoration have been included to achieve European finance. The management team has external technicians in fluvial hydromorphology

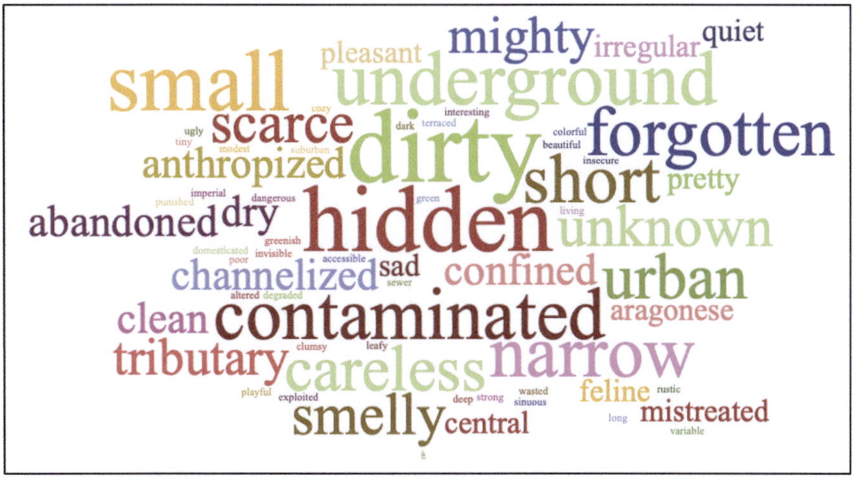

Fig. 13.6 Word cloud about the Huerva River from the first 100 surveys. Prepared by the authors. Software: https://www.jasondavies.com/wordcloud/

and in biodiversity, who will intervene in the control of the work (2024–2025) and the following monitoring, ensuring restoration goals.

Difficulties of acting in Huerva River are very notable due to its current high degradation that implies a negative ecological state. With hydromorphological restoration, promoted from the call, it is intended to improve geodiversity and biodiversity and longitudinal continuity as an ecological corridor, in tune with Ebro and Gállego Rivers, and for Zaragoza to increase its leisure places imbricated with nature (Gallardo 2023).

The area of action is divided in two reaches. The upstream one obtained European finance (1.6 M€ for a total of 9.9 M€) in the 2022 call of *Fundación Biodiversidad*. It includes a storm tank in the widest sector. The downstream one, from the end of the underground reach to the confluence with Ebro River, submitted to the 2023 call, has obtained €3.4 M of European funds, with a total budget of €13.9 M. Therefore, the underground section is not planned, although sediments retained in it can be used for mobilization downstream. More specifically, hydromorphological restoration measures planned include the proposal for a hydrological regime, both ordinary and extraordinary, which will be adjusted to the expansion of the projected channel section in some sectors, for which the removal of anthropic material fillers, old factory works, and defence works will be necessary. For the definition of the flow regime, sediment load and channel slope must also be considered, which will also be subject to restoration. Restoration of bed structure and substrate is projected to be improved through the injection of sediments, which will be naturally redistributed by morphological floods generated with discharge derived from the Imperial Canal, as well as with Huerva floods themselves. Some of the removed riprap blocks will be used to create a diverse longitudinal structure, featuring rapids and backwaters. These deflectors will also help the formation of bars in riverbed. The action also projects the construction of disturbance block ramps to improve longitudinal continuity for fish. Finally, a restoration of riparian forest is projected by eliminating exotic species and planting native. This is intended to improve its structure and longitudinal, vertical and transversal connectivity.

(c) *Gállego River, the Freest*

Gállego River is seen as a less urban river by the population surveyed. The experts highlight the problems of its basin and its upstream peri-urban section: (i) potential risk of contamination by lindane from Sabiñánigo dumps; (ii) problem of discharges from Montañana paper industry, which also has a horizontal bridge at risk; (iii) its excessive regulation and important derivations that reduce and denature its flow, circulating a low ecological flow most days of the year; (iv) loss of braided pattern and strong incision from Urdán weir; (v) lack of purification of Peñaflor rural neighbourhood; (vi) arrival of direct discharges (CSO) of rainwater from the "Transport city" and San Juan de Mozarrifar; (vii) progressive invasion of *Arundo donax*. In the most urban area, rectilinear channelization between bridges and drawing a curve downstream is highlighted, which conditions and simplifies the channel, as well as invasions of DPH, such as some housing states (Fig. 13.7).

Gállego River has been subject to recovery works by the City Council in the last four decades, improving quality of its banks and punctually channel dynamics. Most of the substrate is rubble from the city. First actions of revegetation failed due to incision and phreatic lowering, and it was necessary to apply irrigation to plantations. It also evolved with the species, achieving the final success with autochthonous ones. Riparian corridor achieved is remarkable and what it has been planted is already perfectly integrated with what it has colonized spontaneously. This renaturation has led to the expansion of beavers and otters. With the accompanying project for Expo

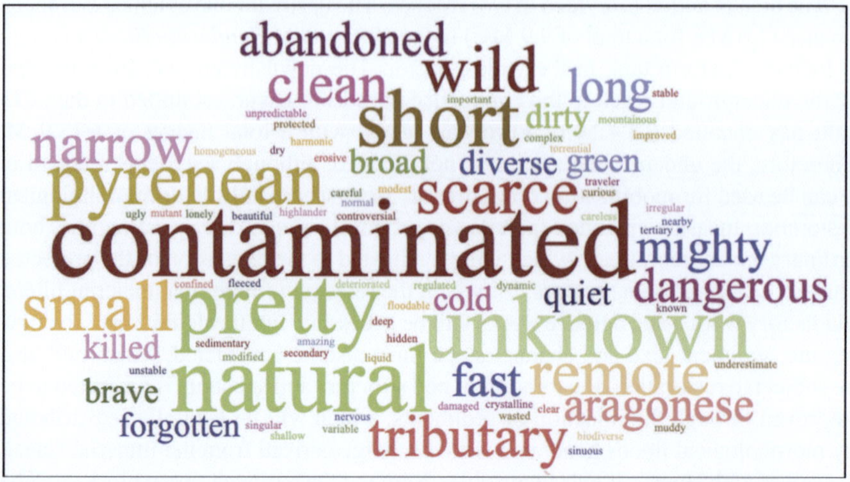

Fig. 13.7 Word cloud about the Gállego River from the first 100 surveys. Prepared by the authors. Software: https://www.jasondavies.com/wordcloud/

Fig. 13.8 Different images of problems in river-city interaction. **a** Invasion of *Azolla filiculoides*, Ebro River (May 2023); **b** Vadorrey pier and weir (with gates open, Google Earth, April 2019); **c** Pavilion-Bridge (Google Earth, April 2019); **d** urban leisure space, Ebro River (2015 flood); **e** rubble bank, Ebro River; **f** dredging (2010) for the passage of pleasure boats; **g** channelized final segment, Huerva River; **h** tree removal, Huerva River (photo: M Mérida); **I** Gállego-Ebro confluence; **j** erosive concave bank, lower Gállego River. Photos by the authors

2008, non-existent previous accessibility was achieved, so that the Gállego riverbanks are used by public leisure activities. Channel dynamics seek greater width in the final urban segment, initiating processes of lateral erosion in the last floods. However, the presence of a sewage collector and a gas pipeline parallel to the channel and the channelization with gabions condition and limit natural dynamics. The City Council attempted a cut-off to solve the erosion in a curve, resulting in failure (Ferrer-Boix et al. 2021).

The colony of *Riparia riparia* that settled on the left bank next to the mouth between 2018 and 2022 is very remarkable. It has disappeared in coincidence with some unfortunate actions in which a flow evacuation corridor has been opened in the last bar of sediment and the island that existed at the confluence with the Ebro has been dredged.

(d) *Problems and Proposals (Citizens' Opinions)*

In the citizen survey, questions were asked about the problems of three rivers and about possible measures to solve them (results in Table 13.3). The problem that most stands out for population in Ebro River is floods, in Gállego River the contamination by lindane (risk in basin headwaters), and in Huerva River the lack of "cleaning" and dredging in-channel. This mantra of the lack of "cleanliness" is repeated, occupying second place in Gállego River and fourth amongst the problems of Ebro River. Along the same lines, population calls for the "cleanliness" of channel and banks as first measure in the three rivers, a very popular measure that, however, no interviewed expert has indicated. The most environmental solutions, which have been advocated mainly by experts, appear in some cases amongst proposals of the citizens, but in the last places of the ranking.

13.4 Discussion, Perspectives, and Challenges

The "urban stream syndrome" (Walsh et al. 2005) is the consistently observed ecological degradation of streams draining urban land. Homogenization, loss of personality of urban rivers, and difficulties of their restoration have been highlighted by many other authors (e.g., Findlay and Taylor 2006; Gurnell et al. 2007; Francis 2012; Vietz et al. 2016; Santasusagna 2016; González-Rojas 2017; Horacio et al. 2018; Durán 2020; Pirchi 2020; Durán et al. 2021; Espinosa 2023) and are also found in Zaragoza, although this city has the advantage of having three very different rivers that can cover also different citizen objectives and ecosystem services, achieving social and environmental connectivity (Kondolf and Pinto 2017; Tort et al. 2020; Zingraff-Hamed 2018).

Urban fluvial planning (acronyms in Table 13.4) has not, and still does not, address the environmental changes associated with climate emergency or fluvial emergency (Ollero 2023). In this context, the forecast of future urban-fluvial behaviour is highly uncertain. The PGOU of 2007 has been sensitive to natural values of municipality, but it does not make it immune from future unsustainable urban developments. Expo 2008

Table 13.3 Ranking of responses of people who completed the survey on the problems and solutions of three urban river areas

Ebro river			
Problems	Num. resp.	Solutions, proposals	Num. resp.
1. Floods, flooding	34	1. To "clean" channel and banks	33
2. Dirt, pollution, waste	29	2. To expand recreational activities	9
3. Defences, narrowing, channelization	15	3. Dikes removal, to unprotect banks	6
4. Lack of dredging and veget. clearance	14	4. To eliminate invasive species	4
5. Invasive species	10	5. Dredging	4
6. Lack of flow, drought	7	6. To renature	4
7. Neglected, wasted river	6	7. Weir removal	3
8. Denatured, anthropized river	5	8. Flood management	2
9. The weir	4	9. Social participation	2
10. The river is a barrier to the city	4	10. More security	2
Huerva River			
Problems	Num. resp.	Solutions, proposals	Num. resp.
1. Lack of dredging and veget. clearance	23	1. To "clean" channel and banks	22
2. Dirt, pollution, waste	15	2. To improve accessibility	5
3. Neglected, wasted river	9	3. To eliminate invasive species	4
4. Underground and channelization	8	4. To decontaminate	4
5. Lack of flow, drought	8	5. To open underground segment	3
6. Low accessibility, disconnection	7	6. Integration in the city	2
7. Bad smells	5	7. Water sewage	2
8. Degraded riparian corridor	5	8. To renature	2
9. Insecurity	3	9. To expand recreational activities	2
10. Flow irregularity	2	10. Flood management	2
Gállego River			
Problems	Num. resp.	Solutions, proposals	Num. resp.
1. Pollution, lindane	29	1. To "clean" channel and banks	18
2. Lack of dredging and veget. clearance	12	2. To decontaminate	8
3. Low accessibility, disconnection	12	3. To improve accessibility	5

(continued)

Table 13.3 (continued)

Ebro river			
Problems	Num. resp.	Solutions, proposals	Num. resp.
4. Neglected, wasted river	5	4. To expand recreational activities	5
5. Little-known	4	5. To renature	3
6. Floods, flooding	4	6. Improve confluence Gállego-Ebro	3
7. Lack of flow, drought	4	7. Room for the river	2
8. Degraded riparian corridor	4	8. Dikes removal	2
9. Incision	3	9. Amount dams removal	2
10. Gravel extractions	1	10. Flood management	1

Table 13.4 Acronyms of the different management plans

Acronym	Spanish	English
DPH	Dominio Público Hidráulico	Public river space (Spanish water law)
PDH	Plan Director del Huerva	Huerva River director plan
PDIVZ	Plan Director de la Infraestructura Verde de Zaragoza	Green infrastructure director plan of Zaragoza
PGOU	Plan General de Ordenación Urbana	General urban management plan of Zaragoza
PGRI	Plan de Gestión del Riesgo de Inundación	Ebro basin flood risk management plan
PHE	Plan Hidrológico del Ebro	Ebro basin hydrological plan
PORN	Plan de Ordenación de los Recursos Naturales	Ebro River natural resources management plan
PRTR	Plan de Recuperación, Transformación y Resiliencia	Recovery and resilience facility NextGenerationEU

and its accompanying riverside plans were positive for Ebro River-city integration, but an opportunity was lost due to the lack of river restoration criteria. Today, urban rivers of Zaragoza continue to lack protection as rivers, they are considered "modified special reaches", which are not included in the PORN Sotos y Galachos del Ebro, nor in Ebro Resilience project, nor in the special programs of the PHE or the PGRI. The PDIVZ does not propose adopting the strategy of zoning a fluvial territory in urban section because it is considered excessively anthropized.

Neither is there an authentic consideration of rivers and banks as part of civic life. This is a disease that is widespread internationally in different contexts (Petts 2006). In Zaragoza, the deficit in environmental education of citizens is notable, with almost no proposals that take river ecosystems as axes. It is essential and urgent that local population appreciates rivers as they are and their diversity. There is also no scientific monitoring of fluvial processes in the city, nor a good connection between science and citizenship that could be supported by the application of urban fluvial health ratings, such as those by Gurnell et al. (2016), Murphy (2020), or Ranta (2021). People surveyed are unaware of the plans and actions in the rivers of Zaragoza. They do not know of any plan or project for the Gállego River. For Huerva River, some people, very few, mention the urban burial as an old negative project, the PDH, and the current rehabilitation project, which is being reported by the local press. For Ebro River, Expo 2008 is mentioned, with positive and negative opinions, dredging, with more positive opinions, embankments, with divided opinions, and the Ebro Resilience project with a positive vision.

In Zaragoza city, no river restoration projects have yet been addressed. The last possibility has focused on Huerva River by political decision, despite its few possibilities. But the great potential as an object of restoration is held by Gállego River, less urbanized, with great hydromorphological and environmental interest, which can still be the representative of fluvial freedom, as well as a laboratory and fluvial learning space (Cuello 2022). Political decisions have had, and still have, a lot of weight and do not meet river needs, but are based on capturing votes. The last municipal political change (2019) implied the distancing of the PDIVZ, which contributed ideas and environmental proposals of maximum interest for the three rivers, including recovery of hydrogeomorphological processes.

The experts also highlight the need to improve city's sanitation system, expanding the separative networks, installing storm tanks, lamination zones, and controlled alleviation points that minimize the impacts of said discharges in the rivers.

A paradigm shift is required in the treatment of urban rivers: less intervention and more adaptation, which should be a universal principle. Several interviewed experts request a river agreement or contract in the city, which integrates aspects of risk prevention (going beyond the city, including rural neighbourhoods and associated with Ebro Resilience project), natural protection, uses of the river space, treatments against exotics and invaders, etc. A rivers-city pact that changes perspective to look at the city from the river, instead of looking at the rivers from the city, sustainably making the health of river system compatible with citizen uses. This initiative should come from the citizens in a participatory manner, not from the administration, without waiting for political consensus, and be based on scientific knowledge and on measures agreed upon in PDIVZ: a greener Zaragoza with more dynamic rivers and permeable connectivity. For this, continuous environmental education is essential, with programs to value sediments and geobiodiverse fluvial functioning, rejecting common errors and myths in the media and society (Ollero 2023), until they consider rivers a necessary part for health and for the identity of the city.

More specifically, for each of the rivers, the following proposals by the experts have been compiled, which are very different from those proposed by the non-expert population (Table 13.3).

For the Ebro River: (i) setback or permeabilization of embankments with bidirectional dampers upstream (Almozara and Juslibol) and downstream (Movera) of the city; (ii) eliminate Vadorrey weir or improve its controlled opening and closing system (lower it with less flow or at night), as well as review the quality of retained water and rehabilitate fish ladder; (iii) remove or recondition piers; (iv) integrate the urban sphere into Ebro Resilience strategy; v) restoration of Almozara rubble bank; (vi) expansion of the river territory in urban reach and upstream and downstream, with the special case of the southern parking of Expo 2008, which can reduce the problem of Pavilion-Bridge (Abdullayeva 2022); (vii) remove ripraps; (viii) complete, connect, and improve the system of collectors and treatment plants; (ix) management plan for rainwater and storm tanks on the left riverside; x) effective protection of peri-urban riparian forests (e.g., Cantalobos) and their nearby orchards; (xi) garbage clean-up campaigns with clear institutional support; (xii) transformation of riverside areas into natural spaces; (xiii) administrative and scientific coordination to monitoring the evolution of channels; (xiv) camouflaged observatories for bird watching; (xv) maintenance and improvement of Expo 2008 elements; (xvi) creation of equipped public beaches in appropriate places; (xvii) recovery of the right span of Piedra Bridge.

For the Huerva River: (i) improve the current rehabilitation process with participation and ideas of the PDH; (ii) improve accessibility to the channel and use it for environmental education; (iii) remove rubbles and garbage from the riverbed and banks and hide collectors; (iv) eliminate all exotic and invasive vegetation; (v) naturalize confluence with Ebro River; (vi) change necessary defences for bioengineered solutions; (vii) also act in the southern upstream reach, giving continuity and room for the river, exercising expropriations, with an urban rethinking, as well as connectivity with the pine forests (PDIVZ); (viii) review current hydrological management, reduce alleviation from the Canal, maintain ecological flows, and generate floods to favour fluvial dynamics; (ix) recovery of an open-air river, at least on Constitución Boulevard, where there is enough space.

For the Gállego River: (i) restoration plan that includes sediment contribution from hanging areas and braided channel regeneration with formative floods, in the entire incision reach from Urdán weir; (ii) return room for the river and eliminate defences, resolving the risk of floodable housing states; (iii) divert the collectors, gas pipeline, cycle paths, and other structures attached to the channel; (iv) resolve the discharge from Montañana paper industry, for example, with a collector to Ebro River; (v) complete the recovery of riverbanks with a new phase of revegetation; (vi) increase environmental flows; (vii) urban and natural redevelopment in final segment to the mouth, dismantling an old factory and junkyards, making the Z-40 permeable and participatory planning; (viii) awareness campaign to increase knowledge and enjoyment of the river, creating a Gállego River centre (e.g., in Torre de los Ajos, Montañana).

As a last reflection and as a conclusion, we propose to respect the fluvial diversity of Zaragoza, not to homogenize it. They are three different rivers that complement each other in the green and blue infrastructure of the city. On this basis, the diagnostic and consultation exercise carried out in this work may constitute the basis for a future global plan or agreement between the city and its rivers.

Acknowledgements To degree students of Geography and Territorial Planning (University of Zaragoza) who have collaborated in carrying out the surveys, and to the experts, scientists, and public managers interviewed: JM Arnal, T Artigas, D Ballarín, JL Briz, O Conde, C Ferrer-Boix, L Manso de Zúñiga, C Marcén, JP Martín Vide, F Pellicer, J San Román, L Soriano, and three people who have opted for anonymity.

References

Abdullayeva Y (2022) De-coding velocity. Reintroducing approach to river control through creation of a performative system for the city of Zaragoza. Master's Programme in Urban Studies and Planning, Aalto University

Benito G, Pérez-González A, Gutiérrez F, Machado MJ (1996) Modelo morfo-sedimentario de evolución fluvial cuaternaria en condiciones de subsidencia kárstica de evaporitas (río Gállego, cuenca del Ebro). Cuad Geol Ibérica 21:395–420

Benito G, Sancho C, Peña-Monné JL, Machado MJ, Rhodes EJ (2010) Large-scale karst subsidence and accelerated fluvial aggradation during MIS6 in NE Spain: climatic and paleohydrological implications. Quatern Sci Rev 29:2694–2704

Cuello A (2022) Los entornos fluviales urbanos como recurso para la educación ambiental. Estudio de casos en los ríos Guadalquivir y Guadalete en Andalucía. PhD Thesis, Universidad de Sevilla

Díaz-Redondo M, Marchamalo M, Egger G, Magdaleno F (2018) Toward floodplain rejuvenation in the middle Ebro River (Spain): from history to action. Geomorphology 317:117–127

Durán F, Pons JJ, Serrano M (2021) River-city recreational interaction: a classification of urban riverfront parks and walks. Urban Forestry and Urban Greening 59. https://doi.org/10.1016/j.ufug.2021.127042

Durán F (2020) Ríos y ciudades: delimitación y análisis del espacio fluvial en España. Estudio del uso público y la recuperación de riberas. Ph.D. Thesis, Universidad de Navarra

Espinosa NP (2023) Living at (in) the edge. River restoration as a new urban design strategy. Ph. D. Thesis, KU Leuven

Ferrer-Boix C, Boix J, Martín-Vide JP, Ollero A (2021) Alluviation of a side-channel by bed material load. Field measurements and modelling. Geomorphology 389. https://doi.org/10.1016/j.geomorph.2021.107801

Ferrer-Boix C (2010) Incisión de ríos por extracción aluvial y retirada de presas. Estudio matemático y experimental. Ph.D. Thesis, Universitat Politècnica de Catalunya

Findlay SJ, Taylor MP (2006) Why rehabilitate urban river systems? Area 38(3):312–325

Francis RA (2012) Positioning urban rivers within urban ecology. Urban Ecosystems 15:285–291

Gallardo JJ (2023) Mejora de la biodiversidad a través de la restauración paisajística (hidromorfológica) y acciones de participación en el entorno del río Huerva (RE-PAPAH). IV Congreso Ibérico de Restauración Fluvial RestauraRíos, Toledo

González-Rojas D (2017) Bases conceptuales y metodológicas para el estudio de los espacios fluviales urbanos. Un estudio de caso en Andalucía. Estudios Geográficos 283:657–679

Gurnell AM, Lee M, Souch C (2007) Urban rivers: hydrology, geomorphology, ecology and opportunities for change. Geogr Compass 1(5):1118–1137

Gurnell AM, Shuker L, Wharton G (2016) Urban river survey manual. Queen Mary University, London

Gutiérrez-Serret RM (2002) Actuaciones urbanas en cauces y riberas: el caso del Ebro en Zaragoza. In De la Cal P, Pellicer F (coord) Ríos y ciudades. Aportaciones para la recuperación de los ríos y riberas de Zaragoza. Zaragoza, Institución Fernando el Católico, pp 159–180

Horacio J, Ruiz-Chacón M, Duarte P, Noguera I, Ollero A (2018) Propuesta de trabajo para la restauración fluvial del río Arga en el ámbito urbano de Iruña-Pamplona. Technical report. Iruñeko Udala

Kondolf GM, Pinto PJ (2017) The social connectivity of urban rivers. Geomorphology 277:182–196

Longares LA, Peña-Monné JL (2013) Aportación a la reconstrucción topográfica de la ciudad romana. In Escudero F, Galve MP (coord) Las cloacas de Caesaraugusta y elementos de urbanismo y topografía de la ciudad antigua. Institución Fernando el Católico, Zaragoza

Magdaleno F (2011) Evolución hidrogeomorfológica del sector central del río Ebro a lo largo del siglo XX. Implicaciones ecológicas para su restauración. Ph.D. Thesis, Universidad Politécnica de Madrid

Marco A (1991) Analisis de la evolucion reciente de la morfología del cauce del Bajo Gállego en las proximidades de Zaragoza: influencia de las actuaciones humanas en su entorno. Acta Geologica Hispanica 26(1):23–33

Marqués LA (2018) Alteraciones hidrogeomorfológicas en el Bajo Gállego a partir del registro instrumental. Ph.D. Thesis, Universidad de Zaragoza

Martín-Vide JP, Ferrer-Boix C, Ollero A (2010) Incision due to gravel mining: modeling a case study from the Gállego River, Spain. Geomorphology 117:261–271

Martín-Vide JP, Fael CMS, Núñez F, Ferrer-Boix C, Santos CAV, Prats A, Chavarrías V (2022) A large bridge pier in an alluvial channel: local scour versus morphological effects and the role of physical models. J Hydraulical Eng 148(8):05022001

Martín-Vide JP (Dir. 2018a) Estudio hidrológico, geomorfológico, hidraúlico y ecológico del bajo Gállego en Zaragoza para su gestión como espacio fluvial dentro del desarrollo urbano sostenible de la ciudad. Research unpublished for the Ayuntamiento de Zaragoza

Martín-Vide JP (Dir. 2018b) Estudio del caudal de inicio del movimiento de fondo en el azud del Ebro en Zaragoza. Research unpublished for the Ayuntamiento de Zaragoza

Mateo J, Sánchez-Navarro JA (2011) Análisis de circulación de crecidas mediante el programa SHEE. Aplicación a la cuenca del río Huerva, España. Revista de la Sociedad Geológica de España 24(3–4): 187–195

Molina P, Jendrzyczkowski L, Berrocal AB, Allende F (2020) The analysis of urban fluvial landscapes in the centre of Spain, their characterization, values and interventions. Sustainability 12. https://doi.org/10.3390/su12114661

Murphy B (2020) Urban stream assessment procedure: a framework for assessing stream health in the urban environment. Watershed Management Conference 2020, 99–107

Ollero A (2010) Channel changes and floodplain management in the meandering middle Ebro River, Spain. Geomorphology 117:247–260

Ollero A, Ibisate A, Granado D, Real de Asua R (2015) Channel responses to global change and local impacts: perspectives and tools for floodplain management (Ebro River and tributaries, NE Spain). In: Hudson PF, Middelkoop H (eds) Geomorphic approaches to integrated floodplain management of lowland fluvial systems in North America and Europe. Springer, New York, pp 27–52

Ollero A, García JH, Ibisate A, Sánchez-Fabre M (2021a) Updated knowledge on floods and risk management in the Middle Ebro River: the "Anthropocene" context and river resilience. Cuadernos De Investigación Geográfica 47:73–94

Ollero A, Ballarín D, García JH, Ibisate A, Mora D, Sánchez-Fabre M (2021b) Diagnóstico fluvial, impactos en cauces y cambio global: aplicaciones del índice hidrogeomorfológico IHG. Geographicalia 73:295–316

Ollero A (Dir. 2005) Estudio hidrológico, geomorfológico, hidráulico y ecológico del bajo Gállego en el término municipal de Zaragoza para su gestión como espacio fluvial. Research unpublished for the Ayuntamiento de Zaragoza

Ollero A (Dir. 2018) Propuesta y justificación de alternativas para anteproyecto demostrativo de generación de espacio de inundabilidad: acción b2. LIFE 12-ENV/ES 000567 Zaragoza Natural

Ollero A, Briz JL (2018) Ciudad y territorio fluvial en Zaragoza: principales retos y estrategias de futuro. International workshop "El agua y los ecosistemas fluviales en la ciudad", Ayuntamiento de Zaragoza, Cámara de Comercio and FNCA.

Ollero A, Sánchez-Fabre M, Marín JM, Fernández D, Ballarín D, Mora D, Montorio R, Beguería S, Zúñiga M (2004) Caracterización hidromorfológica del río Gállego. In: Peña-Monné JL,

Longares LA, Sánchez-Fabre M (eds) Geografía física de Aragón. Aspectos generales y temáticos. Zaragoza, Universidad de Zaragoza & Institución Fernando el Católico, pp 117–129
Ollero A (1992) Los meandros libres del río Ebro (Logroño-La Zaida) geomorfología fluvial, ecogeografía y riesgos. PhD Thesis, Universidad de Zaragoza
Ollero A (2023) Los paisajes fluviales peninsulares en un contexto de cambio hidroclimático ambiental: los retos de la gestión de riesgos y de la restauración. In: Zaragoza MF (coord) El Bajo Segura como enclave hidrológico: territorio, economía y paisaje. València, Ed. Tirant Humanidades, pp 287–318
Pellicer F, Sopena MP (2019) Grandes eventos, huellas del futuro. Las riberas del Ebro y Expo Zaragoza 2008. Zarch 13:62–75
Pellicer F (2015) La recuperación de las riberas del Ebro en Zaragoza. Un efecto perdurable del evento efímero Expo 2008. In: De la Riva J, Ibarra P, Montorio, R, Rodrigues M (eds) Análisis espacial y representación geográfica: innovación y aplicación. Zaragoza, Universidad de Zaragoza & AGE, pp 353–362
Peña-Monné JL, Rubio V, Longares LA, Sánchez-Fabre M (2011) El meandro de la Peña del Cuervo: un ejemplo de la dinámica fluvial actual del bajo Gállego (Depresión del Ebro). Geographicalia 59–60:281–294
Peña-Monné JL, Longares LA, Rubio V, Sampietro MM, Sánchez-Fabre M (2020) Dynamic changes in the lower Gállego river (Ebro basin, NE Spain) and their relationship with anthropic activities and the quaternary substrate. Cuadernos De Investigación Geográfica 46(2):371–393
Petts J (2006) Managing public engagement to optimize learning: reflections from urban river restoration. Hum Ecol Rev 13(2):172–181
Pirchi VN (2020) Dinámica y morfología fluvial de un tramo antropizado: el cauce periurbano norte del arroyo Napostá Grande. Unpublished technical report, Bahía Blanca
Pueyo Á, Climent E, Ollero A, Pellicer F, Peña JL, Sebastián M (2017) L'interaction entre Saragosse et ses cours d'eau: évolution, conflits et perspectives. Sud-Ouest Européen 44:7–23
Ranta E, Vidal-Abarca MR, Calapez AR, Feio MJ (2021) Urban stream assessment system (UsAs): an integrative tool to assess biodiversity, ecosystem functions and services. Ecological Indicators 121. https://doi.org/10.1016/j.ecolind.2020.106980
Santasusagna A (2016) Ciutat i riu. Mig segle de transformacions urbanístiques als espais fluvials de quatre poblacions catalanes (Manlleu i el Ter, Terrassa i les seves rieres, Lleida i el Segre, Sant Adrià de Besòs) i una francesa (Lió, el Roine i el Saona). Ph.D. Thesis, Universitat de Barcelona.
Tort J, Santasusagna A, Rode S, Vadrí MT (2020) Bridging the gap between city and water: a review of urban-river regeneration projects in France and Spain. Science of the Total Environment 700. https://doi.org/10.1016/j.scitotenv.2019.134460
Vietz GJ, Walsh CJ, Fletcher TD (2016) Urban hydrogeomorphology and the urban stream syndrome: treating the symptoms and causes of geomorphic change. Progress in Physicial Geography. https://doi.org/10.1177/0309133315605048
Walsh CJ, Roy AH, Feminella JW, Cottingham PD, Groffman PM, Morgan RP II (2005) The urban stream syndrome: current knowledge and the search for a cure. J N Am Benthol Soc 24(3):706–723
Zingraff-Hamed A (2018) Urban River Restoration: a socio-ecological approach. Ph.D. Thesis, Technische Universität München, Université de Tours

Chapter 14
Change in Perceptions and Social Value of the River Corridors in the Metropolitan Region of Barcelona: The Case of the Besòs Basin

David Pavón, Marta Benages-Albert, Pere Vall-Casas, Xavier Garcia, and Anna Ribas

Abstract In addition to their environmental and landscape potential, river corridors play a strategic role as socio-spatial elements that allow the structuring of the urban systems arranged around them. In addition, they strengthen citizens' connections with the place through daily practices. Following years of intense degradation and abandonment, the growing social use of these corridors has become prevalent in developed countries since the 1980s. The metropolitan region of Barcelona is a privileged territorial exponent of this transformation. This chapter will address the changes in perceptions of and increased social value awarded to these spaces, which are progressively and simultaneously conceived as ecological corridors, heritage spaces and social cohesive meeting points. For this purpose, emphasis will be placed on a project carried out in the Besòs basin and on the strategies implemented for its improvement.

Keyword Public perception · Social value · River corridors · Besòs basin

D. Pavón (✉) · A. Ribas
Department of Geography, SAMBI (Grup de Recerca en Canvi Socioambiental), University of Girona (UdG), Girona, Spain
e-mail: david.pavon@udg.edu

A. Ribas
e-mail: anna.ribas@udg.edu

M. Benages-Albert · P. Vall-Casas
UIC Barcelona School of Architecture, Universitat Internacional de Catalunya, Barcelona, Spain
e-mail: martabenages@uic.es

P. Vall-Casas
e-mail: perevall@uic.es

X. Garcia
Catalan Institute for Water Research (ICRA-CERCA), University of Girona (UdG), Girona, Spain
e-mail: xgarcia@icra.cat

© The Author(s), under exclusive license to Springer Nature Switzerland AG 2024

J. Farguell Pérez and A. Santasusagna Riu (eds.), *Urban and Metropolitan Rivers*, The Urban Book Series, https://doi.org/10.1007/978-3-031-62641-8_14

14.1 The Relevance of the River Corridors and Their Interpretation in the Metropolitan Area of Barcelona

Unsustainable management and land use have severely impacted rivers and streams for decades. These impacts have been driven by urban development, intensive flow regulation, river channelization, point and non-point sources of water pollution and other anthropogenic drivers. As a result of these human alterations in the biodiversity, hydromorphology and physical and chemical characteristics of the water that flows into rivers, their ecological functioning has been severely constrained (Everard and Moggridge 2012). In addition to this deterioration in ecological value, the degradation of riparian landscapes also has a negative impact on socio-cultural values and well-being. For instance, the urban transformation of rural riparian areas jeopardizes the public's sense of connection to their local rivers, a connection that is normally strong in traditional rural communities (Benages-Albert et al. 2015). As a result of these and other adverse effects, rivers in many areas have a negative image amongst authorities and the public due to their unpleasant appearance, disconnection from the community and/or the problems they cause the population (e.g., mosquitoes, smell, etc.) (Garcia and Pargament 2015).

In recent years, the authorities in charge of rivers and streams have begun to realize the potential of these degraded ecosystems as a unique opportunity for recovering a diverse and broad suite of ecosystem services (Everard and Moggridge 2012; Lundy and Wade 2011). Within this context, river rehabilitation refers to significant improvements in ecosystem functioning and biodiversity when it comes to rivers and streams (Findlay and Taylor 2006). Despite growing interest, several planning and management issues have discouraged the widespread initiation and implementation of river rehabilitation plans (Zaugg 2002). In addition to the high technical and ecological complexity and the high costs of such interventions, one main reason for doubt is that river rehabilitation schemes can cause controversy due to conflicting interests and expectations, often resulting in a lack of local public support or even resistance (Buijs 2009; Junker et al. 2007; Vining et al. 2000). It is therefore critical to understand the factors that influence local public acceptance. For instance, information regarding public perceptions and preferences regarding changes in sensitive landscapes may be important in meeting demands and improving acceptability (Casado-Arzuaga et al. 2013; Ryan 1998). Furthermore, ensuring that the preferences of the general public are recognized may encourage the public to value these changes and thus become watchful caretakers of rehabilitated landscapes (Grêt-Regamey et al. 2016; Nassauer 2004). For the abovementioned reasons, researchers have employed case-specific analyses to investigate which factors may condition public reactions to environmental changes derived from river rehabilitation projects.

River corridors play an essential role in the structuring of metropolitan regions. These geographical spaces often accommodate important communication routes and extensive urbanized continuums, with complex mixtures of infrastructures, urban fabrics and open spaces. Their management requires a renewed territorial vision, capable of assuming the functioning and morphology of the contemporary metropolitan city, beyond administrative boundaries and the current regulatory framework (Font 2004). Likewise, the complexity of the riverside landscape requires approaches that adopt a global perspective to combine simultaneous analyses on multiple scales aimed at integrating the structure of each level into a coherent whole (Panareda 2009).

The strategic value of river corridors is particularly significant in the case of the Barcelona metropolitan region where, from the 1950s onwards, the Llobregat and Besòs valleys became axes of territorial development (Barcelona Provincial Urban Planning Commission 1963; Vendrell and Presmanes 1993). The traditional agricultural use of water resources was progressively replaced with supplying the population and growing industrial activities, whilst agricultural activity, still relevant in some river areas, inevitably receded. The case of the port and the airport, in the Llobregat delta, and the CIM Vallès goods exchange centre, in the Caldes Stream, are illustrations of this dynamic. The metropolitan rivers have currently become road and rail communication axes between the central city and the successive metropolitan crowns, i.e. axes of urbanization formed by the aggregation of residential and industrial fabrics and metropolitan service nodes (Font et al. 1999; Vecslir 2007), as well as drainage and energy supply channels (García and Godé 2006).

14.1.1 Towards a Change in the State of Metropolitan River Environments: From Degradation to Environmental and Social Recovery

From the 1980s onwards, the overexploitation of water resources and degradation of ecosystems caused by anthropic pressure (Prat et al. 1982–1983; Prat and Rieradevall 1992) led to the progressive acquisition of a new environmental consciousness in technical and political circles. However, the scientific community was already warning about the pollution of rivers and aquifers due to deficient management of water resources back in the 1970s (Díaz and Queralt 1970; Planas et al. 1976; Queralt 1974; Canton et al. 1975). At the end of that decade, the horizon of European integration and consequent obligation to seriously address wastewater treatment led to the diagnosis of the ecological state of Catalan rivers being assigned to the University of Barcelona's Department of Ecology (Margalef and Prat 1979; Prat 1979). In addition, in parallel with this process of environmental awareness induced "from above", the social pressure "from below", triggered by alarming situations such as epidemic outbreaks and floods, played a decisive role in legitimizing political decisions aimed at improving water management. Gradually, rivers ceased to be perceived as simple water channels circulating through a residual space, and commitments were made to achieve their complex and integrated environmental recovery (Simon 1994; Vendrell 1994; AADD 1994). Within this context, the progressive revaluation of the riverside landscape allowed it to overcome the oblivion to which it had been relegated in previous decades (Panareda 2009). What became known as "river parks" emerged as a leisure space and strategic instrument for metropolitan regeneration (Torra et al. 2008). At the beginning of the twenty-first century, river corridors are now part of the system of metropolitan public spaces and play an important social and economic role beyond their strictly ecological function (Benages and Vall 2014).

In terms of their social role, rivers and streams are widely used as spaces for leisure and reconnecting with nature (Batlle 2011; Vallerani 2012). River courses, together with forests and immediate agricultural spaces, make up the everyday landscape of many citizens, foster cohesion and local identity and compensate for the lack of conventional public spaces in nearby residential areas (Benages 2011). For this reason, a commitment to responsible social use and a civic reinterpretation of the water system (Vall 2010; Vall et al. 2011), following the pioneering example of the Boston Park System in the second half of the nineteenth century (Zaitzevsky 1982), is essential for the physical and social integration of the metropolitan region (Torra et al. 2008; Generalitat de Catalunya 2010b). In recent times, this model has been revised taking into account new demands (Platt 2006; Forman 2004), including public health taking on special relevance, given that the direct influence of the environmental improvement of rivers and the increase in their accessibility can foster healthy lifestyles and, consequently, improved health amongst citizens and a reduction in medical expenses (Nilsson and Nielsen 2006; Nogué et al. 2008; Rydin et al. 2012).

In terms of their economic role, rivers and streams also have an important function in terms of configuring the productive corridors, a characteristic of the metropolitan arch of Barcelona, which take advantage of the pleasant topography of the fluvial plains and the privileged connection with the central city. The corridors of Granollers and Sabadell, organized respectively around the Congost and Ripoll rivers, exemplify the recurrent presence of these productive spaces distinguished by their dynamism (Trullén 2003). Within this context, improving the quality of the river area helps facilitate the progressive consolidation of productive activities with higher added value (Morris 2003) and, in short, increases the competitiveness of the metropolitan region within the framework of a new economy. In the context of the Besòs basin and its tributaries, the Caldes stream corridor exemplifies this trend, with new premises being integrated in the IT and pharmaceutical sectors, and especially non-directly productive units dedicated to directional uses, attracted by quality industrial land equipped with large riverside parks (Generalitat de Catalunya 1987, 1988). Thus, several studies have suggested that in the Barcelona metropolitan region, factors such as the environmental quality of the space, together with accessibility to high value-added services and the socio-business prestige of the locations, have become decisive for the location of economic activities linked to knowledge and innovation (Pérez and Marmolejo 2008).

The aforementioned economic and social interest in the river corridors in the Barcelona metropolitan region, added to their importance as road and environmental axes, make it advisable to approach their management from an integrated urban perspective. This approach has been widely accepted in political circles (Llop 2008).

The current Metropolitan Territorial Plan, in force since 2010 (Autonomous Government of Catalonia 2010b), has led to the development of urban master plans aimed at reconciling ecological, economic and social objectives set around various fluvial axes. These plans have been conceived as instruments for technical, political and social agreement (Nel·lo 2006). For this reason, effectively managing citizen participation acquires special importance. In this regard, it is generally accepted that a solid social base will reinforce the technical content of the plan and allow for its correct implementation (Kaplan et al. 1998).

14.2 The Case of the Besòs Basin and the Enhancement of Its River Corridors

14.2.1 Key Elements in Improving Its Fluvial Environments

The Besòs basin, enclosed between the Pre-Coastal and Coastal Mountain ranges, mostly runs through the Vallès depression and is made up of six main river courses: the Mogent, the Congost, the Tenes, the Caldes stream, the Ripoll and the Besòs (Fig. 14.1). Over two million inhabitants live in the urban systems articulated around these river courses, demonstrating the social dimension of this water system, for which plans are in place to make them a metropolitan system of public spaces (Fig. 14.1). A detailed analysis was conducted of the Besòs basin at the end of the 1940s by the geographer Josep M. Puchades, in his work *"El río Besós: estudio monográfico de hidrología fluvial"*—The River Besós: a monographic study of river hydrology—(Puchades 1948). The author called for a first systemic vision of the fluvial environment in order to address negative impacts deriving from its overexploitation. In line with this seminal contribution, the environmental and social revaluation of the Besòs basin has been high on the agendas of local governments and numerous civil entities over the last three decades (Gordi 2008; Torra et al. 2008). Throughout this time, the approach to river regeneration has evolved, broadly speaking, from an initial, sectorial and parametric vision, basically oriented towards monitoring the quantity and quality of water, to the current, more unitary and complex vision, which recognizes the integrity of the river landscape and necessary citizen involvement in maintaining it. The European directives on water and landscape (Council of Europe 2000a and 2000b) have played a key role in this evolution, as analysed below. The creation of the Consortium for the Defence of the Besòs River Basin in 1988 is essential in this regard. Its lines of action will be low-level sanitation (municipal sewers), high-level sanitation (control of discharges, collectors, pumping and wastewater treatment plants—or WWTPs), the improvement of the fluvial environment, environmental promotion and education and support for consortium entities.

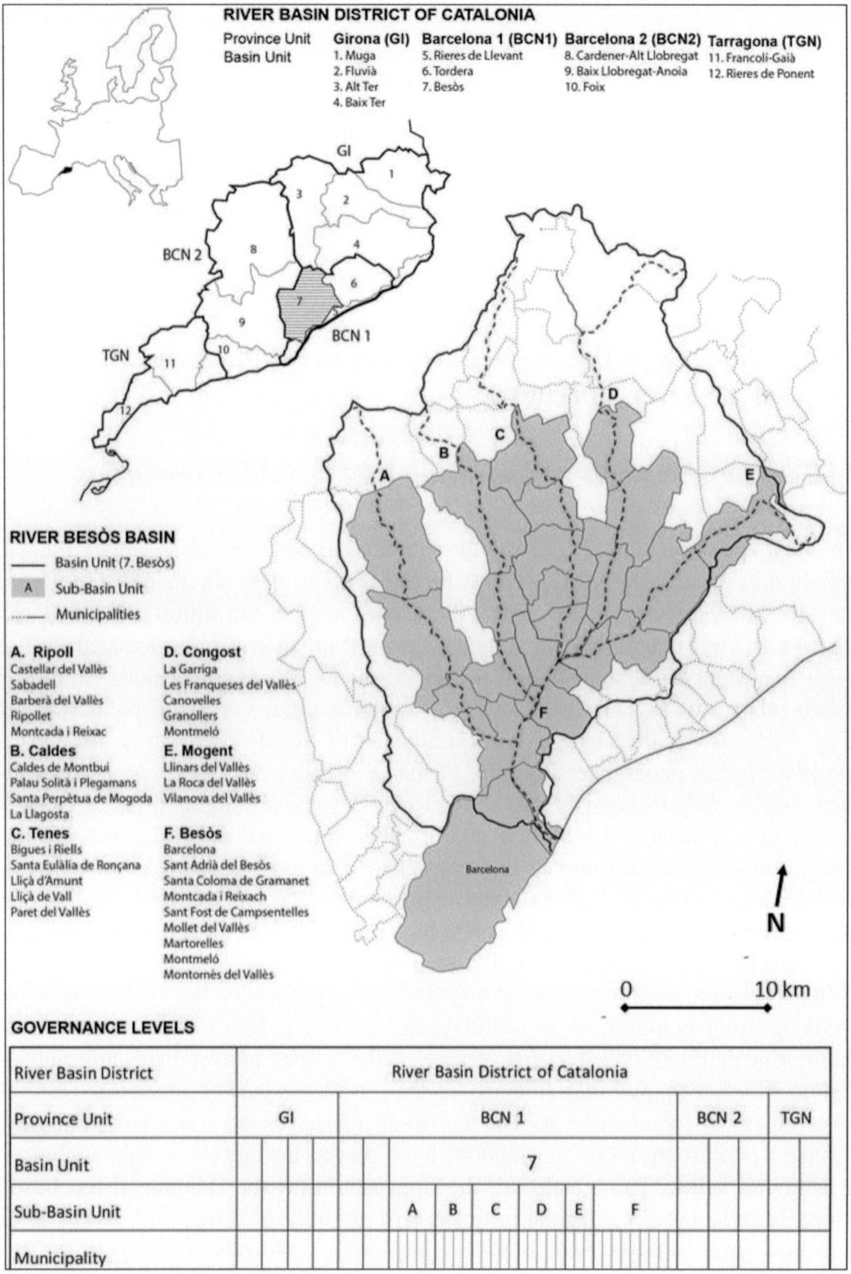

Fig. 14.1 Location of the Besòs river basin within the river basin district of Catalonia. *Source* Authors (2017)

As wastewater sanitation actions take effect and the water quality of the rivers improves, the physical and social dimensions of the fluvial space can be progressively integrated. Under the normative framework of the *Pla Director dels Espais Fluvials de la Conca del Riu Besòs* (Master Plan for Fluvial Areas of the Besòs River Basin 2000), projects for environmental recovery and the arrangement of fluvial paths are drafted for each of the sub-basins. At the same time, environmental education and river stewardship initiatives are being initiated. These initiatives exemplify the paradigm shift in water management that is definitively consolidating the European Water Framework Directive (Council of Europe 2000a). This new regulatory framework obliges local authorities, also in the case of Catalonia, to standardize priorities and intervention instruments under sustainable development criteria that integrate institutions and citizens through participatory processes.

Furthermore, in the early years of the twenty-first century, the new landscape culture became a relevant issue in Catalan territorial policies (Busquets 2005; Tarroja and Camagni 2006; Cortina 2010). Following the guidelines established by the European Landscape Convention (Council of Europe 2000b), the Autonomous Government of Catalonia passed Law 8/2005, on the protection, management and planning of landscape in the region, and created the Landscape Observatory of Catalonia, which set the drafting of the landscape catalogues in motion with the aim of incorporating landscape objectives into territorial planning. This new regulatory framework establishes an integrated vision that recognizes the multiple natural and cultural values of the landscape, whilst arguing for public participation as a tool to involve citizens and make them jointly responsible in management and planning. In the specific case of metropolitan river corridors, the Barcelona Metropolitan Region Landscape Catalogue supports the continuity and quality of river environments by allocating compatible social and natural uses, as well as emphasizing the importance of awarding social value to rivers as "reference points for the creation and recreation of the place's identity".

14.2.2 Key Strategies for the Revitalisation of River Corridors

The progressive assumption of the new water and landscape cultures detailed in the previous paragraphs entails the implementation of revaluation processes for metropolitan river landscapes, conceived simultaneously as ecological corridors, heritage spaces and social cohesive elements. Within this context, the tasks of strict physical repair are enriched with more far-reaching proposals that integrate improvement of the natural environment with bringing citizens closer to the rivers and streams, authentic triggers of their own territorial identity. With regard to the Besòs basin, this new generation of initiatives aimed at the parallel support of the ecological and social values of fluvial spaces contains the following four main work lines: environmental education; environmental volunteering; historical heritage pedagogy and promoting leisure. These are common strategies in contemporary processes aimed at regenerating metropolitan river spaces, as illustrated by the exemplary case of Emscher Park (Government of North-Rhine Westphalia 1989).

(a) *Environmental education*

Environmental education has been included in the recovery of the Besòs basin. During the project, the local authorities have implemented significant educational and dissemination programmes aimed at citizens. It has done so through exhibitions and publications such as *Viu el riu: Idees, propostes i paisatges per a la conca del riu Besòs* (Live the river: ideas, proposals and landscapes for the Besòs river basin—AADD 1994), *Ciutat i riu dins l'àmbit metropolità: La conca del Besòs* (City and river in the metropolitan area: the Besòs basin—AADD 1999) and *El paisatge fluvial a la conca del Besòs: Ahir, avui..., i demà?* (The river landscape in the Besòs basin: yesterday, today... and tomorrow?—Gordi 2005). Over the years, these efforts to disseminate the environmental values of the basin as a whole have acquired a more technical and retrospective character, with evaluations of the work carried out and meetings such as the Conference for the Restoration of Fluvial Spaces in the River Besòs Basin (2009).

Progressively during the 1990s, citizens belonging to civic organizations promoted numerous environmental education initiatives aimed at adults and families through conferences and nature discovery walks. These were on a local scale and often sponsored by the local authorities. Fairly active local organizations such as *l'Associació per a la Defensa i l'Estudi de la Natura* (Association for the Defence and Study of Nature—Sabadell 1982), *l'Associació Amics del Sender* (the Friends of the Footpath Association—Caldes de Montbui 1997), *l'Associació Amics del Ripoll* (the Friends of Ripoll Association—Sabadell 2000), *l'Associació de difusió i conservació del patrimoni natural LESTES* (the Association for the Dissemination and Conservation of Natural Heritage *LESTES*—Santa Perpètua de Mogoda 2001) and *el Grup d'Ornitologia del Tenes* (the Tenes Ornithology Group—Santa Eulàlia de Ronçana 2007) are an illustration of the commitment shown by civic society around the Besòs basin. All these environmental education and promotion initiatives shared the same underlying objective: to raise awareness of the natural dimension of the river environment that must be recovered in order to respect and preserve it as a common asset (Fig. 14.2).

(b) *Environmental Volunteering*

This is a very effective way of raising awareness through the direct involvement of citizens in improving the river environment. Once a certain degree of environmental awareness had been achieved, thanks to the efforts of the 1990s, environmental volunteering programmes aimed at recovering river courses began to become more widespread. In Catalonia, the *Projecte Rius* (River Project), promoted since 1997 by the Habitats Association in collaboration with the University of Barcelona's Department of Ecology and the Ter Defence Group, has become an innovative social awareness-raising initiative that promotes three types of actions: inspections of the ecological state of rivers; actions to demand the cleaning of watercourses and the

Fig. 14.2 Awareness-raising walk (Stream Day), 20th of November 2016 (Caldes de Montbui) *Source* Authors (2016)

adoption of river sections. This programme is implemented on a regional scale through local volunteer networks in charge of carrying out actions in all basins, with the technical support and advice of the entity promoting the activity. Alongside this regional initiative, on a more local scale, civic organizations have received collaboration from the local authorities to set up volunteer programmes with citizens, schools and private companies, organizing tree plantings, riverbed restorations and river environment clean-up days. This is the case with the ADENC project *Fes reviure el Ripoll!* (Bring the Ripoll River back to life!), which promotes citizen participation through an in situ environmental education exercise on a stretch of the Ripoll River in Sabadell.

These programmes have been shown to be very effective in increasing a feeling of community, self-esteem, commitment and identification with the natural environment (Kaplan et al. 1998). In this respect, environmental volunteering contains a high civic potential that activates effective processes of citizen ownership through actions deployed on the environment (Vidal i Pol 2005). Over the last thirty years, this approach has led to a proliferation of civic organizations and initiatives aimed at promoting environmental volunteering as a dual process of learning and community service through specific activities.

(c) *Interventions in Fluvial Historical Heritage*

Thirdly, the recovery of historical heritage linked to water has become an opportunity to bring the population closer to the rivers and the metropolitan streams, and to educate them in this respect. In general terms, although the industrial and agricultural heritage of the Besòs has deteriorated due to urban overpressure in recent decades, a certain variety of heritage elements remains, especially in the sub-basins, such as irrigation systems, water mines, mills and ice wells (Dantí 2010). The recovery and dissemination of these elements allow for a cultural interpretation of the river, complementary to the naturalistic vision.

However, the reference initiative where cultural discourse around the river is most consolidated is without doubt the Ripoll River Fluvial Park in Sabadell, with four itineraries dedicated to river heritage (Fernández and Prat 2004) and a plan to recover the pre-industrial heritage of Ripoll (Sabadell City Council 2006). There are also two relevant cases located in municipalities in the upper reaches of the rivers, further from Barcelona's area of influence. In Caldes de Montbui, the rehabilitation of the Riera promenade and the subsequent landscape restoration of the river environment aims to articulate the various heritage elements linked to the water cycle (Roman baths, washing places, vegetable gardens, flour mill, etc.) (Caldes de Montbui Town Council 2007). Along the same lines, between the municipalities of La Garriga, Figaró-Montmany and Tagamanent, the Congost Itinerary covers part of the old royal road and marks the most important local elements of natural and cultural heritage. In all of these cases, the link between the historical materials and the river or stream that accommodates them and gives them meaning is recognized, as well as offering a supportive interpretation of the fluvial landscape through itineraries aimed at discovering natural and cultural heritage.

(d) *Promoting Leisure*

Lastly, it is worth highlighting initiatives related to the promotion of leisure. In order to reach a wider range of citizens and simultaneously promote the physical improvement of the river environment and respectful social use, the local authorities have opted to create recreational spaces along specific highly frequented stretches, compatible with the protection of other riverside spaces. Within this context, the river park provides a response to the growing demand for places for citizen leisure, and the improvement of paths along rivers and streams is particularly necessary to link the river parks with the nearby urban systems. The precedents for this strategy in the Besòs basin are to be found around the Ripoll River, in the city of Sabadell (Vidal 1999), and in the final stretch of the Besòs (Alarcón 1999), probably due to the greater urban pressure exerted on the river in these areas. Following the example of these pioneering cases, an increasing number of riverside municipalities in the Besòs basin are developing itineraries based on the riverside path construction projects drawn up by the Consortium for the Defence of the Besòs River Basin. These include, amongst others, the Ronda Verda de Palau-solità i Plegamans (2011), the Riera de Caldes de Montbui walk (2012) and the Santa Eulàlia de Ronçana Riverside Path (2012). These itineraries allow for a new repertoire of oft-frequented peripheral open areas to be added to the existing central urban open spaces.

14.3 Participated Actions and Lived-In Space: A Change of Perspective in Interventions

The retrospective analysis discussed in the previous paragraphs on the most relevant actions undertaken over the last forty years to reverse the degradation of rivers and encourage citizen ownership in the context of the Besòs basin allows us to gauge the importance of the efforts made by the local authorities to date. Copious investments have been earmarked for the construction and management of sanitation infrastructures, the prevention of risks derived from flooding, the environmental recovery of river areas and the repair of paths and leisure spaces. Likewise, resources have been used to fund an emergency citizen education programme related to the river environment, combining environmental education and volunteering with the promotion of cultural and sporting leisure. These initiatives have made a significant contribution to turning the river area into a space frequented by many, and a gradual increase in the number of individuals committed to its care and valuation as an area of daily enjoyment full of aesthetic and affective values.

Generally speaking, we can state that the initiatives fostering the social use of watercourses presented in the preceding paragraphs have been induced "from above" and based mainly on an exhaustive technical knowledge of the natural and cultural resources available. Issues such as water quantity and quality, fauna and flora and the environmental impacts caused by urban overpressure have been recurrent arguments in the pedagogy promoted around river spaces and aimed at fostering responsible ecological behaviour. Although urban planning initiatives, the didactic signposting of itineraries, publications, information days and various awareness-raising activities have been used to describe how the river space works, very little has been done to explore "how it is experienced". In short, the link between citizens and the river has been built primarily on measurable technical content, whilst experiential content has been relegated to second place. This imbalance constitutes a significant obstacle in progress towards the participatory regeneration of metropolitan river corridors, and highlights the need for a clearer focus on experience as a starting point. More specifically, a better knowledge of the river environment as a lived-in space has two significant and closely interrelated methodological advantages: firstly, it offers a means of accessing relevant invisible data related to the connection between citizens and the river space; and, at the same time, it facilitates communication between citizens and local managers throughout the regenerative process.

Within this context, the geo-ethnographic diagnosis (Benages and Vall 2014) aimed at discovering the lived-in space deserves special attention. In recent years, a number of studies have focused on analysing river rehabilitation processes driven by self-organized ("bottom-up") groups. These studies have highlighted important benefits for river space management, such as the co-production of knowledge (Van Buuren 2013) and more inclusive decision-making (Robinson et al. 2015). The key factors for the success of these processes have also been analysed (citizen's organizational capacity, types of leadership and institutional support) (Igalla et al. 2019). However, little knowledge is available on how and under what conditions bottom-up processes can culminate in the formation of citizen-based river groups. This lack of knowledge limits the ability of local authorities to foster the self-organization of local groups and sustain these effectively over time. In order to advance in this field and provide support for the collaborative management of urban rivers, the pilot participatory process "*Viu la riera*!" (Long live the stream!) was designed and implemented.

14.3.1 The Pilot Project Viu La Riera! (Long Live the Stream!) as an Example

The pilot participatory process *Viu la riera!* (Long live the stream!) began in 2016, aimed at activating the involvement of river communities in caring for the lower and middle sections of the Caldes stream, a tributary of the Besòs river. This 19-km section of the Caldes stream has become a civic corridor with intense social use that connects the low-density residential and industrial fabrics of four municipalities, covering a population of 71,928 inhabitants (2022): from north to south, Caldes de Montbui, Palau-Solità i Plegamans, Santa Perpetua de Mogoda and La Llagosta (Vall et al. 2019).

(a) *Design and Development*

Viu la riera! was implemented between May 2016 and December 2017 with two main objectives: (1) to assess local residents' landscape values and preferences for improvement (Garcia et al. 2017); and (2) to promote the creation of a citizen-based group committed to caring for the Caldes Stream (Vall et al. 2021).

The participatory process combined GIS technology for public participation (SIGPP) with face-to-face participatory methods, such as focus groups, leisure activities, direct observations, face-to-face workshops and presentations by experts. On the *Viu la riera!* website (www.viulariera.org), participatory activities were announced, photos and videos of the river landscape were disseminated that had been provided by the public and the SIGPP tool was made available (Fig. 14.3). In total, six face-to-face sessions and three online actions were employed to gather public opinion and promote decision-making and citizen self-organization.

The participatory process was divided into three phases. In the first phase, "Opina" (*Give your opinion*), positive and negative landscape values and preferences for improvement were identified. In the second phase, "Decidim" (*We decide*), improvement actions were agreed based on a vote on the results of the previous phase. In addition, experts on controversial issues were invited to give informative talks aimed at overcoming preconceived opinions and finding viable solutions. Finally, in the "Mulla't" (*Get involved*) phase, support was provided for citizen self-organization to carry out some of the selected actions.

A total of 327 people participated in the process, including a wide range of profiles: from people interested in enjoying experiences related to the natural environment, in this case the river, but not very likely to participate in volunteer activities, to people

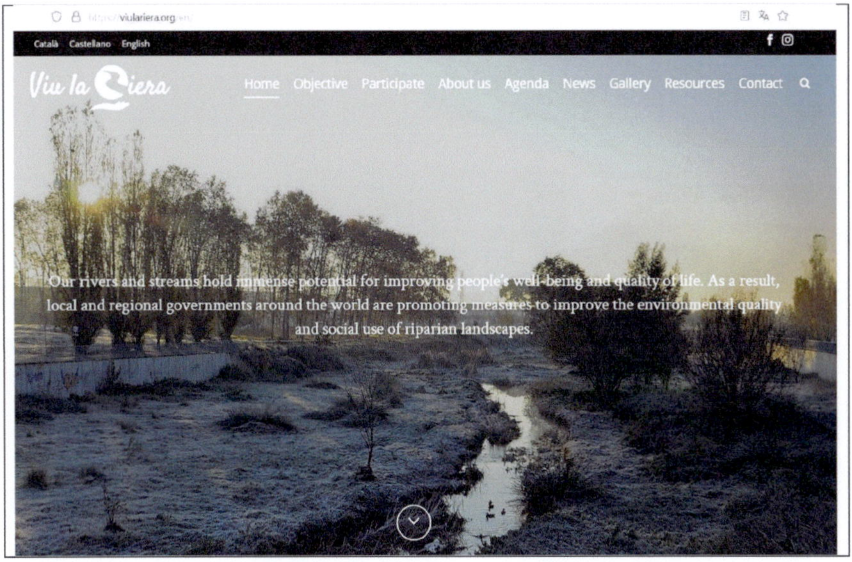

Fig. 14.3 Website header of project *Viu la Riera!*. *Source* Authors (2017)

with a history of pro-environmental behaviour aimed at caring for and safe-keeping nature (Alonso-Monasterio, Alonso-Monasterio and Viñals 2015).

(b) *Citizens' Perception of River Areas at the Heart of the Debate*

The contents addressed during the process diverged from those currently at the centre of the debates engaged in by the Catalan Water Agency (ACA), which are oriented towards ecology and the management of water resources (pollution, wastewater, hydro-morphological and biological quality, water supply). The pilot project participants perceived the river as a green environment with strong links to social and ecological dimensions. The socio-ecological approach to identifying landscape values and preferences for improvement proved crucial for the active involvement of citizens. For example, the overlapping of positive and negative landscape values, such as naturalness and flood risk, both related to the presence of trees in the riverbed, was not perceived as a source of conflict but as an opportunity for a more nuanced understanding of the river.

The face-to-face *Decidim* workshops allowed for expression of the different, equally reasonable, opinions gathered in *the Opina* phase. Specifically, the expert talks provided a technical basis that helped build consensus amongst the mobilization drivers (Fig. 14.4). Although some of the predominant proposals for improvement were not immediately translated into practical actions and will require further details and support, the pilot project reinforced and aligned the participants' willingness to contribute to this regard. Thus, a mixed socio-ecological approach, attentive to the diversity of meanings that the stream has for participants, was the starting point for forming the citizens' group.

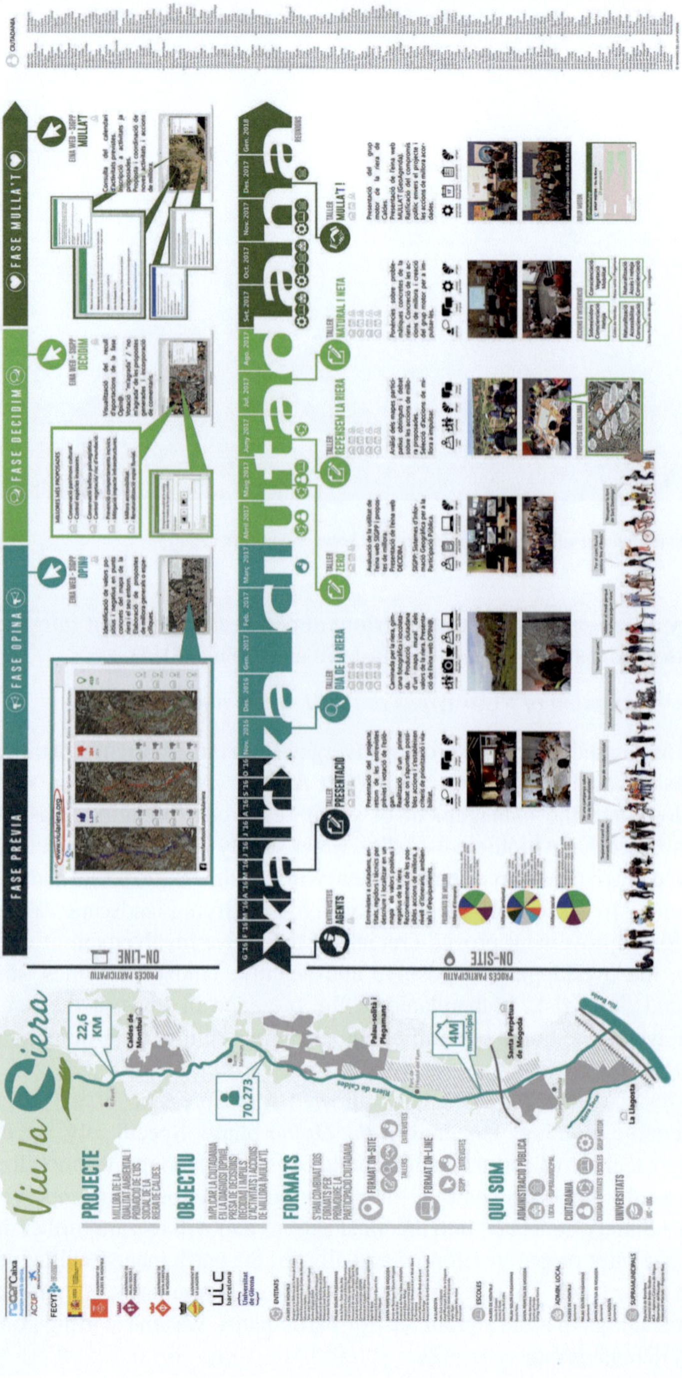

Fig. 14.4 Itinerant panel explaining the phases of *Viu la Riera!* Project. *Source* Authors (2017)

Fig. 14.5 Participative workshop with citizens of the basin to share knowledge and learnings obtained during the project 'Viu la Riera!' 17th of May 2017. *Source* Authors (2017)

14.4 Concluding Remarks

After forty years of policies aimed at recovering Catalan rivers, the river corridors of the Besòs basin have become true metropolitan public spaces. Different instruments are needed to highlight their collective value and increase the involvement of the population in caring for them (Fig. 8.5). Within this context, geo-ethnographic diagnosis can provide the necessary experiential knowledge to feed the ongoing participatory regeneration processes. This qualitative research methodology is particularly appropriate at the present time, when growing social interest in the environmental and landscape values of metropolitan river areas converges with the economic savings of co-responsible management.

Either way, the convenience of incorporating experiential knowledge within the participatory territorial project of rivers and streams should be understood in conjunction with the policies deployed in favour of new landscape and water cultures to date (Fig. 14.5).

References

AADD (1994) Àrea: Revista de debats territorials, 4. Diputació de Barcelona
AADD (1999) «Ciutat i riu». Notes, 13. Centre d'Estudis Molletans

Alarcón A (1999) La recuperació del riu Besòs a la desembocadura. Notes 13:67–77

Alonso-Monasterio M, Alonso-Monasterio P, Viñals JM (2015) 'Natusers' motivations and attitudes in urban green corridors: challenges and opportunities. Case Study of the Parc Fluvial del Turia (Spain). Boletín de la Asociación de Geógrafos Españoles, 68(2):369–383

Batlle E (2011) El jardín de la metrópoli: Del paisaje romántico al espacio libre para una sociedad sostenible. Gustavo Gili

Benages M, Di Masso A, Porcel S, Pol E, Vall-Casas P (2015) Revisiting the appropriation of space in metropolitan river corridors. J Environ Psychol 42:1–15

Benages M, Vall P (2014) Vers la recuperació dels corredors fluvials metropolitans. El cas de la conca del Besòs a la regió metropolitana de Barcelona. Documents d'Anàlisi Geogràfica, 60(1):5–30

Benages M (2011) Re-pensar la urbanidad en zonas residenciales de baja densidad desde la mirada comunitaria. In Vall P, Mendoza C, Cuéllar A, Carracedo O, Benages M (eds) LAU II. Regeneración de territorios intermedios. Repensar Encamp. Universitat Internacional de Catalunya, 82–87

Buijs AE (2009) Public support for river restoration. A mixed-method study into local residents' support for and framing of river management and ecological restoration in the Dutch Floodplains. J Environ Manage 90(8):2680–2689

Busquets, J. (2005). Per una nova cultura del paisatge. Escola Catalana. Del paisatge i les persones, 417, 6–9. Òmnium Cultural

Cantó J, Guardiola J, Salvatella N (1975) Evolución de la polución del agua del río Llobregat. Agua 91:15–24

Casado-Arzuaga I, Madariaga I, Onaindia M (2013) Perception, demand and user contribution to ecosystem services in the Bilbao metropolitan greenbelt. J Environ Manage 129:33–43

Comisión Provincial de Urbanismo de Barcelona (1963) Plan Provincial de Barcelona. Col·legi d'Arquitectes de Catalunya, Arxiu Històric, Manuel Baldrich Tibau, H 110A/7/102–113: Estructura urbanística i àrees de desenvolupament preferent

Consell d'Europa (2000a) Directiva Marc de l'Aigua. 2000/60/CE de 23 d'octubre de 2000

Consell d'Europa (2000b) Conveni Europeu del Paisatge [on-line]. <http://www.magrama.gob.es/es/desarrollo-rural/temas/desarrollo-territorial/09047122800d2b59_tcm7–26223.pdf> [Consulta: 16 maig 2012]

Consorci per a la Defensa de la Conca del Riu Besòs (2000). Pla Director dels Espais Fluvials de la Conca del Riu Besòs

Cortina A (2010) Nova cultura del territori i ètica del paisatge. Generalitat de Catalunya. Documents de Recerca, 17

Dantí, J. (Coord.) (2010). L'Aigua i el patrimoni històric industrial a la conca del Besòs. Granollers: Consorci per a la Defensa de la Conca del Riu Besòs

Díaz E, Queralt E (1970) Estudio de la polución de la cuenca del Pirineo Oriental. Documentos De Investigación Hidrológica 9:269–304

Everard M, Moggridge HL (2012) Rediscovering the value of urban rivers. Urban Ecosyst 15(2):293–314

Fernández L, Prat C (Coords) (2004) Quatre itineraris pel Ripoll a Sabadell: Guia del patrimoni fluvial. Amics del Ripoll

Findlay SJ, Taylor MP (2006) Why rehabilitate urban river systems? Area 38(3):312–325

Font A, Llop C, Vilanova JM (1999) La construcció del territori metropolità: Morfogènesi de la regió urbana de Barcelona. Mancomunitat de Municipis de l'Àrea Metropolitana de Barcelona

Font A (ed) (2004) L'explosió de la ciutat: Transformacions territorials recents en les regions urbanes de l'Europa Meridional. Col·legi d'Arquitectes de Catalunya/Fòrum Universal de les Cultures de Barcelona

Forman RTT (2004) Mosaico territorial para la región metropolitana de Barcelona. Gustavo Gili

Garcia X, Pargament D (2015) Rehabilitating rivers and enhancing ecosystem services in a water-scarcity context: the Yarqon river. Int J Water Resour Dev 31(1):73–87

Garcia X, Benages M, Pavón D, Ribas A, Garcia J, Vall P (2017) Public participation GIS for assessing landscape values and improvement preferences in urban stream corridors. Appl Geogr 87:184–196
Garcia X, Benages M, Buchecker M, Vall P (2020) River rehabilitation: preference factors and public participation implications. J Environ Planning Manage 63(9):1528–1549
García E, Godé Ll (2006) La recuperación del Baix Llobregat. Evolución histórica. Congreso Nacional del Medio Ambiente, 8, Desarrollo rural y conservación de la naturaleza, Agua. Vitoria-Gasteiz
Generalitat de Catalunya (1987) Actuacions Industrials de l'Institut Català del Sòl
Generalitat de Catalunya (1988) Actuacions Industrials de l'Institut Català del Sòl
Generalitat de Catalunya (2010b) Pla Territorial Metropolità de Barcelona (PTMB). Generalitat de Catalunya, núm. 5627, 12 de maig
Gordi J (2008) Els paisatges fluvials del Besós. Notes 23:105–128
Gordi, J. (Dir.) (2005) El paisatge fluvial a la conca del Besòs. Ahir, avui..., i demà? Consorci per a la Defensa de la Conca del Riu Besòs
Government of North-Rhine Westphalia (1989) Internationale Bauausstellung Emscher Park: Workshop for the Future of Old Industrial Areas. Memorandum on Content and Organization. Gelsenkirchen: Gesellschaft Internationale Bauausstellung
Grêt-Regamey A, Weibel B, Vollmer D, Burlando P, Girot C (2016) River rehabilitation as an opportunity for ecological landscape design. Sustain Cities Soc 20:142–146
Igalla M, Edelenbos J, van Meerkerk I (2019) Citizens in action, what do they accomplish? a systematic literature review of citizen initiatives, their main characteristics, outcomes, and factors. VOLUNTAS: Int J Voluntary Nonprofit Organ 30:1176–1194
Junker B, Buchecker M, Müller-Böker U (2007) Objectives of public participation: which actors should be involved in the decision making for river restorations? Water Resour Res 43(W10438)
Kaplan R, Kaplan S, Ryan RL (1998) With people in mind: design and management of everyday nature. Island Press
Llop C (2008) Paisatges metropolitans: policentrisme, dilatacions, multiperifèries i microperifèries: del paisatge clixé al paisatge calidoscopi. Papers 47:9–13
Lundy L, Wade R (2011) Integrating sciences to sustain urban ecosystem services. Progress Phys Geogr: Earth Environ 35(5):653–669
Margalef R, Prat N (1979) La Limnologia. Quaderns D'ecologia Aplicada 4:9–23
Morris N (2003) Health, well-being and open space. Literature review. Openspace Research Centre. Edinburgh College of Art. Heriot Watt University
Nassauer JI (2004) Monitoring the success of metropolitan wetland restorations: cultural sustainability and ecological function. Wetlands 24(4):756–765
Nel·lo O (2006) Els plans directors urbanístics: Una nova generació de plans. Espais, 52:3–11
Nilsson K, Nielsen AB (2006) Urban forestry for human health and wellbeing. Danish Center for Forest, Landscape and Planning
Nogué J, Puigbert L, Bretcha G (eds) (2008) Paisatge i salut. Observatori del Paisatge de Catalunya & Generalitat de Catalunya
Panareda JM (2009) Evolución en la percepción del paisaje de ribera. Boletín De La AGE 51:305–324
Pérez Cl, Marmolejo C (2008) La localización intrametropolitana de las actividades de la innovación: un análisis para la región metropolitana de Barcelona. Scripta Nova, XII, 270 (153)
Planas D, Vidal A, Folch R (1976) Problemàtica de les aigües continentals: Llibre blanc de la gestió de la natura als Països Catalans. Mem. Inst. Cat. Hist. Nat., 9. Barcino, 6–109
Platt RH (ed) (2006) The humane metropolis: people and nature in the 21st-century city. Lincoln Institute of Land Policy
Prat N (1979) La Xarxa Hidrogràfica. Quaderns D'ecologia Aplicada 4:87–107
Prat N, Rieradevall M (1992) La degradació del riu Besòs. Lauro: Revista del Museu de Granollers, 15–19

Prat N, Puig MÀ, González Gl (1982–1983) Predicció i control de la qualitat de les aigües dels rius Besòs i Llobregat. Diputació de Barcelona. Servei de Medi Ambient
Puchades JM (1948) El río Besós: estudio monográfico de hidrología fluvial. Miscelánea Almera. Diputació de Barcelona, 197–354
Queralt A (1974) La contaminación de las aguas. CAU 25:83–110
Robinson CJ, Bark RH, Garrick D, Pollino CA (2015) Sustaining local values through river basin governance: community-based initiatives in Australia's Murray-darling basin. J Environ Planning Manage 58(12):2212–2227
Ryan RL (1998) Local perceptions and values for a midwestern river corridor. Landsc Urban Plan 42(2–4):225–237
Rydin Y et al (2012) Shaping cities for health: complexity and the planning of urban environments in the 21st century. The Lancet 379:2079–2108
Simon A (1994) Per una política metropolitana d'espais lliures. Papers: regió metropolitana de Barcelona: Territori, estratègies, planejament, 20, 9–16
Tarroja A, Camagni R (Coords) (2006) Una nueva cultura del territorio: Criterios sociales y ambientales en las políticas y el gobierno del territorio. Diputació de Barcelona
Torra R, Farrero A, Ténez V (2008) La recuperació dels paisatges fluvials metropolitans: El projecte de recuperació ambiental i paisatgística del riu Llobregat a la comarca del Baix Llobregat. Papers 47:44–53
Trullén J (2003) Economia de l'arc tecnològic de la regió metropolitana de Barcelona: Delimitació de pols i corredors, estructura econòmica i indicadors d'economia del coneixement. Diputació de Barcelona. Sèrie Elements de Debat Territorial, 18
Vall P (2010) Territorios intermedios en la región metropolitana de Barcelona. Identidad y reciclaje. Ciudad y territorio. Estudios Territoriales, XIII 164:267–283
Vall P, Koschinsky J, Mendoza C (2011) Retrofitting suburbia through pre-urban patterns: introducing a European perspective. Urban Design Int 16:171–187
Vall P, Benages M, Elinbaum P, Garcia X, Mendoza C, Cuéllar AR (2019) From metropolitan rivers to civic corridors: assessing the evolution of the suburban landscape. Landsc Res 44(8):1014–1030
Vall P, Benages M, Garcia X, Cuéllar A, Pavon D, Ribas A (2021) Promoting citizen-based river groups at subbasin scale for multi-level river governance in the Barcelona Metropolitan Region. Local Environ 26(2):1–17
Vallerani F (2012) Franges hidràuliques, entre angoixes geogràfiques i estratègies de supervivència: El cas de la terra ferma de Venècia. A: Nogué J, Puigbert L, Bretcha G, Losantos A (eds) Franges: Els paisatges de la perifèria. Observatori del Paisatge de Catalunya, 229–252
Van Buuren A (2013) Knowledge for water governance: trends, limits, and challenges. Int J Water Governance 1:157–175
Vecslir L (2007) Paisajes De La Nueva Centralidad. Urban 12:34–55
Vendrell J (1994) Realitzacions i propostes metropolitanes en els espais públics: parcs, rius i platges. Papers 20:71–88
Vendrell J, Presmanes S (1993) La recuperación de los ámbitos fluviales metropolitanos de Barcelona. OP. Revista del Colegio de Ingenieros de Caminos, Canales y Puertos [on-line], 26
Vidal P (1999) El projecte del Parc Fluvial del Ripoll. Sabadell. Notes 13:91–116
Vidal T, Pol E (2005) La apropiación del espacio: Una propuesta teórica para comprender la vinculación entre las personas y los lugares. Anuario De Psicología 36(3):281–297
Vining J, Tyler E, Kweon BS (2000) Public values, opinions, and emotions in restoration controversies. In Restoring nature: perspectives from the social sciences and humanities, Gobster PH, Hull RB (eds). Island Press, 143–161
Zaitzevsky C (1982) Frederick Law Olmsted and the Boston Park System. Harvard University Press
Zaugg M (2002) More space for running waters: negotiating institutional change in the Swiss flood protection regime. GeoJournal 58(4):275–284

Chapter 15
An Educational Approach to River Restoration in the City

Agustín Cuello Gijón

Abstract The restoration of river systems in cities has a new role in the twenty-first century, in the context of resilient cities, climate change, the water transition or nature-based solutions. In these fluvial and urban transformation processes, education must have a relevant presence, as a necessary strategy to improve knowledge and consideration of rivers amongst citizens, thus enriching their participation in decision-making and commitments for action. This chapter will attempt to analyse aspects of river restoration that can facilitate the development of critical and participatory education in the city, characterising the intervention models that are best suited to teaching and learning processes in schools and citizens. Finally, with a practical nature, lines of action are proposed at a constructive and educational level.

Keywords River restoration · Civic education · Environmental education · Learnig city · Learning river

15.1 Introduction

The role of rivers in the historical and urban development of cities is well known, and it is used in formal teaching when discussing the great river civilisations such as the Egyptian or Mesopotamian. The most populous urban agglomerations on the planet, Shanghai, Karachi, New York, Cairo, etc., were born from rivers, and the largest European cities such as London, Paris, Rome or Lisbon, as well. In the Iberian Peninsula, a dense river network has historically structured the settlement system and as a consequence, although only 6% of the river courses are urban, 38 of the 47 provincial capitals of peninsular Spain are river-based, which represents 88.94% of

A. Cuello Gijón (✉)
Fundación Nueva Cultura del Agua (FNCA), Education Commission, Cádiz, Spain
e-mail: agustin.cuellogijon@mail.uca.es

© The Author(s), under exclusive license to Springer Nature Switzerland AG 2024
J. Farguell Pérez and A. Santasusagna Riu (eds.), *Urban and Metropolitan Rivers*, The Urban Book Series, https://doi.org/10.1007/978-3-031-62641-8_15

the capital's population.[1] The urban morphologies of river cities are due to the river, which defines the lines of defence against floods and a large part of the landmarks and communication axes through bridges and riverbanks. Cities depend on rivers to a greater or lesser extent, for their supply, food, energy, transport or waste disposal, which has led to a difficult but enriching coexistence.

The relationship between human settlements and their rivers has evolved unevenly, according to scales, cultures and geographical conditions, in the vast majority of cases reaching the total occupation of the river space, altering its dynamics and polluting the waters. In the Iberian basins, until the middle of the twentieth century, the physical connection between the cities and their rivers was maintained with a certain degree of naturality with fishing, milling in their mills, laundry washing or the keeping of livestock on the riverbanks being frequent. In later years, the regulation of rivers for agricultural, energy or supply uses and the implementation of structural flood defence measures, favoured urban growth towards the riverbanks and flood plains. Under these circumstances, the river environments of many cities reached a critical state of degradation, with concreted riverbeds and banks, unhealthy waters, simplification of landscapes and loss of ecosystem health. The dictatorship of the city over the natural environment led to the imprisonment of watercourses, the concealment of Sewer Rivers under traffic routes or diversion outside the city. With the rivers fragmented and expelled from their primitive routes in favour of the growth of the city (Cataldo Costa 2011), the population forgot about them, transforming the river spaces into pestilential and marginalised places. The city became a traumatic anecdote in the life of the river.

The situation in many cities in the last quarter of the last century was alarming, due to the state of the rivers and the lack of effectiveness of structural measures against floods, in addition to the increasing social pressure. All this led to the adoption of a new paradigm in the treatment of rivers, both in the hydrological and in the urban and environmental spheres (Santasusagna and Tort 2013). This new relationship between the city and the river, already advanced in many European countries, has led to the implementation of numerous and varied projects for recovery, refurbishment, environmental improvement, etc., with the intention of integrating the river into the city (ENRR 2022). This set of works, which we will call fluvial-urban interventions because they have an impact on the interface between the fluvial and urban spheres, has worked on urban, hydraulic, sanitary and environmental aspects, offering the city a new image of the fluvial space, healthier, more functional and public, revaluing its banks and facilitating the emergence of important economic synergies (Osuna 2014). However, there are well-founded doubts that citizens' relationships with their rivers have improved, that river city transformation operations have reached the educational sphere to enrich learning or environmental awareness or that water management has changed. On the contrary, the cognitive and affective distancing of citizens from their rivers remains, with social rejection of rivers being evident every time there are floods (Cuello 2018) or in the face of urban planning measures that limit the occupation

[1] According to population data from the Spanish National Institute of Statistics as of 1 January 2020.

of flood beds. The interventions have lacked significant participatory processes and have lacked the appropriate educational projection despite the enormous potential they have in this sense, by generating a large amount of information, creating new resources and recovering others, and even building facilities with enormous didactic possibilities.

The pedagogical interest of rivers is widely documented and validated by the educational community, with numerous experiences of their use in natural spaces and rural areas. The use of rivers in the city is less common, although their educational potential is much greater, as they integrate the river's own resources, those of the city and those generated by the relationship between the two systems: the river-urban interface. River restoration interventions further enrich educational resources and constitute a set of opportunities to improve teaching and learning processes, not only for students but also for teachers and citizens. At present, the known experiences are not good examples of this possibility because the necessary conditions that favour the encounter between river restoration and school education or citizenship education do not exist. To this end, river-urban interventions should have specific characteristics aimed to educational purposes, and education should also be improved to use the actions for learning purposes. The vocation and educational use of river restoration works would facilitate a better appreciation of rivers in the city, would reinforce the cognitive and affective connection with nature, culture and river heritage and would help to form more active citizens in the defence of their rivers. Finally, interventions in urban rivers can be a powerful resource for the development of environmental education in favour of rivers.

15.2 River-Urban Interventions

A historical review of urban transformations indicates that urban planning and engineering interventions in the fluvial-urban sphere have been constant, with the objective of minimising risk and increasing available land, in many cases with poor results, but with fatal consequences for the river and its environment. At present, river interventions are returning to the economic and environmental agenda of cities with a different focus, integrating the river into the urban system, valuing its ecosocial role as well as its hydrological and urban planning role. The paradigm of urban sustainability and the need to adjust water systems to social demands and climate change require new approaches to the role of rivers in the city, where nature-based solutions such as green infrastructures and the naturalisation of watercourses are presented as the most effective and efficient ways of tackling the environmental, economic and social problems posed by cities.

The need and interest in the incorporation of nature-based measures in river restoration have led countries such as Spain (Ministry for Ecological Transition and Demographic Challenge) to assign part of the European Union-NextGeneration funds to facilitate local administrations and other social and academic entities, to develop river restoration projects in which this type of measures are key to obtaining such aid. In this context, projects should contribute to urban and river renaturation, as well as mitigate flood risks, increase green infrastructure and the connectivity of green and blue spaces, whilst increasing biodiversity in urban environments and its conservation. The lines of intervention proposed for nature-based river restoration include, amongst others, the recovery of copses and riverside woodlands, the enlargement of the fluvial space, the recessing of motes, the recovery of old riverbed branches, the renaturalisation of the riverbed and plains in urban areas for flood lamination, the elimination of transversal barriers, permeabilisation of obstacles, control of invasive exotic species, stabilisation of banks with bioengineering techniques and the removal of concrete and breakwaters. The idea is to integrate the river into the city, overcoming the concept of an obstacle or border to give it a structuring and organising function of the territory, incorporating natural and landscape elements into the city's planning (Andrés and Masiá 2010).

In the social context, governance and participatory processes are also timidly incorporated into water management instruments at different scales, involving the participation of social, professional and scientific actors to contribute knowledge and experience, raising public awareness of river restoration, transfer of acquired knowledge to other places with similar problems, as well as in gender mainstreaming to ensure the balanced participation of women and men at all levels: from decision-making to working teams.

From a viewpoint of urban and environmental sustainability, the proposed river and urban planning interventions based on " natural solutions must go beyond those that have predominated in the past, not only those of a structuralist and hydraulic type, which have proved to be ineffective and have had a very high environmental and social cost, but also those euphemistically calledriver restoration", based on cosmetic landscaping, land speculation or hydraulic adaptation. The new paradigm of intervention must approach real restoration, restore the functionality of the ecosystem, recover the space and freedom necessary for the river to adjust the processes of erosion and sedimentation to the conditions of circulation at any given time, allowing its evolution and self-regulation (Fernández Yuste 2012).

The difficulty lies in the existing obstructions and confinement in most urban scenarios, where it is only possible to develop specific elements or areas (Heinz 2007). The competition for space between urban uses, the use of water resources and the need to preserve river dynamics, make it difficult for a clearly artificial and rigid artefact such as the city to coexist with a fully natural, changing and flexible one such as the river (Ureña et al. 1999).

From the socio-cultural viewpoint, the objective must be directed towards the recovery of the identity and heritage value of the river by the citizens, the improvement of river knowledge amongst students, the change of attitude towards degradation and the construction of ideas, propositions and actions in favour of the vitality of the river. To achieve these objectives, it is necessary to incorporate specific educational measures in the whole process of river intervention, turning them into educational resources in themselves and generating others that can be used by the educational community and the general public. In addition to the hydraulic, urban planning and

Fig. 15.1 Rivers for citizen learning, Guadaira River, Sevilla (España) (*Source* Author)

environmental objectives of river and urban interventions, socio-cultural and educational objectives must be incorporated at the same level of detail and specificity (Fig. 15.1).

15.3 Educational Potential of the Changing River Environment

The climatic changes in which we are immersed predict rivers with lower water inflows, more intermittent flows, greater disconnection of their aquifers, sporadic and extreme floods and long periods of low water, an evolution towards more Mediterranean models such as wadis. Faced with this situation, in addition to restoration based on natural processes, a new approach to water management is needed in which educational measures are relevant, measures that address the problem of widespread ignorance and the low value that rivers have amongst the population. In this objective, the school population should be the target group, taking advantage of the educational potential of river and urban areas, especially those that have been the object of interventions with different characteristics and impact (Cataldo 2011; Cuello 2022).

There are numerous and diverse explications that justify the lack of knowledge and the lack of appreciation of the river in the city and that serve to guide possible solutions. As a general framework, the social perception of water has been commodified and the tyranny of the city has been imposed on the river space, reducing the sensory contact of citizens and young people, disassociating the impacts of the city from the river under attack (Naredo and Rueda 2010); on the other hand, the easy access to water in homes, offices or business masks the dependence that our daily life has on the river or aquifers. In terms of interventions, structural measures have led to physical and visual obstacles which, together with motorised transport, have contributed to the oblivion of the river to the point of making it disappear from everyday reality. In the projects, the usual practice is to ignore the educational aspects, tiptoeing over the need for participation and limiting information to mere publicity. Technical language, difficulties in accessing documents or the preservation of the works for security, reasons are also major obstacles, not to mention the lack of maintenance and abandonment of the works once they have been completed. In the educational field, the contents are usually far removed from the local reality, and in the specific case of the urban river reality and its problems, the treatment is scarce and can be improved (Ladrera 2016, Cuello and García Pérez 2019); school activities out the door have decreased notably, so that the students' knowledge of their river, the dependence on the city or the transformations that may be taking place is almost nil. It should be added that the generation that lived near the river and its resources is disappearing and with it a river culture of enormous heritage wealth. The recovery of the river for citizens requires its contextualisation in the urban and socio-environmental environment, as well as the development of river literacy strategies of a certain intensity and permanence, starting with the educational system (Fig. 15.2).

Fig. 15.2 Accessibility is key to a river of learning, Guadalete River, Arcos de la Frontera (España) (*Source* Author)

The framework for a literacy strategy must be situated in a process of water transition[2] in the river area, understood as a process of social change in favour of a new river culture in the city, based on sustainable and participatory water management, which harmonises urban development with the demands of climate change and with the natural recovery of the river's dynamics. With this idea in mind, educational action must generate knowledge and skills to critically and creatively address the problems posed by the river-urban reality and take advantage of it as a set of opportunities, considering the geographical, historical, economic, heritage and environmental prominence of rivers in our country (Aguilar and Del Moral 2008; CEDEX 2017). The aim is to place the river at the centre of a teaching and learning process that begins with the recognition of the river: a river flows through the city, to which it owes its history and development. The successive phases must recognise its characteristics in the context of the city and its territory, upstream and downstream (basin), assess its state based on the desirable river model, the values and services it provides or could provide and analyse the actions that should be taken or must be taken to achieve the desired river. In this educational process, river-urban interventions offer exceptional opportunities to provide resources of a very diverse nature, even more so if they incorporate this option in all their phases, from the study of the situation and data collection, the discussion and choice of solutions, designs, execution, to the evaluation of results.

The river in the city is a space with its own educational value as it facilitates the development of curricular educational content and the use of interdisciplinary methodologies, contributing to education and scientific training in a multiple dimension of development, knowledge and skills, attitudes and values (sensations, feelings) and action in favour of the socio-natural environment. The natural or forced transformations that the river has suffered in its relationship with the city enrich the educational capacity of the river, regardless of their environmental result, as in any case they facilitate the construction of knowledge, whether about the natural character of the river ecosystem or the problems caused by inappropriate actions. One of the didactic strengths of the urban river environment is the possibility offered by socio-environmental problems, to organise teaching and learning processes and the application of knowledge around them.

In any case, it is a question of reorienting the learning contents towards the river and urban reality, focusing educational activity on the integrated treatment of the river elements that are traditionally dealt with in a disjointed manner and with a disciplinary vision. The changing fluvial-urban environment has the enormous capacity to offer complex themes and problems that can be approached from a critical and globalised point of view, from the local to the territorial level and at different time scales. The contents of school work must incorporate urban management options based on nature, on the limits and change of activities, the options of degrowth and the

[2] The concept is a specification of Ecological Transition and is equivalent, for the purposes of this text, to hydrological transition. It is a current of thought, management and intervention in the territory in favour of the socio-environmental sustainability of water and water systems in the new scenarios of climate crisis.

technical and social fight against climate change in cities, without abandoning the emotional, sensitive, artistic aspects of enjoying the river in all its magnitude.

15.4 An Educational Dimension of River Restoration

It is clear that the objectives of river-urban interventions are fundamentally environmental, hydrological and urbanistic, although others of an economic, health or landscape nature can be derived from them. Also, as we have been saying, we defend the educational utility and therefore the need to incorporate, specifically and explicitly, actions in this direction. In this section we put forward proposals to reinforce the educational dimension of the river-urban interventions, essentially in the school sector, but with a civic projection, thus favouring the amplitude of the learning space, in the physical, cognitive and emotional spheres.

Assuming the above implies making changes in the conception of the intervention project and of the teams that design, direct and execute the works. It seems obvious that there is a need to incorporate professional profiles from the fields of social knowledge, education, sociology or social anthropology to complement the traditional fields of engineering, hydrology or architecture and to design and develop the connections between work and citizenship, work and school, as well as the ethnographic or identity aspects of the interventions.

In general terms, it should be considered that each river and its relationship with the city is a specific and unique case, and therefore requires a singular and detailed study, analysing specific problems and solutions (Conde 2011; Osuna 2014). Recovering areas of floods, meanders, flows, heritage, etc., together with the regeneration of urban environments, marginal neighbourhoods, degraded industrial areas, agricultural areas, infrastructures, providing the city with new open spaces, leisure and recreational areas, sports activities, etc., are actions in which the educational viewpoint must be incorporated into the design, execution and monitoring.

Intervening in the river must also include the social look and, by inclusion, the viewpoint educational. The improvement of interventions in this sense involves using parameters of a social nature in the analysis and evaluation of the different intervention models, such as the degree of public acceptance, the educational use that has been made or the level of recovery of river culture. In many cases, during the development of an intervention, there are opportunities for social activation that we can take advantage of to design participation strategies in educational terms, intentionally promoting knowledge of the actions, diagnosis and discussion of the problems to be addressed, the solutions adopted, the processes and techniques to be carried out, the expected results, periodic evaluations, etc.

Clearly state the intended river model and argue the solutions with similar examples in other basins or localities. Make accessible and understandable the information related to the diagnosis and definition of problems, the study of solutions, evaluation of alternatives, impact assessment, modelling, tests, etc. Teach and explain the works in real time, carry out dissemination activities in situ, with the necessary safety

Fig. 15.3 Landscape analysis with future teachers, Guadalquivir River, Córdoba (España) (*Source* Author)

measures, in order to show the work processes, materials, construction innovations, the role of science and technology, effects on the environment, etc. Develop support programmes for educational centres on the basis of collaboration agreements with provincial or local educational institutions, with the aim of advising or assisting teams of teachers who wish to incorporate the work processes into their school work (Fig. 15.3).

The publication of specific adapted documents, plans, videos, etc., which can be consulted online or on a specific website, will facilitate the use of information in the classroom and incorporate it into the teaching and learning processes in the classroom. It is very useful to have graphic documentation of a landscape nature, of the evolution of the river over the course of an intervention in order to be able to carry out diachronic studies, drawings and models of the future, designs of situations of the evolution of the work according to phases. It is advisable to prepare educational guides of the urban river sections, incorporating new elements generated or recovered with the works, their characteristics and details for learning. The companies carrying out the works have all the means and facilities to carry out this type of work and it would only be a matter of incorporating them into the technical specifications of the contract.

Better quality information improves participation, which is as necessary as it is ineffective because it is limited to legal requirements. An intervention that favours educational processes must incorporate, from the beginning, a planned process of participation that involves citizens and students in the different phases of the intervention. Public participation processes should serve to acquire reasoned criteria of opinion that facilitate progress towards a new vision of rivers. They should be based on technical proposals, not closed ones, that allow for the substantial contribution of people and are opportunities for collective learning and citizenship (González del Tánago 2007). They must also be flexible so that they can adapt and react to new circumstances.

15.5 Project Decisions and Elements With an Educational Approach

The improvement of communication and the creation of new accesses are a constant in river restorations, in the effort to open the river to the city, an opening that must also be cognitive and sensorial. An accessible river drastically improves its educational use. All the transit routes, such as the opening of paths, footbridges and pedestrianised bridges, spaces to be and contemplate, access to heritage sites or equipment, etc., are a clear improvement for the planning of recognition and study activities outside the classroom. The recovery of paths that link the city with its periphery through the riverbanks, even connecting riverside towns of a certain proximity, facilitates the carrying out of comparative studies between urban and peripheral river sections. Bridges stand out as privileged observation points over the riverbed and riverbanks, as constructive elements through which to analyse the history of certain technological elements, design, materials, etc., as well as the evolution of urban growth and the communications system.

Trails should, whenever possible, recover traditional paths and tracks, diversify the routes with diverse and attractive shapes, pavements and elements that motivate learning and facilitate the interpretation of the space through which they circulate; incorporate modules and dynamic information that help to recognise processes and realities of the environment, as well as viewpoints and views along the route; take into account links with bus stops, rest areas and safety areas with vehicle access for maintenance and support.

Accessibility to the sheet of water must be ensured. Physical contact with the water is essential as well as walking along the shore or close to it for a stretch in which different substrates, biotopes, ecosystems, etc. can be appreciated. Security measures, transit limitations due to environmental or heritage issues, risk of occupation, etc., must be made compatible with contact with the river in spaces and/or times that have been prepared or defined for this purpose; facilitate river transport tours for educational purposes with the reconstruction of abandoned structures or the construction of new piers.

In terms of educational installations, preference is given to the recovery of hydraulic engineering buildings such as mills, mini-hydroelectric power stations, ironworks, paper or textile factories (Fig. 15.4). The keys are sobriety, functionality and valuing the building as an exceptional container, as well as the elaboration of an exhibition argument that integrates constructive elements, technical processes linked to its former function, ethnographic, historical and cultural aspects, all in a simple way and with a professional and well-founded didactic base. The installations must have viable educational programmes, connected with other installations and with a stable and constantly updated team. Equipment linked to dams and reservoirs is also important, as it facilitates understanding of the city's dependence on the river for its supply, energy consumption or agricultural production, aspects that are difficult for the urban population to perceive (De la Lastra, Barrionuevo and Delgado 2016). Create or fit out multi-purpose spaces for meetings, conversation, debate, workshops for scientific, artistic or technological creation and any activity that generates learning. Environmental issues should permeate all educational activity in these installations.

Incorporate the landscape dimension into the design and execution of the actions, considering spaces for looking, enjoying the scene, observing in peace, painting, photographing, etc. Enhance sections with observation capacity, especially viewpoints and panoramic landmarks with elements that facilitate landscape interpretation (Pardo García 2015). Support, through urban, environmental or cultural planning, measures to protect the fluvial landscape and for places of privileged observation, geomorphological landmarks or geomorphsites (Reynard, Pica and Coratza

Fig. 15.4 Interpreting in a flour mill, Water Museum, Benamahoma (España) (*Source* Author)

2017) referring to places of geomorphological interest with unique views. Incorporate diversity in the landscape, alternations of open and closed areas, connections with city parks, creation of bands and corridors, varied treatments of light, colour, textures, etc. Produce didactic resources to facilitate knowledge and interpretation of the landscape.

The signposting and interpretation elements must be studied in detail, in terms of design and materials, correcting classic errors of poor location, inadequate information, lack of maintenance, overlapping of actions, etc. Establish maintenance and material replacement programmes to ensure the permanence of the actions over time. The lack of maintenance and the consequent vandalism is one of the biggest problems affecting the interventions once they have been completed, and the worst educational example. The lack of maintenance reinforces the ideas against river restoration, feeding rejectionist attitudes and increasing mistrust of future actions on the river.

Flood risk is often an argument used by part of the population to oppose river restoration models that propose river renaturation measures and other nature-based solutions. Intervention projects should address these situations with social, discussion, participation and education strategies that soften the conflicts generated by negative perceptions of the river. The problems that have led to historical flooding need to be highlighted and new solutions argued for. An interesting activity is to commemorate historical events in this sense, to recover traces and elements of memory.

15.6 A School Looking at the River

The deficiencies in the technical and political sphere of river-urban interventions, due to their disconnection from the citizenry and the school, have their parallel in the educational sphere due to their disconnection from the immediate reality, in our case the city's river reality. The incorporation of educational measures in river-urban interventions requires an educational system capable of using and improving them. Both systems must evolve in a joint and coordinated way, creating spaces for meeting, discussion and mutual use.

At present, there is a worrying distancing of educational activity from the problems of the environment, and those related to the river are no exception. The situation responds to diverse and intertwined issues, some related to the management system of schools and the administration, others to the curriculum, teaching materials or teachers' perceptions and others to pressures from society itself. The state of the environment also has an influence, the health of the rivers and the resources available, and this is where the above proposals come in.

An urban school that looks at its river must get as close and as good as possible to it, use it as a place to work, a laboratory of experiences, a space for acquiring data, sensations and perceptions. It must incorporate into its educational aims reflection and debate on the reality of the river in relation to the city and citizenship, as well as the generation of commitments and strategies that enable action to be taken in adverse environmental situations in order to transform and improve them. A school that looks at the river must take advantage of its integrative capacity, focus the curriculum on this capacity with holistic didactic activities and take into account the passage of time. It must seek opportunities to propose attractive projects that have an impact on the improvement of the environment, taking advantage of the resources available and the situations offered by the city. It is necessary to confront the school with the daily, uncertain, changing and complex reality, attending to the construction of useful and practical knowledge that allows it to understand and interpret it and generate greater analytical capacity (Rivas 2004).

One option with great didactic potential is to work on socio-environmental problems through school research processes, which facilitates the application of knowledge based on the analysis of a reality, assessing solutions and designing interventions. From this viewpoint, river restorations can fill school research processes with content, providing information and methodology duly adapted to the learning conditions of schoolchildren.

In this context, improving the treatment and presence of urban rivers in textbooks is an unavoidable necessity, which would facilitate a more rigorous knowledge of the river reality. Textbooks should substantially improve the treatment of water and river issues, facilitate the construction of more useful knowledge schemes and greater scientific soundness to understand current problems and counteract established discourses (Cuello and García Pérez 2019). It is worrying that the school textbooks do not mention the main European Directives on water, nor the river restoration programmes that have been developed in Spain over the last forty years; nor do they mention the activity of social and academic movements in favour of river restoration and the development of a new water culture. Research in this field highlights the need to give visibility to the river as an element of urban identity, valuing the socio-environmental services that it offers, as well as to focus on river restoration with nature-based measures as an appropriate response to climate change.

River restoration has the responsibility of improving the concept of the river that is often generated in schools, reinforced by inadequate information in school textbooks. The recovery of the naturality of the river space returns sediments, riverbank vegetation, dead trunks and branches, fauna, flow fluctuations and the irregularity of the riverbed to the river, overcoming the textbook definitions of river-channel (Antoranz and Martínez Gil 2003; Ladrera et al. 2020; Ollero 2017).

The extensions of river restoration beyond the urban space, by means of riverside paths or intervening in areas close to the city, make it possible to extend the spaces for observation and study, establishing connections between the urban stretch of the river and the geographical context in which the city is located, relating geomorphological processes, riverside environments or land use, etc., that are present in different parts of the river. The aim is to provoke learning situations that are somewhat more elaborate than mere observation, to facilitate the construction of relationships, peripheral problems, comparisons, approaching the river in a complex way.

In the context of a fluvial-urban intervention, the school that looks at the river incorporates participation activities in its teaching and learning processes, in order to be able to respond to the calls that the team managing the restoration will make. This type of activity probably has no decision-making capacity, but it does help to learn the techniques and tools of participation, to reinforce knowledge and argumentation and, perhaps, to contribute interesting ideas on a specific aspect of the work.

The monitoring of a river restoration project as an educational resource is in any case an excellent opportunity for the development of environmental education in the classroom and, by extension, in the whole school. Environmental education is defined as education for critical social action in favour of sustainability in a broad sense and can help to achieve a change in the perception of rivers in the city, what we have called the water transition. Education in the school environment should aim to construct a concept of the river in which all of its elements interact: water, solids and other drags, biological diversity, etc., considering the need for sufficient space for the mobility of the river, as well as for water exchanges in depth and laterally, recovering its evocative and playful capacity (Martínez Gil 2010; Ollero 2012). With this objective in mind, the educational measures incorporated into river restoration must facilitate this learning process, through appropriate and sustained actions over time.

References

Aguilar Alba M, Moral Ituarte LD (2008) Evolución 's aportaciones en embalses de cabecera del Guadalquivir: relación con las tendencias climáticas recientes y repercusión en la planificación hidrológica. In Congreso Ibérico Sobre Gestión y Planificación del Agua (8.2008. Vitoria), 50–60. Fundación Nueva Cultura del Agua

Andrés Mateo C, Masiá González Ll (2010) El Miño, configurador de paisajes en su recorrido urbano. En Cornejo C, Morán J, Prada J (Coords.) Ciudad, territorio y paisaje: Reflexiones para un debate multidisciplinar, pp 343–353

Antoranz M, Martínez Gil J (2003) El agua y el sistema educativo español. En Del Moral L, Arrojo P (Coords) La directiva marco del agua: realidades y futuros. Fundación Nueva Cultura del Agua, 385–424

Cataldo Costa R (2011) Parques fluviais na revitalização de rios e córregos urbanos. Doctoral Dissertação. Universidade Federal do Rio Grande–FURG

CEDEX (2017) Evaluación del impacto del cambio climático en los recursos hídricos y sequías en España. Centro de Estudios Hidrográficos. Memoria

Conde O et al (2011) Perspectivas en la búsqueda de soluciones de mitigación al riesgo de inundación en la ribera alta aragonesa del Ebro. I Congreso sobre gestión y restauración de ríos. León

Cuello Gijón A, García Pérez FF (2019) ¿Ayudan los libros de texto a comprender la realidad fluvial de la ciudad? Revista de Humanidades, UNED, 37, 209–234. Disponible en http://revistas.uned.es/index.php/rdh/article/view/22895

Cuello Gijón A (2018) Las inundaciones del invierno 2009–2010 en la prensa, un recurso educativo para las ciencias sociales. Revista de Investigación en Didáctica de las Ciencias Sociales REIDICS. 2:70–87. Universidad de Extremadura. https://mascvuex.unex.es/revistas/index.php/reidics/article/view/3039

Cuello Gijón A (2022) Los entornos fluviales urbanos como recurso para la educación ambiental. Estudio de casos en los ríos Guadalquivir y Guadalete en Andalucía. (Tesis Doctoral). Universidad de Sevilla. https://idus.us.es/handle/11441/140111

ENRR (2022) Estrategia Nacional de Restauración de Ríos 2022–2030. Ministerio para la Transición Ecológica y Reto Demográfico. Secretaría de Estado de Medio Ambiente. https://www.miteco.gob.es/content/dam/miteco/images/es/borrador-enrr_tcm30-547863.pdf

Fernández Yuste JA (2012) Principios básicos de la restauración de ríos en entornos urbanos. El caso de la rehabilitación del río Huécar a su paso por Cuenca. En: XXXIX Congreso Nacional de Parques y Jardines Públicos. León, España. http://oa.upm.es/20911/

González del Tánago M (2007) Mirando al futuro, la concreción espacial: restauración fluvial. Jornada de Seguimiento de la implementación de la Directiva Marco del Agua en España, Madrid

Heinz Patt H (2007) Experiencias de rehabilitación y restauración de tramos urbanos (Review of the Development of Urban Rivers in Germany). En Restauración de Ríos II Seminario Internacional, Madrid

Jaso León C, Bastida Colomina G, Ibarra Murillo J (2002) Valoración de obras de restauración fluvial en Navarra: criterios de evaluación. Ecosistemas, vol 11, 1 http://hdl.handle.net/10045/9840

Ladrera Fernández R, Rodríguez-Lozano P, Verkaik I, Díez JR (2020) What do students know about rivers and their management? analysis by educational stages and territories. Sustainability 12(20):8719

Ladrera Fernández R, Prat N (2016) Las políticas europeas y el consenso científico en materia de gestión y conservación de aguas no llegan a la escuela. Fundación Nueva Cultura del Agua (Ed.), IX Congrés Ibèric de Gestió i Planificació de l'Aigua. 637–648 Universidad de Valencia

La de Lastra Valdor I, Barrionuevo Ferrer A, Delgado López C (2016) La dinamización social de los embalses como lugares públicos del territorio. Una cuestión pendiente. Los embalses de Minilla y Gergal en Sevilla. En Actas del IX Congrés Ibèric de Gestió i Planificació de l'Aigua. Universidad de Valencia y Fundación Nueva Cultura del Agua, 525–538

Martínez Gil J (2010) La experiencia fluviofeliz. Edit. Fundación Nueva Cultura del Agua

Naredo JM, Rueda S (2010) La ciudad sostenible: resumen y conclusiones. En Ciudades para un futuro sostenible. http://habitat.aq.upm.es/cs/p2/a010.html

Ollero Ojeda A (2012) Territorios fluviales sostenibles. Iraungarritasunari buruzko III. Jardunaldiak Arabako Campusean. Universidad de Álava. En http://docplayer.es/39774157-Territorios-fluviales-sostenibles.html

Ollero Ojeda A (2017) Hidrogeomorfología y geodiversidad: el patrimonio fluvial. Centro de Documentación del Agua y el Medio Ambiente (CDAMAZ). Ayuntamiento de Zaragoza

Osuna Pérez F (2014) Córdoba y el Guadalquivir: construcción de un ideario de futuro. Secretariado de Publicaciones, Universidad de Sevilla

Pardo García SM (2015) Las vistas panorámicas de núcleos urbanos: propuesta para su análisis y aplicación al caso de Andalucía. (Tesis doctoral). Universidad de Málaga

Reynard E, Pica A, Coratza P (2017) Urban geomorphological heritage. An overview. Quaestiones geographicae, 36(3)

Rivas P (2004) La formación docente, realidad y retos en la sociedad del conocimiento Educere, 24:57–62

Santasusagna Riu A, Tort Donada J (2013) A propósito de la interfaz ciudad-río. Retos y oportunidades de los espacios fluviales urbanos. En VIII Congresso Ibérico de Gestão e Planeamento da Água, 565–575. Fundación Nueva Cultura del Agua

Ureña JM, Ascorbe A, Canteras JC, Garmendia C, García Codrón JC, Liaño A, Sainz Borda A (1999) Ordenación de las Áreas Fluviales en las Ciudades: un enfoque metodológico. OP: revista del Colegio de Ingenieros de Caminos. Canales y Puertos 46:4–17

Chapter 16
Reconstructing a City's Evolving Physical Environment Through Its Hydronymy: Barcelona as a Case Study

Joan Tort Donada and César López Leiva

Abstract Current research on the urban growth of Barcelona emphasizes the relevance of its physical environment and, in particular, of its hydrographic network, in the construction, development and transformation of this Catalan city. Barcelona's hydrography can be considered a paradigm of the "Mediterranean city": that is, an urban core in which physical environmental factors, above all those related to *water* (i.e., a coastal site influenced by a series of regionally significant river courses and an internal drainage dependent on a network, in the case of Catalonia, of *rieras* and *torrents*—ephemeral gully streams), have been fundamental in its growth. In this framework, we conduct a study of social perceptions of the importance of the "hydrographic environment" in the configuration of the city, focusing our analysis on the thematic thread provided by the city's *hydronymy* (that is, the names given to these streams and rivers).

Keywords Barcelona · Mediterranean environment · Urbanization process · Intermittent streams · Hydronymy

16.1 Introduction

By taking both a prospective and analytical approach but, also, by adopting a historical-evolutionary perspective, we seek to reflect on how the physical environment of the city of Barcelona is perceived—that is, as a space heavily conditioned over the centuries by its hydrographic network or, more strictly speaking,

J. Tort Donada (✉)
Department of Geography. GRAM (Grup de Rercerca Ambiental Mediterrània) and Water Research Institute (IdRA), University of Barcelona (UB), Barcelona, Spain
e-mail: jtort@ub.edu

C. López Leiva
School of Forest Engineering and Natural Resources (ETSI), Polytechnic University of Madrid (UPM), Madrid, Spain
e-mail: cesar.lopez@upm.es

© The Author(s), under exclusive license to Springer Nature Switzerland AG 2024

J. Farguell Pérez and A. Santasusagna Riu (eds.), *Urban and Metropolitan Rivers*, The Urban Book Series, https://doi.org/10.1007/978-3-031-62641-8_16

micro-network—and based on the premise that at the heart of this reflection lies what we identify as the key element of the perception of any inhabited geographical environment: its *toponymy*.

The choice of the city of Barcelona as the focus for such a study is by no means accidental. Barcelona, in many ways, is a paradigmatic example of a *Mediterranean* city and not simply because it constitutes one of the most important urban areas on the western coast of the Mediterranean Sea, but, also, because its *Mediterraneanity*, that essential quality associated with its geographical location and the city's latitudinal coordinate in the northern hemisphere, is explicitly manifest in its hydrography. Or, more simply stated, and as learned scholars of Barcelona's physical environment have stressed time and again (see, for example, Vila and Casassas 1974; Casassas 1974; Casassas and Cuixart 1983; Casassas and Riba 1992; Riba 1992 and 1993; Riba and Colombo 2009), any "geographical reading" of the city of Barcelona needs to take into account the constant interaction between its *natural* and its *built environment* and, within this framework, sight should not be lost of the conditioning role played by its hydrographic network—a network which, despite operating basically at the micro-scale, is characterized by a set of parameters that typify those of Mediterranean hydrology, that is, very low rates of discharge (in stark contrast with the transcendence of the network's flood episodes), its intermittent and irregular nature and the distributional randomness of its surface waters. The combination of these parameters means that the element of *water* in the city of Barcelona, as seen from the perspective of its use and exploitation, is a classic example of what economists refer to as a "scarce good", while from the perspective of urban and environmental planning it is a factor of the greatest significance, characterized by the complexities and hazards of its management.

The rest of this article, following on from this introduction, is organized as follows. The second section outlines the methodology we adopt and identifies the sources we draw on, with a particular emphasis on the relevance of the city's toponymy as a methodological tool. In the third section, we describe the general characteristics of our chosen area of study—that is, the city of Barcelona and its surrounding area. The fourth section, entitled *Analysis and interpretation of Barcelona's physical environment and its modern evolution, based on a selection of hydronyms*, represents the bulk of our research findings. Finally, we complete the discussion with a series of conclusions, which make up the fifth and last section.

16.2 Methodology and Sources. The Use of *Toponymy* in the Present Study

The main thread around which the present study is developed is that of Barcelona's toponymy or, more specifically, fifteen names from the city's geographical area linked to water (that is, *hydronyms*), and which we consider significant for interpreting the evolution of Barcelona's physical environment (that is, the area historically known as the *Pla de Barcelona* or the Barcelona Plain) and which highlight the importance of *water* as a determinant of maximum relevance. These fifteen names can be divided between four "estructural" hydronyms: *Mediterrània* [represented on the map by the number [1], *Besòs* [2], *Llobregat* [3] and *Rec Comtal* [4] and eleven "referential" hydronyms which, from east to west and from north to south, are: the *riera d'Horta* [5], *la Llacuna* [6], *el Bogatell* [7], *el Merdançar* [8], the *torrent de l'Olla* [9], the *riera d'en Malla-riera de Vallcaraca* [10], the *riera del Pi* [11], *la Rambla* [12], *el Cagalell* [13], the *riera de Magòria* [14] and the *riera Blanca* [15]. The distinction we draw between "structural" toponymy—understood as that which alludes to elements of the territory with an organizational or structural value at the general scale—and "referential" toponymy, which operates at the micro-scale and has a more specific value, has been used by the authors previously (Tort and Santasusagna 2023; López-Leiva 2016; López-Leiva and Tort 2018; Tort 2013; Tort 2014; Tort and Sancho 2014; and, albeit with a complementary character, within a similar analysis, Tort 2020a, 2020b; and Tort and Membrado 2018) and is used here because we consider it particularly apt for the purposes of the present study, above all, because it allows us to capture the spatial/physical significance of the toponyms operating at two distinct scales.

Mention should be made at the outset of the numerous sources (see the bibliography for details) related to Barcelona and its geographical environment, which allude directly to the fundamental questions we raise in our analysis, and, above all, to those that in recent years have specifically concerned themselves with the transformations of the urban environment as a result of the historical process (both ancient and modern) of urbanization. The works of Lluís Casassas, Pau Vila and Oriol Riba alluded to earlier—above all Riba and Colombo (2009)—are of particular relevance, as are those of Jaume Olivé (1993) Francesc Carreras i Candi (1918), Pierre Vilar (1964), Duran Sanpere (1972) and Josep Moran et al. (1980) insofar as they provide a general contextualization of the issues addressed by focusing on the city's historical geography.[1] As for more specifically toponymic and onomastic sources, mention must be made of the two major lexicographic and onomastic corpora that Joan Coromines dedicated to the Catalan language (1983–1991 and 1989–1997; and discussed in Tort 2020a, 2020b), and the work of reference that preceded them (Balari 1899), as well as the works that concerned themselves with aspects of a theoretical, conceptual or methodological nature related to place names and their spatial or geographical meaning. Indeed, the authors of the present study have, over the last few years, developed a common line of research centred on the exploration, at different levels, of the questions raised, and the results of these endeavours—collected, to some extent, in the studies included in the bibliography—have been an obligatory starting point and an indispensable basis for reflection and discussion for the present study.

We complement the above with what might be referred to as the bibliography of the sector (e.g., urban planning, geology, etc.), which has been used in a complementary manner and that corresponds essentially to the specific contributions made by

[1] Readers should consult the detailed bibliography at the end of this article.

Ildefons Cerdà (1991[1859], Maria Àngels Marquès (1984) and Josep M. Puchades (1948).

We should also mention here the map of Barcelona and its surrounding area that accompanies this article. Our purpose in drawing it has been to capture in a single cartographic document the basic spatial information (especially, as regards the hydrographic network) discussed throughout the text. The map shows the fifteen hydronyms selected for analysis (numbered from 1 to 15), a numbering that is also employed in the text to facilitate the reader's task.

16.3 The Geographical Context: *Pla De Barcelona*

The geographical area in which we focus our study is not so much the city of Barcelona, understood as a strictly urban entity, but rather the sum of the city and its immediate surrounding area. Specifically, we refer to a territory known historically as the *Pla de Barcelona*,[2] strategically located in the north-eastern corner of the Iberian Peninsula, and which is delimited very clearly by the four major natural features from which it is made up: the Mediterranean shoreline to the south-east; the mountains of the Serra de Collserola—with the crowning peak of Tibidabo (512 m) that overlooks the city—parallel to the shoreline, on the inland side; the final stretch of the Besòs river, in the north-east; and, the final stretch of the Llobregat river, lying behind the tectonic relief feature of Montjuïc (185 m) overlooking the city harbour, to the south-west (Fig. 16.1).

In practice, this consideration of the aforementioned geographical features at the regional scale of Catalonia highlights an important fact: namely, the high degree of centrality of the Barcelona area and its deep integration within the immediate hinterland and surrounding lands. An extract from the work of historian Pierre Vilar (1984), describing the location of Barcelona, is particularly illustrative in this regard:

> It is evident that its location was not only highly advantageous, but ultimately also decisive. Here, Barcelona was able to crystallize coastal life, establish itself as a single unit, allowing it to demand from the countryside what it might then convey to the sea. Along the unbroken but narrow shore, which forms the Maresme as far as Tordera, emerges an abrupt expansion, an advance of the plain, influenced by the small dip at the foot of an isolated horst [Montjuïc] and by the work of two rivers, which on this occasion are neither *rambles* or *rieres* [ephemeral streams or gullies], but which spring from the heartland of Catalonia and, before that, the Pyrenees [...], rivers that delimit [...], before the mountainous amphitheatre of Tibidabo, a space that it now invades, and which one day it dreams of occupying as it establishes itself as one of the most powerful urban organisms in the Mediterranean. (Vilar 1964, I: 298)

Physically, the central area of the *Pla de Barcelona*—coinciding with the present-day conurbation of Barcelona—occupies a roughly 6-km wide coastal plain, extending from the foot of the Serra de Collserola to the coast that slopes uninterruptedly to the sea. The plain itself is easily confused on both sides with the low-lying lands of

[2] Name documented as early as the tenth century, according to Moran (1982).

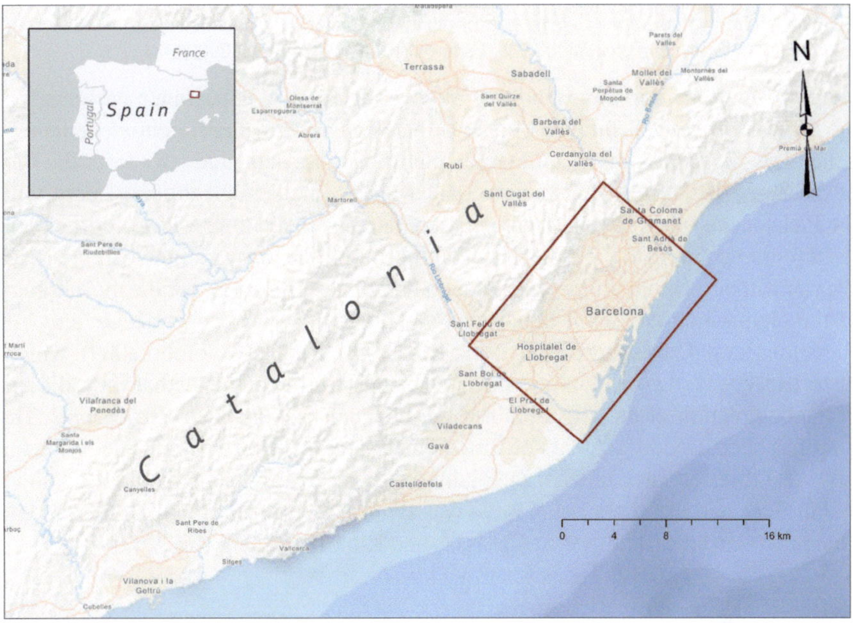

Fig. 16.1 Location of the studied area, in the context of the Iberian Peninsula and Catalonia (*Source* Authors)

the deltas of the Llobregat, to the south-west, occupying some 90 km^2, and those of the Besòs, to the north-east, occupying some 16 km^2. Both rivers cross the Serra forming narrow passes—Montcada in the case of the Besòs and Martorell in that of the Llobregat—separated by a distance of less than twenty kilometres and which have historically channelled the communication infrastructure between the coast and inland Catalonia. However, within this general framework, it is important to stress—above and beyond, that is, the role played by the two rivers in delimiting the Barcelona area—the relevance of Barcelona's *minor* hydrographic network: the less visible, less obvious stream network, to the extent that it has been profoundly altered, and even made invisible, by the transformation and expansion, over time, of the *built* city. However, the apparent invisibility of the minor hydrographic network in Barcelona today does not mean we should ignore its relevance; rather, just the opposite: a detailed understanding should enable us to appreciate the significance of the changes (from rural to urban) undergone by the morphology of Barcelona from the Middle Ages down to the present day. The extract we reproduce below is eloquent in this regard:

> The territory of Barcelona in the late Middle Ages (11th century) maintained its rural nature, unchanged, in terms of its relief and crops, and, above all, in the flow of its waters and its small lakes and marshlands (...). The *rieres* of Magòria, Creu d'en Malla, Malla, and the *torrents* of Olla, Vidalet and Pregon all flowed directly into the city. What's more, we should not forget the intermittent streams that flowed down from Montjuïc feeding into the Cagalell

> lake. All freely flowed crossing the territory that would later be surrounded by the mediaeval wall, work on which began in the 13th century. (Riba and Colombo 2009: 48)

Here, it should be borne in mind that the overall physical appearance presented by the *Pla de Barcelona* is far removed from that of a uniform, undifferentiated plain. In general, we are speaking of an inclined plain that descends gradually from the foot of the Serra de Collserola to the sea (and which coincides, for the most part, with the Pleistocene substrate), and which is separated from the lowest levels of the plain itself (corresponding to the most recent Holocene substrate) by a slight, barely visible escarpment or tectonic prominence, but one of great significance for the morphology of the city, which crosses it transversally. This is the so-called *graó barceloní* or Barcelona step (Casassas & Riba 1992; Riba 1993; Riba & Colombo 2009), which, historically, was of critical importance in the construction of the mediaeval walls and in the completion of the city's main water channel, the so-called *Rec Comtal*. The *graó* has been used by scholars of Barcelona to present what they consider to be the two great differentiated units of the plain: the lower Besòs deltaic plain, also known as *Pla Baix*, and the upper plain, known as *Pla Alt*, identified also as the Pleistocene *Samontà* (Carreras Candi 1918; Casassas and Riba 1992) (Fig. 16.2).

16.4 Analysis and Interpretation of Barcelona's Physical Environment and Its Modern Evolution, Based on a Selection of Hydronyms

16.4.1 First Level of Analysis: The Consideration of Four "Structural Hydronyms" in the General Perception of the Barcelona Territory

As discussed, we have selected four hydronyms from the geographical environment of Barcelona that might be considered as having a structural value throughout the history of the collective perception of this geographical environment (and which must therefore be understood within the framework of a scale that, taking the city of Barcelona as a reference, can be deemed as being general in nature): they are *Mediterrània* [Mediterranean], *Besòs*, *Llobregat* and *Rec Comtal*. The four names have in common the fact that they allude to the element of *water* as a constituent and that they allow us to establish—considered from the perspective of their significance, in space and time—correlations of sufficient relevance for the understanding of the Barcelona environment in terms of its evolution.

[1] ***Mediterrània*** [Mediterranean]. As noted in describing Barcelona's geography, the city's location on the western shores of the Mediterranean was a prime factor in appraising Barcelona's potential as an "urban centre" capable of structuring the territory at different scales. Here, we focus more specifically on the strict meaning of the toponym in relation to the city and should stress, from the outset,

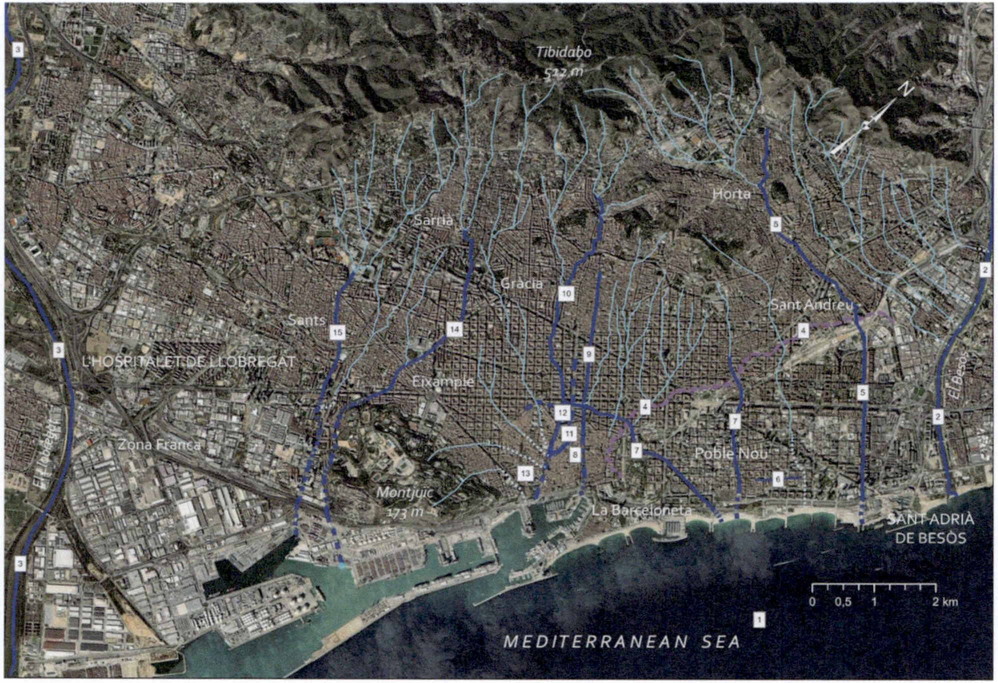

Fig. 16.2 General map of the surface hydrographic network of the city of Barcelona and its surroundings. As a reference we have taken the hydrographic network established by Olivé (1993) and adopted by Riba and Colombo (2009), on which we have highlighted the fifteen elements selected for our study and identified through their respective names, that is: four *structural hydronyms* (**1**. Mediterranean; **2**. Besòs; **3**. Llobregat; **4**. Rec Comtal) and eleven *referential hydronyms* (**5**. Riera d'Horta; **6**. La Llacuna; **7**. El Bogatell; **8**. El Merdançar; **9**. Torrent de l'Olla; **10**. Riera d'en Malla-Riera de Vallcarca; **11**. Riera del Pi; **12**. La Rambla; **13**. El Cagalell; **14**. Riera de Magòria; **15**. Riera Blanca). The courses represented by intermittent lines are only assumptions (*Source* Own elaboration based on MAXAR orthophotography July 2023)

a fact that might often not be that obvious: the *mar* [sea] (by way of synonymy, the *Mediterrània*), whether explicitly or implicitly, constitutes something of an undoubted and persistent virtual nature in the collective perception of the Barcelona space: and this perception has a toponymic translation in the sense that the name typically used to allude to this "dimension" of the city has not been the learned, more cultured form, but, simply, *la mar* or, alternatively, *la marina* [coast/shore]. Indeed, in relation to the latter, Coromines (*DECat*, V: 461) stresses that references to the "area along the coast" are the most deeply rooted popular expression in Catalan since the origins of the language, both in Barcelona as well as everywhere else in the language domain.

Yet, it should be borne in mind that Barcelona's *marina*—apart from being the usual way the city's residents allude to the city's maritime façade—has never in terms of its geography been looked upon with any degree of favour. On the contrary, the material qualities of this seafront for the development of a port that might be considered proportional to the power of Barcelona's urban reality have never been—and not by a long shot—the most suitable, given the accumulation of sand carried by the sea currents (from the mouth of the Besòs). Pierre Vilar, who has discussed the historical dimension of the problem, sums it up as follows (1964: 305): "Barcelona, a magnificent urban site, has been, like so many other Mediterranean ports, a mediocre maritime place".

[2] ***Besòs***. In the north-east corner, the rectangle that forms the *Pla de Barcelona* is limited by the final stretch of the Besòs river—which here takes the form of a delta: an area of less than 20 km^2, adjacent to the city (in the conurbation formed by Barcelona, Sant Adrià, Santa Coloma and Badalona), which, throughout the last century, has been completely urbanized. Yet, the river basin lies, in the main, outside Barcelona's geographical area. It is organized, in terms of its hydrography, like a fan—with five tributaries that converge in the Montcada gorge (the *congost de Montcada*: one of the most significant place names in the neighbourhood north of Barcelona). In fact, the name Besòs refers only to the river course downstream of the confluence of the Mogent (Fig. 16.3).
and the gorge (Puchades 1948). It should be stressed that we are speaking of a river with great hydrological irregularity throughout the year and throughout its history, with marked contrasts between prolonged periods of severe drought and massive, sudden and catastrophic floods: the so-called *besossades*. These are conditioned by the torrential nature and the rainfall regime of the Mediterranean climate. Indeed, this river, with all its idiosyncrasies, and despite its modesty, has been a territorial element of decisive importance in the general process of physical construction of the city:

> Historically, the Besòs, upon reaching the Delta, has diffused forming several distribution channels, which today no longer exist. The best-documented branch (...) is the one on the right bank of the river, and which was known as *Besòs Vell* to distinguish it from the present-day branch or *Besòs Nou* (...). *Besòs Vell* has disappeared today, but its course (...) served as the boundary for the parishes of Sant Adrià and Sant Martí de Provençals. (Riba and Colombo (2009: 158–160)

Fig. 16.3 **a** The Besòs as it enters the *Pla de Barcelona*, between Barcelona and Santa Coloma de Gramenet. At the bottom, the Montcada gorge: an obligatory passage for the main communication routes in the north of the city (*Source* Wikipedia Commons. Author: Amadalvarez (2004). b The Llobregat as it flows under the Mercabarna bridge, on the border between the districts of Barcelona, L'Hospitalet and El Prat de Llobregat) (*Source* Creative Commons. Author: Pere_prlpz 2020)

Its relevance has not diminished in modern times, despite the intensive urban occupation that has taken place and its physical degradation (until the great coastal renovation works of the 1990s and 2000s). In the words of the mentioned authors (2009: 163): "The urbanization of the Besòs riverbed and the Mar Vella [or Bella] beach took place in one of the most abandoned sectors of the big city (…). The anthropic degradation with which the Besòs river was afflicted in those years at the end of the century was more than evident".

[3] ***Llobregat***. In the south-west corner, the conventional delimitation of the Barcelona quadrangle corresponds to the course of another river, the Llobregat, the importance of which needs to be appraised, given, above all, the role it plays as a fundamental "connection" between the capital and the inland and northern regions of Catalonia. Its final stretch, as in the case of Besòs (albeit on a much greater scale of development), also coincides with a delta, physically separated from Barcelona, and from its central plane, by a raised fault block: the Montjuïc *horst*, which has been, at all times, of strategic importance for the city. At the beginning of the 1960s, Vilar characterized it as "a convenient field of urban extension: an easy place for contemporary industry, for railway stations, for the airfield, and free port projects. It is also the city's vegetable garden" (Vilar 1964: 302). Almost half a century later, and in the midst of the expansion on this "field of urban extension" of the communication, port and logistical infrastructure of the last two decades, Riba and Colombo (2009: 228) presented this geographical space as finding itself at the end of a cycle:

> The Llobregat Delta has gradually lost its natural functionality and the agricultural character that still characterized it in the 1930s. Today, many factors of degradation impact its environment: the lack of detrital flows, the flood defence dikes, the major works creating the port and airport, the deviation of the riverbed and the creation of industrial and service zones and the adjacent neighbourhoods of El Prat and L'Hospitalet, the motorway network and the new railway lines of the *Generalitat*, RENFE and the TGV, have led to a movement to protect what is left in an 'agricultural park'.

[4] ***Rec Comtal***. We complete this section of references to the hydronyms that might be considered as having structured or organized Barcelona's geographical environment by focusing on a key element in the hydraulic infrastructure of the city's history: the Rec Comtal. An irrigation channel documented since the tenth century, which runs over the *Pla de Barcelona* for twelve kilometres, from the Besòs water catchment to the lowest sector of Ciutat Vella, and which is extraordinarily representative of the agricultural and industrial development of the Barcelona area throughout its history and, moreover, which is perfectly adapted to the conditions of its topography (Martín 1999). From a geological perspective, Riba and Colombo (2009: 105) describe it as follows (Fig. 16.4).

> The channel known as Rec Comtal carried Barcelona's water supply from Montcada and Sant Andreu to Ciutat Vella. It was built on the surface of the Barcelonès Samontà, with a very gentle slope (of less than 1/1000), following, apparently, a route very similar to that of an ancient Roman road. The raised position of the river, close to the platform [Pleistocene],

Fig. 16.4 The so-called Montcada Mine, starting point of the Rec Comtal, in the municipality of Montcada i Reixac (*Source* Wikipedia Commons. Author: Pere_prlpz 2015)

> allowed the installation of water mills and dikes for the irrigation of the delta (...) and the supply of water to the city.

16.4.2 Second Level of Analysis: The Eleven "Referential Hydronyms" Selected as Being Representative of Barcelona's "Internal" Hydrographic Network

The eleven names we analyse and interpret below have been selected primarily for their representativeness (in relation, that is, to the geographical area of Barcelona) and because they have, within the framework of this study, an essentially "referential" character. Indeed, they can be considered as concrete markers or referents of the territory at a scale that coincides with the compact urban fabric of Barcelona: the sum of the mediaeval city—comprising the Barri Gòtic and Raval—and the modern city—comprising the 19th-century expansion or *Eixample* and the developments of the last century and a half, as well as the surrounding neighbourhoods that make up the periphery of Barcelona's urban fabric.

The order in which we have opted to analyse these hydronyms adheres, as mentioned, to a certain spatial order, that is, from north to south and from east to west. However, establishing their precise location is far from straightforward given that these hydrographic elements (be they phenomena of a point or linear nature) have undergone significant deviations or alterations through the ages, "fossilized", moreover, by large-scale urban planning projects. Here, we should also stress the arguments forwarded by scholars of Barcelona's physical environment regarding the "mobile" nature of these water courses, which has resulted in considerable changes in their location over time, attributable to both man-made diversions and natural avulsions (Travesset 1994).

[5] ***Riera d'Horta***. The *riera d'Horta* is considered one of the most important river courses between the Besòs and the Llobregat, and its basin, which drains a large part of the sunny *adret* slopes of the Serra de Collserola, is urbanized (except in its highest reaches) practically in its entirety. The *riera* drains an extensive sector of the eastern half of the *Pla de Barcelona* (coinciding, to a large extent, with the old district of Horta: a territory traditionally known for its abundance of water and its exploitation in the rural economy). Its final stretch coincides with the Rambla de Prim, where it has, in general, played a very important role in the morphological development of Poblenou. Geologists highlight its raised river course, in the form of a ridge, which extends between Sant Andreu and the sea, and the function it has served hydrologically, acting as a "partition" or "hinge" on the Poblenou plain. In this sense, its course has acted, historically, as a kind of "natural dike" blocking the passage of "the lateral flood waters of the Besòs towards the SW" (Riba and Colombo 2009: 146). The speed of the formation of the riera d'Horta compared to that of the Besòs Vell has meant that the "geomorphological capture" of one course of the other has not occurred. On the other hand, the aforementioned authors consider that the basin of the riera d'Horta, unlike most of the courses of the primitive network of the Collserola Samontà, remained intact until the nineteenth century.

[6] ***La Llacuna***. Coromines tells us (*OnCat*, V: 26) that the term *llacuna* [lagoon] corresponds to a "small lake", and that the term identifies the Barcelona neighbourhood referred to by this name, "established in the nineteenth century in a place that had been flooded by the sea waters". Located to the east of Bogatell (and documented for the first time in 1057, according to Moran 1980), it was a good example of the endorheic, closed basins and lagoons characteristic of Barcelona's *Pla Baix*. However, as Riba and Colombo (2009: 127–128) point out, these features were far from easy to map because their boundaries varied greatly depending on the season of the year and the rainfall regime: "No map after the seventeenth century records these expanses of water or the wetlands that surrounded them (…). These lagoons were very small in extent, ephemeral in nature, with their shallow, seasonal waters". These two authors further venture that the *llacuna* that survived longest was located between the confluence of the streets of Almogàvers and Pallars with Pere IV, and that its persistence is attributable to the "fact that it was fed with waters from different tributaries, including the Bogatell torrent". Among the factors that led to its disappearance were the numerous wells exploited intensively during the years

of nascent industrialization. This caused a drop in the piezometric surface and allowed the infiltration of rainwater and the loss of many wetlands. This desiccation was irreversible. Throughout the nineteenth century, according to the aforementioned authors, the *llacunes* disappeared.

[7] ***El Bogatell***. Balari records this Barcelona hydronym, as a derivative of the word *buc* (for its etymology see below) and discusses it extensively in the chapter of his work dedicated to "drainage channels" (Balari 1964 [1899]: 201–204). Specifically, he provides the following description: "The channel that, altering the natural course of the riera d'en Malla (…), collects the waters from Tibidabo and carries them to the sea is known as *Bogatell.* (…) Before this canal was built, the city of Barcelona was criss-crossed by torrents, still remembered as *rieres*—a name that some of its streets bear to this day. The alteration in the course by means of the *Bogatell* became necessary when the mediaeval walls, which were to surround the entire city until their demolition in 1854, were built".[3] Coromines, likewise, records this Barcelona hydronym but with the spelling *Bugatell* (*OnCat*, III: 135–136, i) and points out, in agreement with Balari, that the name is a diminutive of *buc* and that, etymologically, it is equivalent to the idea of an "inner cavity"—that is, of a tree, building or object, in which it has an important structural function (*DECat*, II: 312–315). Riba and Colombo point out (2009: 99) that the new canalization also collected water from the Olla and Vidalet torrents, and converged, near Portal Nou [sector of the present-day Arc del Triomf], with the waters of the Pregon torrent and the *riera* del Camí d'Horta". One of the streams involved in this task, apparently, was the Bogatell. The canal then headed right to the sea along *carrer* Taulat, the other side of the Old Cemetery. "This first channel was named *Bogatell Vell.* With the construction of the Citadel [1716–1725] it underwent an initial alteration; in the middle of the nineteenth century, the last artificial section, sloping down and rectilinear to the sea, was excavated (…) [and named] *Bogatell Nou*" (Riba and Colombo 2009: 99–101).

[8] ***El Merdançar***

Also known as *Areny del Merdançar*, this drain or sewer formed in the lower part of the city where another drain, known as the sewer of *carrer* Ample, converged with the Rec Comtal. From here, it continued downstream along the *carrers* del Rec and d'Ocata, until reaching the current outlet of the wastewater system in Barceloneta. Coromines points out that its name, also spelled *Merdançà*, and mentioned as early as 1029 ("torrente quam dicunt Merdancianum"), is based on the root MERD, applied to a remarkable number of rivers, all of them derived from MERDA [shit], "so that the waste accumulates around the towns" (*OnCat*, V: 257–258). Balari reports many other toponymic examples formed from this same root in Barcelona and in different parts of Catalonia (Balari 1964 [1899]: 149–151). In the case of the Merdançar sewer, a note introduced by Riba and Colombo (2009: 43–44) is worth highlighting:

---

[3] The italics are from Balari.

"All historians accept that the clean waters of the Rec Comtal became putrid as they crossed the city. This polluted urban stretch received the expressive name of *Merdançar*".

[9] ***Torrent de l'Olla***. We have selected this hydronym of a river course, located in a strategic sector of the *Pla de Barcelona* (approximately on the bisector that runs NW to SE, and which cuts the plain in roughly two equal halves, to the east and west), considering it highly representative of the many torrents that cross what we could call the "central nucleus" of the Barcelona conurbation. If the *riera d'Horta* was, in a certain sense, the characteristic hydrographic element of the neighbourhood of Horta, something similar might be said about the *torrent de l'Olla* in relation to the neighbourhood of Gràcia. Joan Lafarga describes it as follows: "The torrent de l'Olla runs straight, like a line drawn on a piece of paper, from above Vallcarca, crossing Gràcia from top to bottom, before continuing along the route occupied today by *carrer* Roger de Flor" (Lafarga 2001: 28). The conversion of the stream bed into an urbanized street occurred quite late: Carreras Candi (1918: 987) places it around 1883; however, this same author, documents the name (torrent de l'Olla or de Llepa-olles) to the thirteenth and fourteenth centuries (1918: 305). The toponym, whose etymology is probably related to *gorg* (Coromines, *DECat*, VI: 53–54, article *olla*), is one of the four street names formed with "torrent" in Gràcia (there is also *carrer* del Torrent de les Flors, de Can Mariner and de Can Vidalet). Mention should also be made of the *carrer* de la Riera de Sant Miquel and of the fact that the riera d'en Malla became the *avinguda* del Príncep d'Astúries in 1935. According to Riba and Colombo (2009: 59), at the end of the eighteenth century the torrent de l'Olla had already become a sewer, receiving the name of Sant Joan. It originated in Jonqueres, flowed along the *riera* of that same name (today, in part, Via Laietana), and in its final stretch it joined the waters of Rec Comtal forming the Merdançar.

[10] ***Riera d'en Malla-Riera de Vallcarca***. The headwaters of this long, complex river are known as the *riera d'en Malla*, while its lower course is known as the *riera de Vallcarca*. In its mountain course, as Riba and Colombo (2009: 52) remind us, it has been subject to many human interventions. Its basin is made up of the *torrent de Can Gomis* (today *carrer* Esteve Terrades) and the *torrent de les* Arenes (formerly the *avinguda* Hospital Militar), which come together under the Vallcarca bridge. A little further downstream, on the left bank, the course received the *torrent* de Farigola and, from there, took the name of *Riera de Vallcarca* (or *d'Arija*). It was here that the city's drinking water was first obtained: "Much of the drinking water that was carried by this *riera*, and that which originated from the springs, wells and mines in the old municipalities of Sant Gervasi, Gràcia and Horta, was collected and channelled directly to Ciutat Vella to supply the city with potable water", according to the aforementioned authors. The genealogy of the name (documented as *Valle Carcara*, 1124 and *Valle Carchara*, 1179) has, according to Coromines (*OnCat*, VII: 404–405), an interesting geographical dimension: "Vallcarca, the well-known hollow, covered for almost a hundred years by the small village of Barcelona, which

descends from the massif of Muntanya Pelada-Sant Josep-Carmel, towards the heights of Gràcia, between Josepets and the ring roads". Etymologically, Coromines writes, the toponym would appear to be related to CARCAR, a vulgar variant of the Latin CARCER: "prison", "enclosed place". Balari (1964 [1899]: 143), who notes that *Vallcarca* is an abbreviated form of *Vallcàrcara*, draws a similar conclusion: it is a name, he says, that in its widest sense is equivalent to a "narrow or closed passage similar to a prison".

[11] ***Riera del Pi***. The hydronym *riera* del Pi is included here as being representative of the intensely anthropized hydrology of the Ciutat Vella (especially, following the construction of the mediaeval walled enclosure). In fact, the river course originally flowed down from Collserola, between the riera d'en Malla and the torrent de l'Olla, and, before the walls, entered the city in a direction that today would be equivalent to the Portal de l'Ángel–Cucurulla–*carrer* Pi thalweg, then continuing towards the Rambla along *carrer* Cardenal Casañas—which, in the words of Riba and Colombo (2009: 57), "was the riera del Pi until the nineteenth century". The virtual nature of the name today, despite the loss of the original hydrographic reference, is probably related to the fact that the old river course crosses the neighbourhood known as Santa Maria del Pi, one of the best known in Ciutat Vella today. We should add, however, that there are elements of the present-day morphology of this urban sector that are of great interest as evidence of the old riverbed and of the processes of flooding and sedimentation that occurred there: for example, the meandering profile of *carrer* Cardenal Casañas or the "topographical anomaly of the Boqueria" (or the inversion of the slope between the central riverbed of the Rambla and its adjacent streets or former tributaries), which Casassas and Riba explain in great detail (1992: 16–19).

[12] ***La Rambla***. Within the *Ciutat Vella* (the name used to refer to the mediaeval centre of Barcelona, encircled with a city wall from the 13th–fourteenth centuries until 1854, when the demolition of the walls was decreed), the *Rambla*, the old riverbed of the final stretch of the Riera d'en Malla, and which would became, over time, the city's promenade par excellence, forms the dividing line between the two large sectors of the mediaeval city: the so-called *Barri Gòtic*, which grew up on its east side, and the *Raval*, which extends in its western corner. Having lost its functionality as a river course in the fourteenth century, once the walled perimeter was completed, the riverbed of the Rambla was used as a road and, later, developed as a pedestrianized street or promenade. Indeed, today, its physical environment is entirely urbanized, and only certain elements (such as the sustained slope, the over-elevation of the riverbed due to the historical process of sedimentation and the meandering tendency of the final stretch, before reaching the Columbus monument) bear testimony to its former role as a river course—which, on the other hand, is irrefutable evidence, as Casassas and Riba (1992: 16–17) emphasize, of the two bridges—excavated by archaeologists—that crossed it: one in front of the gateway to the Boqueria market and one in front of the iron gateway of Portaferrissa. The name, *rambla*, which Coromines points out is of Arabic etymology, being

derived from *raml*, "sandy riverbed" (*DECat*, VII: 79), is not documented until 1443 (Duran Sanpere 1972). It should be stressed, however, that the walling of the mediaeval city of Barcelona meant that, between the 14th and the middle of the nineteenth centuries, the city was protected from the dangers of flooding and that its only problems in relation to the hydrographic network were of an internal nature: the evacuation of rainwater and sewage accumulated within the walls—the result of the *impoldering* of the walled city, to cite Riba and Colombo (2009: 43) (Fig. 16.5).

[13] ***El Cagalell***. This hydronym serves as another example of a name with a direct relation to the drainage of urban waste and detritus. The name, in common with other comparable examples forwarded by Balari (1964 [1899]: 189–190), can be associated with the Latin verb CACARE, "to defecate". In discussing this name, Riba and Colombo (2009: 37) report: "To the north of the coastal strip of *carrer* Nou [that is, between Montjuïc and the Rambla] *Cagalell Vell* was formed, the most important wetland near the Rambla, with *carrer* de Sant Pau as its main axis". To the south of this cordon lies a series of depressions that the aforementioned authors propose referring to as *Cagalells Nous*. Both this and the former "have persisted, until the definitive urbanization of the Raval, as a swampy, unsanitary land (…). The flood-prone nature of Cagalell Vell persists to this day". The authors mention, in this regard, the floods of 1974 and 1995, and stress that the problem is attributable to the poor drainage provided by the old sewers. The first known document referring to *Cagalell Vell* dates from 1023. Balari quotes a document from 1104: "in territorio barchinone…, in stagnum quod vocatur cagalel" (1899: 189–190).

[14] ***Riera de Magòria***. The *riera* de Magòria, with its headwaters located on the central-western slope of the mountains of Serra de Collserola, crossed the district of Sarrià and used to serve as the boundary for the districts of Corts, Sants and Hostafrancs. It entered Ciutat Vella from the Raval side, until 1755, that is, when its waters were redirected towards the Llobregat (Riba and Colombo 2009: 51). Coromines, in *OnCat* (V: 131–133), informs us: "Riera de Magòria. On the western outskirts of Barcelona. Formerly the name of an extensive area, from where this stream originates, located perhaps around the present-day neighbourhood of Les Corts de Sarrià". The name has been documented from ancient times: "in termino de Mogoria" (1002); "in territorio Barchinone, ad ipsas Cortes, in loco qui vocatur Mogoria" (1066). Coromines hypothesizes that the toponym "of pre-Roman appearance and Indo-European flavour—Celtic or Sorothaptic—(…), rather than Ibero-Basque", derived possibly from Magauria or Mogauria, which might mean a 'large stream' (*OnCat*, V: 132).

[15] ***Riera Blanca***. Of all the watercourses making up the hydrological network of Barcelona's micro-urban environment, the *riera* Blanca is the one located furthest to the west of what is strictly the city's nucleus. Its course used to drain the south-east slope of the hill of Sant Pere Màrtir (in the SW sector of Serra de Collserola), discharging to the west of Montjuïc, in what is today the Zona Franca. Its name, which can be conceptualized as a chromato-toponym (that

Fig. 16.5 View of the final stretch of the Rambla, from the Columbus monument. Note how the slightly undulating shape of the promenade reproduces the meandering profile of the old riverbed (*Source* Wikimedia Commons. Author: Ralf Roletschek 2015)

is, one that alludes to the colour of the referent), can, most likely, be related to other toponyms in this part of the *Pla de Barcelona* (including *Pedralbes* and *Collblanc*) where chalk forms, dolomites and other whitish rocks outcrop at the level of the geological substrate. The *riera* collected the waters of the *torrent d'Escuder* and the *torrent de Sants*. Today, this watercourse is canalized and circulates beneath the city streets. Its lower course runs under the road identified as *carrer* de la Riera Blanca and constitutes the current boundary between the municipalities of L'Hospitalet de Llobregat and Barcelona (in mediaeval times, between the districts of Santa Maria de Sants and Santa Eulàlia de Provençana). This virtual border (totally at odds with the urban continuum affected, which clearly constitutes a single unit) was made all too evident—generating a political conflict that the media seized upon—during the COVID-19 pandemic in May 2020, when organizing the de-escalation of the lockdown by health regions.[4]

16.5 Discussion and Conclusions

In 1883, Jacint Verdaguer—one of the founding poets of modern Catalan literature—in his well-known ode, *A Barcelona*, celebrates the capital of Catalonia, invoking its urban vitality, at a time when it was undergoing a phase of great transformation ("from river to river it now extends"), and evoking its timeless link with the Mediterranean ("of the sea you remain queen/your sceptre its trident").[5]

This highly expressive image serves as an ideal point of reference for the final discussion of our article. By means of this simple allusion to the two rivers that delimit Barcelona's geographical space and to the sea front which, in one sense or another, has constantly conditioned the history of the city, Verdaguer, a veritable "poet-geographer",[6] conveys the idea of a city *in movement* in which the interaction with its surrounding environment provides us with the key to its *raison d'être*.

Here, within a more objective framework of analysis, but adopting the same points of reference (not only at the broad scale but also at the micro-scale), we have undertaken a study whose primary interest has been Barcelona's toponymy (or more specifically, its hydronymy). In practice, if on the one hand our consideration of the hydronyms that we have conceptualized as being *structural* in nature has allowed us to provide a context for the general geographical environment in which the city is sited, the eleven *referential* hydronyms, on the other, have made it possible for us to devise and conduct an internal survey of its urban space—that is, of the "built environment". Indeed, the analysis conducted at the micro-scale level has striven, at all times, to illustrate the interaction between the urban space and the natural environment, something that the toponyms, thanks to their intrinsic quality as geographical names, highlight by their very nature. Moreover, in undertaking this analysis, we have been able to draw on a significant number of bibliographical contributions which, from the most local to the more general scale, have examined in depth the geography of Barcelona over the course of the last five or six decades.

[4] https://elpais.com/cat/2020/05/06/catalunya/1588755347_118945.html last consulted on 25/8/2023.

[5] See the reference in the bibliography (Verdaguer, 1883). The ode, *A Barcelona* (and, above all, the metaphorical allusion to its location stretching "from river to river") has been used on multiple occasions in recent decades by Barcelona's public institutions—above all, since the 1992 Olympic Games—as a way to "illustrate" the progressive configuration of Barcelona as a metropolitan city.

[6] To cite Josep M. de Casacuberta, 1953: 95.

One conclusion we believe the preceding toponymic analysis has enabled us to draw is that of the transcendence of *water* as an element in the construction of the city throughout its history, at both of the scales at which we have conducted our study. Generally speaking, *water*, in the area of Barcelona, is an element that is not especially visible; yet, it is one that has been of fundamental importance at every moment in that history, whatever the scale of analysis and regardless of the sector of the city under consideration. It is clearly a critical element within the hydrographic network: a network that has just two water courses of any specific relevance and which can indeed be qualified as *rivers*, the Besòs and Llobregat, and, yet, paradoxically, both (especially the latter) are located in a somewhat peripheral location with respect to the "central urban agglomeration". However, the history of Barcelona would not be the same without the (fairly distinct) roles played by each. In the case of the Besòs, for example, despite its limited discharge, the fact that its aquifer has, since the tenth century, provided a continuous supply of water to the city via the Rec Comtal is illustrative of a singular expression of the aforementioned paradox: that is, the water is there, but it cannot be seen. Moreover, its materialization as a resource (or managing this resource in an ordered, rational manner compatible with the needs and interests of citizens) has always required a major effort, one that is especially evident throughout the city's history. Yet, this conclusion is one that to a large extent can be extended, in all probability, to many other cities of comparable dimensions sited along the Mediterranean coast.

However, we have shown that it is precisely at this micro-scale, where the specific characteristics of the Barcelona environment in relation to water and the *complexity* of the hydrographic network are most clearly apparent. Indeed, this complexity is evident (both today and when adopting a historical perspective) in the situation and functioning of the city's "hydrological system" bounded, to the east, by the final stretch of the Besòs, and, to the west, by the Llobregat. This complexity also emerges clearly in the case of the Riera d'Horta—an independent basin, and one of great relevance within the north-east quadrant of the Pla de Barcelona, yet, at the same time, indivisible from that of the Besòs as far as its geomorphology is concerned—as well as, on the other side of the *Pla*, in the cases of the Riera Blanca and Riera de Magòria (also constituting an independent basin, but with different dynamics with regards to their "urban" history: the former exemplifying an urbanized river course, but one that has not been diverted, and which serves today—despite all logic—as an inter-municipal boundary between Barcelona and l'Hospitalet; the latter exemplifying a course that was diverted in the direction of Llobregat back in the eighteenth century and which, until that date, was one of the many examples of Barcelona's *interior* river courses which, following the construction of the mediaeval walls, caused a serious obstacle for the drainage of water to the sea—especially in times of flooding). A similar complexity, albeit more intense in space and time (owing to the number of small inland river basins affected) is evident in relation to the Bogatell, a transversal *canal* which, since the construction of the walls, has diverted the waters of several *torrents* that flowed, from Collserola, directly into the sea, towards the plain that lies to the east of the walled city and that today corresponds to the neighbourhood of Poblenou; and, in any case, a channel that is inseparable, in its lowest reaches, from the wetlands and areas of lakes that characterized certain sectors of Barcelona's coastal plain before urbanization, as best typified by La Llacuna. It is precisely in these lowest sectors of the plain where we find other places in the city of a similar nature and which, historically, have acted as sewers or accumulation tanks for the city's detritus and wastewater: places like Cagalell and Merdançar, albeit located in different sites on the plain, very much speak for themselves. Finally, in the central sector of Barcelona's "hydrological system", we find what we could characterize as its own internal subsystem, given the specificities and interactions it presents: we refer to the area identified by the following hydronyms: torrent de l'Olla, riera d'en Malla-riera de Vallcarca, riera del Pi and La Rambla, river courses that, as we have seen above, have had a marked historical dynamic intensified by the process of Barcelona's urbanization. It is not for nothing that we speak of the "central core" of ancient and mediaeval Barcelona (with its three circuits of walls: the early Roman, the thirteenth century and the fourteenth century): a broad area, with a high density of human occupation lying adjacent to the coast. In other words, a series of ingredients that allows us to understand the special importance that its hydrological dynamics and processes of sedimentation have had throughout its history: a set of factors that have very directly conditioned, in this part of the city, its urban morphology, and that have their most highly symbolic manifestation in the Rambla (understood both as a river course and as an urbanized promenade).

In short, we believe we have demonstrated here that Barcelona is a highly relevant example of a Mediterranean city in which, as is often the case, the water factor and the hydrological network in general have been, and are (and, predictably, will continue to be) fundamental and of marked strategic significance.

Acknowledgements This paper has been prepared as part of the Research Project PID2021-126922NB-C21, supported by the Ministerio de Economía y Competitividad, Government of Spain (MINECO/FEDER, UE), and within the research group GRAM (Grup de Recerca Ambiental Mediterrània), supported by the Generalitat de Catalunya (2021SGR00859).

The authors express their gratitude to Iain K. Robinson for his linguistic advice, in the complex exercise of finding the most exact equivalence possible between the terms in English, Catalan and Spanish that form the empirical basis of this research.

References

Balari J (1899) Orígenes históricos de Cataluña. Establecimiento Tipográfico de Hijos de Jaime Jepús, Barcelona

Carreras Candi F (1918) Geografia General de Catalunya. Vol. III: La ciutat de Barcelona. Barcelona: Establiment editorial d'Albert Martin

Casacuberta JM (1953) Excursions i sojorns de Jacint Verdaguer a les contrades pirinenques. Barcino, Barcelona

Casassas L, Cuixart M (1983) Barcelona. Gran Geografia Comarcal de Catalunya, vol 18. Fundació Enciclopèdia Catalana, Barcelona, pp 80–129

Casassas L, Riba O (1992) Morfologia de la Rambla barcelonina. Treballs De La Societat Catalana De Geografia 33–34:9–29

Casassas L (1974) El Barcelonès i Barcelona Ciutat. In: Solé-Sabarís. L. (dir.) Geografia de Catalunya. Vol. III. Barcelona: Aedos: 641–686

Cerdà I (1991) Teoría de la construcción de las ciudades aplicada al proyecto de reforma y ensanche de Barcelona. Madrid: Instituto Nacional de la Administración Pública-Ajuntament de Barcelona.[Original work: 1859]

Coromines J (1983–1991). Diccionari etimològic i complementari de la llengua catalana (9 vol). Barcelona: Curial-Caixa de Pensions. Abbreviation: DECat

Coromines J (1989–1997). Onomasticon Cataloniae (8 vols). Barcelona: Curial-Caixa de Pensions. Abbreviation: OnCat

Duran Sanpere A (1972) Barcelona i la seva història: la formació d'una gran ciutat. Curial, Barcelona

Lafarga J (2001) Gràcia: de rural a urbana. Història d'un territori. Barcelona: Taller d'Història de Gràcia

López-Leiva C, Tort J (2018) Toponyms related to plants in transitional vegetation areas: how diversity is conveyed by place names. Onomastica Uralica 11:117–130

López-Leiva C (2016) Onomástica, ecología y territorio. La toponimia de La Rioja como indicador biogeográfico y de la dinámica del paisaje forestal. Ph. Thesis. Madrid: Universidad Politécnica de Madrid 2016. https://oa.upm.es/39636/1/CESAR_LOPEZ_LEIVA.pdf

Marquès MA (1984) Les formacions quaternàries del delta del Llobregat. Institut d'Estudis Catalans, Barcelona

Martín, M. (1999). El Rec Comtal (1822–1879). La lluita per l'aigua a la Barcelona del segle XIX. Barcelona: Fundació Salvador Vives i Casajuana

Moran J et al (1982) El Barcelonès. Gran Geografia Comarcal de Catalunya, vol 8. Fundació Enciclopèdia Catalana, Barcelona, pp 11–337

Moran J (1980) Notes de toponímia antiga del Pla de Barcelona. In: Estudis de llengua i literatura catalanes-I (Homenatge a Josep M. de Casacuberta). Barcelona: Publicacions de l'Abadia de Montserrat: 103–115

Olivé J (1993) Les rieres del Pla de Barcelona a mitjan segle XIX. III Congrés d'Història del Pla de Barcelona, vol 2. Institut Municipal d'Història-Ajuntament de Barcelona, Barcelona, pp 399–408

Puchades JM (1948) El río Besòs. Estudio monográfico de hidrología fluvial. In: Miscelánea Almera. Barcelona: Diputación Provincial de Barcelona, 195–354

Riba O (1992) La Rambla de Barcelona: passeig i riera. Muntanya 781:97–100

Riba O (1993) Assaig sobre la geomorfologia medieval de la Ciutat Vella de Barcelona. III Congrés d'Història del Pla de Barcelona, vol 1. Institut Municipal d'Història-Ajuntament de Barcelona, Barcelona, pp 171–176

Riba, O., Colombo F (2009) Barcelona: la Ciutat Vella i el Poble Nou. Assaig de geologia urbana. Barcelona: Institut d'Estudis Catalans-Reial Acadèmia de Ciències i Arts de Barcelona

Tort J (2013) Onomastics in the public space of Barcelona. A compared study between the Old City and the Eixample District. In: Felecan O, Bughesiu A (eds) Onomastics in contemporary public space. Cambridge Scholars Publishing, London, pp 85–101

Tort J (2020a) Names and naming in the Iberian Peninsula. Joan Coromines' intercultural approach in Onomastics. In: Felecan O, Buguesiu A (eds) Names and naming: multicultural aspects. Palgrave Macmillan, Cham, Switzerland, pp 295–314

Tort J, Membrado JC (2018) Urban toponymy as a tool for interpretating the physical environment. A case study: Barcelona's medieval old town. Onomastica Uralica 11:217–229

Tort J, Sancho A (2014) Toponyms as landscape indicators. In Tort J, Montagut M (eds). Els noms en la vida quotidiana. Actes del XXIV Congrés Internacional d'ICOS sobre Ciències Onomàstiques/ Names in daily life. Proceedings of the XXIV International Congress of Onomastic Sciences. Barcelona: Generalitat de Catalunya: 1987–2016

Tort J, Santasusagna A (2023) El paisatge toponímic de Sant Llorenç del Munt i Serra de l'Obac. In Paül V, Arocena ME, García-Abad JJ, Pintó J, Tort J (eds) Geografia, paisatge i vegetació. Estudis en homenatge a Josep M. Panareda. Santiago de Compostela: Universidade de Santiago de Compostela [in press]

Tort J (2014) Microtoponymy as a key for geographical description. In Galkowski, A., Gliwa, R. (eds.). Microtoponimia i macrotoponimia. Problematyka wstępna. Lodz: Wdawnictwo Uniwersytetu Lodzkiego: 89–103

Tort J (2020) On the connection between the physical environment and the toponym. Geography's contribution to clarifying the problem. In Balode, L., Zschieschang, C. (eds.) Onomastikas pētījumi II / Onomastic Investigations II. Starptautiskās zinātniskās konferences "Onomastikas pētījumi" (Rīga, 2018. gada 10–12. maijs) rakstu krājums / Proceedings of the International Scientific Conference 'Onomastic Investigations' (Rīga, May 10–12, 2018). Rīga: LU Latviešu valodas institūts: 286–299

Travesset M (1994) La xarxa hidrogràfica del Pla de Barcelona entre la riera de Magòria i la riera d'Horta. Finestrelles 6:57–69

Verdaguer MJ (1883) A Barcelona. Oda. Premiada en la XXV Festa dels Jochs Florals. Barcelona: Estampa Espanyola

Vila P, Casassas L (1974) Barcelona i la seva rodalia al llarg del temps. Aedos, Barcelona

Vilar P (1964) El medi natural. In: Catalunya dins l'Espanya moderna. Vol. I. Barcelona: Edicions 62:167–432

Epilogue

Having completed sixteen chapters, it is time to ask ourselves what we want the urban rivers of the future to be like, especially in the context of the Anthropocene and the climatic, ecological, and social crisis that is occurring on our planet. There are several reasons that support the need to integrate river ecosystems into the functions of the city, but in the current scientific debate we identify two very clear perspectives when intervening in urban rivers: *greening* and *renaturalization*. It is not a merely conceptual debate. Each perspective offers a variety of different urban and territorial ecology solutions and policies.

In the first case (*greening*), rivers adopt a fundamental role in the system of urban green spaces, since they are considered as new "parks" to which citizens have the right to know and explore. Urban rivers are seen as opportunities for economic and tourism development and are related to an increase in the quality of life of the citizens. On the other hand, from the perspective of *renaturalization*, the priority is not the sociability of the river space, but the restoration of its ecological functioning and structure, either by recovering pre-existing forms or carrying out biodiversity measures. They are two perspectives that at first glance may seem complementary, but the reality is that they advocate different conceptions of integration.

Greening seems to enjoy great social acceptance and a clearly positive perception regarding the river space, because it encourages its use as a healthy environment, which provides environmental, educational and leisure values. *Greening* urban or urbanized river sections seems to be an economic investment in which the results can be seen in the short term, and that satisfies citizens and administrations. Rivers are redesigned by planners; they are perceived as pleasing to the eye and a feeling of "ordered green" is offered to the citizen. The greening policy values the river, moving it away from the conceptions of marginal garbage dumps that river systems have had for centuries, and especially during the second half of the twentieth century. However, this conception has a great drawback: it forgets the river as an active subject. It does not consider that the river flows, modifies and adapts to space, transporting sediment, creating, destroying, or mobilizing sedimentary deposits. The river, as a

© The Editor(s) (if applicable) and The Author(s), under exclusive license to Springer Nature Switzerland AG 2024

J. Farguell Pérez and A. Santasusagna Riu (eds.), *Urban and Metropolitan Rivers*, The Urban Book Series, https://doi.org/10.1007/978-3-031-62641-8

"mobile ecosystem", needs space to adapt and develop fully. This contradiction can lead to problems of coexistence between the river and its inhabitants during times of flooding or with undesirable consequences in the urbanized riverbed.

On the other hand, *renaturalization* (also known as environmental river restoration) focuses on improving the river ecosystem in all its dimensions. This policy focuses, for example, on the improvement of water quality, the recovery of the natural river regime and the recovery of hydro-geomorphological processes. In the latter, it is worth highlighting the recovery of sediment dynamics and transport, and allowing the river to self-adjust, creating and destroying sedimentary islands at will, in what has been called "river self-healing". There are numerous examples of river restoration that include the recovery of river processes, especially in places with an Anglo-Saxon river tradition such as the United Kingdom, the United States or Australia. One of the characteristics of the *renaturalization* policy is that, to be effective, it must impose a limit on the frequency and availability of urban river space for citizens. What in previous decades was a major challenge (bringing the river closer to citizens, to demonstrate its ecological value), is now seen as an impediment to its environmental progress.

Pioneering research on urban rivers reveals the impossibility of recovering river ecosystems to the previous conditions of urbanization or modification due to the magnitude of the modifications suffered at certain points. Therefore, in the dilemma between *greening* or *renaturalizing*, it is necessary to clearly establish the objective of what you intend to do. Depending on the objectives set, appropriate measures must be taken. The success or failure of the actions carried out will depend, ultimately, on the objectives set at the beginning. And it is also important that scientific, technical, and political discourses always accompany this process and are sincere about what they want to put into practice.

Aside from the present dilemma, in the twenty-first century, society demands an increase in green spaces in cities, whether natural or apparent, and river axes can help to expand compact urban areas with an absence of public space. Everything suggests that this trend will increase over the coming decades, especially in Western countries. The way to face these new challenges and check whether the interventions are successes or failures will be based on appropriate monitoring and the analysis of their consequences in the short and medium term, something forgotten in many projects, and which has led to failed actions.

In short, the final objective of this book has been to provide ideas, data and reflections on recent interventions and policies carried out in southwestern Europe (Spain, France, and Portugal). These experiences have their origin in a global process, which is closely related to cases and examples already developed in Anglo-Saxon countries. The coming decades will be key for the development of healthy and renaturalized urban river spaces and the accumulation of experiences and their analysis will be key to resolving which initiatives are best based on the objectives set and thought for each case.

Joaquim Farguell Pérez and Albert Santasusagna Riu
Barcelona, Catalonia (Spain)
March 2024

Index

© The Editor(s) (if applicable) and The Author(s), under exclusive license to Springer Nature Switzerland AG 2024

J. Farguell Pérez and A. Santasusagna Riu (eds.), *Urban and Metropolitan Rivers*, The Urban Book Series, https://doi.org/10.1007/978-3-031-62641-8

MIX
Papier aus verantwortungsvollen Quellen
Paper from responsible sources
FSC® C105338
FSC
www.fsc.org

If you have any concerns about our products,
you can contact us on
ProductSafety@springernature.com

In case Publisher is established outside the EU,
the EU authorized representative is:
Springer Nature Customer Service Center GmbH
Europaplatz 3, 69115 Heidelberg, Germany

Printed by Libri Plureos GmbH
in Hamburg, Germany